PHYSICS RESEARCH AND TECHNOLOGY

# NN AND 3N INTERACTIONS

# PHYSICS RESEARCH AND TECHNOLOGY

Additional books in this series can be found on Nova's website
under the Series tab.

Additional e-books in this series can be found on Nova's website
under the e-book tab.

PHYSICS RESEARCH AND TECHNOLOGY

# NN AND 3N INTERACTIONS

LEONID BLOKHINTSEV
AND
IGOR STRAKOVSKY
EDITORS

nova publishers
*New York*

For permission to use material from this book please contact us:
Telephone 631-231-7269; Fax 631-231-8175
Web Site: http://www.novapublishers.com



**Library of Congress Cataloging-in-Publication Data**

ISBN: 978-1-63321-053-0

*Published by Nova Science Publishers, Inc. † New York*

# CONTENTS

# PREFACE

The study of the nucleon-nucleon interaction has a long history, going back to 1932 (just after the discovery of the neutron by James Chadwick). The force is responsible for binding protons and neutrons into atomic nuclei but, unlike the cases of Coulomb or Newtonian gravitational laws, the form of the fundamental *NN* strong interaction and its constants are *a priori* unknown. The subject is nevertheless important because the energy associated with the binding of protons and neutrons to form atomic nuclei is the energy used in nuclear power. The authors of the present volume deliver accounts of current research on various aspects of the forces between two or more nucleons and how they relate to experiments in nuclear physics. General overviews, illustrated by appropriate examples, are given of both the theoretical and experimental knowledge of nuclear physics that are prerequisites for the understanding of the strong interaction.

A description of how nuclear forces emerge from low-energy QCD via chiral effective field theory is presented in Chapter 1. Over the past two decades, this approach has evolved into a powerful tool to derive nuclear two- and many-body forces in a systematic and model-independent way. Nowadays, chiral nuclear forces are used in most *ab initio* nuclear structure calculations.

Chapter 2 is devoted to a description of the existing data on the deuteron charge exchange on hydrogen, $dp \to \{pp\}_s n$, that leads to a final *pp* system at low excitation energy. In particular it is shown that the proton analyzing power, the deuteron tensor analyzing powers, and the deuteron-proton spin correlations can contribute useful information on the spin dependence of near-backward neutron-proton elastic scattering in the GeV region. At energies where the *np* database is rich, and the partial wave solutions are reliable, the theoretical description works very well, which gives confidence in using the approach at higher energies.

The possibility of constructing a nucleon-nucleon interaction that includes the one-pion-exchange force plus a potential of exponential form is considered in Chapter 3. This leads to node of a wave function only in the *S*-state and allows one to describe well the low energy *np* characteristics, the elastic scattering phase shifts up to 500 MeV, the energy of the deuteron, form factors, and the $e^2$H polarization.

A short overview of the Covariant Spectator Theory and its applications to the study of two- and three-nucleon forces is provided in Chapter 4. The basic ideas are illustrated using a simple model, where two scalar "nucleons" exchange a scalar meson. Recent results obtained by adjusting the free parameters in this model in order to give high-precision fits to the two-nucleon scattering data are discussed. These fits automatically give the correct three-body binding energy without the introduction of any additional parameters.

It is well known that the short-distance behavior of the *NN* interaction cannot be determined uniquely from the analysis of experimental elastic phase shifts. At the same time, it is not known *a priori* whether the two-nucleon interaction in the nuclear medium is the same as that in free space, especially when two nucleons approach each other at distances of the order of the nucleon radius. The study of the short-range two-nucleon dynamics (short-range correlations) in nuclear medium, which is a key topic in nuclear physics, is overviewed in Chapter 5. Recent progress in the experimental investigation of short-range correlations appears to agree with models of the *NN* interaction that exhibit strong short-range repulsion and intermediate-range tensor attraction.

Despite the progress made in understanding the *NN* interaction at long distances, based on effective field theories, the behavior at short distances remains as elusive as ever. One of the most fascinating properties of the short-range interaction is its repulsive nature, which is responsible for the stability of strongly interacting matter. Chapter 6 reviews the current status of the understanding of the QCD dynamics of *NN* interactions at short distances, highlights outstanding questions, and outlines the theoretical foundation of the QCD description of hard *NN* processes.

The deuteron breakup reaction $pd \rightarrow \{pp\}_s n$, where $\{pp\}_s$ is a fast proton pair emerging in the forward direction at small excitation energy $E_{pp} \leq 3$ MeV, is discussed in Chapter 7. The spin-singlet *pp* final state $^1S_0$ dominates under these conditions. The dynamics of this reaction are considered and its relation to backward elastic proton-deuteron scattering discussed.

Chapter 8 deals with a comprehensive overview of *ab initio* results for binding energies and spectra of light nuclei obtained with the JISP16 *NN* interaction. The calculations were performed by means of the No-Core Full Configuration approach, which extends the No-Core Shell Model to infinite basis space through extrapolation.

Finally in Chapter 9, the book presents an overview of the strong interactions of hadron beams with nucleons bound within a complex nucleus. These need not be the same as those measured in the scattering in free space; the strong nuclear fields may induce any of a wide range of modifications of the nucleons, their interactions, and even the properties of the incoming hadron. Quasi-free scattering, which is related as closely as possible to free scattering, is the simplest way to investigate such changes. The conditions for interpreting data within such a simple scattering mechanism are very demanding, and their validity must be demonstrated. The methods and a few examples are given in this chapter. In addition, the special role of spin observables, $\Delta-$ isobars in nuclei, and "swollen nucleons" are reviewed.

Overall, this new book brings together leading research on nuclear physics from around the world.

In: NN and 3N Interactions
Editors: Leonid Blokhintsev and Igor Strakovsky
ISBN: 978-1-63321-053-0


*Chapter 1*

# NUCLEAR FORCES FROM CHIRAL EFFECTIVE FIELD THEORY

*R. Machleidt**
Department of Physics, University of Idaho,
Moscow, Idaho, US

**Abstract**

We review how nuclear forces emerge from low-energy QCD via chiral effective field theory. During the past two decades, this approach has evolved into a powerful tool to derive nuclear two- and many-body forces in a systematic and model-independent way. Nowadays, in most *ab initio* nuclear structure calculations, chiral nuclear forces are used. Therefore, this is a good time for a review of the field. Our presentation is intended for the non-specialist.

**PACS:** 13.75.Cs, 12.39.Fe, 21.30.-x, 21.45.Ff

**Keywords:** Low-energy QCD, effective field theory, chiral perturbation theory, nucleon-nucleon interaction, nuclear many-body forces

## 1. Historical Perspective and Preview

The development of a proper theory of nuclear forces has occupied the minds of some of the brightest physicists for eight decades and has been one of the main topics of physics research in the 20th century. The original idea was that the force is caused by the exchange of lighter particles (than nucleons) known as mesons, and this idea gave rise to the birth of a new sub-field of modern physics, namely, (elementary) particle physics. The modern perception of the nuclear force is that it is a residual interaction (similar to the van der Waals force between neutral atoms) of the even stronger force between quarks, which is mediated by the exchange of gluons and holds the quarks together inside a nucleon.

*E-mail address: machleid@uidaho.edu

### 1.1. Early History

After the discovery of the neutron in 1932, it was clear that the atomic nucleus is made up from protons and neutrons. In such a system, electromagnetic forces cannot be the reason why the constituents of the nucleus are sticking together. Therefore, the concept of a new strong nuclear interaction was introduced. In 1935, the first theory for this new force was developed by the Japanese physicist Yukawa [1], who suggested that the nucleons would exchange particles between each other and this mechanism would create the force. Yukawa constructed his theory in analogy to the theory of the electromagnetic interaction where the exchange of a (massless) photon is the cause of the force. However, in the case of the nuclear force, Yukawa assumed that the "force-makers" (which were eventually called "mesons") carry a mass of a fraction of the nucleon mass. This would limit the effect of the force to a finite range, since the uncertainty principal allows massive particles to travel only a finite distance. The meson predicted by Yukawa was finally found in 1947 in cosmic ray and in 1948 in the laboratory and called the pion. Yukawa was awarded the Nobel Prize in 1949. In the 1950's and 60's more mesons were found in accelerator experiments and the meson theory of nuclear forces was extended to include many mesons. These models became known as one-boson-exchange models, which is a reference to the fact that the different mesons are exchanged singly in this model. The one-boson-exchange model is very successful in explaining essentially all properties of the nucleon-nucleon interaction at low energies [2, 3, 4, 5, 6]. In the 1970's and 80's, also meson models were developed that went beyond the simple single-particle exchange mechanism. These models included, in particular, the explicit exchange of two pions with all its complications. Well-known representatives of the latter kind are the Paris [7] and the Bonn potential [8].

Since these meson models were quantitatively very successful, it appeared that they were the solution of the nuclear force problem. However, with the discovery (in the 1970's) that the fundamental theory of strong interactions is quantum chromo-dynamics (QCD) and not meson theory, all "meson theories" had to be viewed as models, and the attempts to derive a proper theory of the nuclear force had to start all over again.

### 1.2. QCD and the Nuclear Force

The problem with a derivation of nuclear forces from QCD is two-fold. First, each nucleon consists of three quarks such that the system of two nucleons is already a six-body problem. Second, the force between quarks, which is created by the exchange of gluons, has the feature of being very strong at the low energy-scale that is characteristic of nuclear physics. This extraordinary strength makes it difficult to find "converging" mathematical solutions. Therefore, during the first round of new attempts, QCD-inspired quark models became popular. The positive aspect of these models is that they try to explain nucleon structure (which consists of three quarks) and nucleon-nucleon interactions (six-quark systems) on an equal footing. Some of the gross features of the two-nucleon force, like the "hard core" are explained successfully in such models. However, from a critical point of view, it must be noted that these quark-based approaches are yet another set of models and not a theory. Alternatively, one may try to solve the six-quark problem with brute computing power, by putting the six-quark system on a four dimensional lattice of discrete points which represents the three dimensions of space and one dimension of time. This method has become known as lattice QCD and is making progress. However, such calculations are computationally very expensive and cannot be used as a standard nuclear physics tool.

### 1.3. Chiral Effective Field Theory

Around 1990, a major breakthrough occurred when the nobel laureate Steven Weinberg applied the concept of an effective field theory (EFT) to low-energy QCD [9, 10]. He simply wrote down the most general theory that is consistent with all the properties of low-energy QCD, since that would make this theory identical to low-energy QCD. A particularly important property is the so-called chiral symmetry, which is "spontaneously" broken. Massless particles observe chiral symmetry, which means that their spin and momentum are either parallel ("right-handed") or anti-parallel ("left-handed") and remain so forever. Since the quarks, which nucleons are made of ("up" and "down" quarks), are almost mass-less, approximate chiral symmetry is a given. Naively, this symmetry should have the consequence that one finds in nature mesons of the same mass, but with positive and negative parity. However, this is not the case and such failure is termed a "spontaneous" breaking of the symmetry. According to a theorem first proven by Goldstone, the spontaneous breaking of a symmetry creates a particle, here, the pion. Thus, the pion becomes the main player in the production of the nuclear force. The interaction of pions with nucleons is weak as compared to the interaction of gluons with quarks. Therefore, pion-nucleon processes can be calculated without problem. Moreover, this effective field theory can be expanded in powers of momentum over "scale", where scale denotes the "chiral symmetry breaking scale" which is about 1 GeV. This scheme is also known as chiral perturbation theory (ChPT) and allows to calculate the various terms that make up the nuclear force systematically power by power, or order by order. Another advantage of the chiral EFT approach is its ability to generate not only the force between two nucleons, but also many-nucleon forces, on the same footing [11]. In modern theoretical nuclear physics, the chiral EFT approach is becoming increasingly popular and is applied with great success [12].

## 2. Effective Field Theory for Low-Energy QCD

Quantum chromodynamics (QCD) is the theory of strong interactions. It deals with quarks, gluons and their interactions and is part of the Standard Model of Particle Physics. QCD is a non-Abelian gauge field theory with color $SU(3)$ the underlying gauge group. The non-Abelian nature of the theory has dramatic consequences. While the interaction between colored objects is weak at short distances or high momentum transfer ("asymptotic freedom"); it is strong at long distances ( $\gtrsim 1$ fm) or low energies, leading to the confinement of quarks into colorless objects, the hadrons. Consequently, QCD allows for a perturbative analysis at large energies, whereas it is highly non-perturbative in the low-energy regime. Nuclear physics resides at low energies and the force between nucleons is a residual color interaction similar to the van der Waals force between neutral molecules. Therefore, in terms of quarks and gluons, the nuclear force is a very complicated problem that, nevertheless, can be attacked with brute computing power on a discretized, Euclidean space-time lattice (known as lattice QCD). In a recent study [13], the neutron-proton scattering lengths in the singlet and triplet $S$-waves have been determined in fully dynamical lattice QCD, with a smallest pion mass of 354 MeV. This result is then extrapolated to the physical pion mass with the help of chiral perturbation theory. The pion mass of 354 MeV is still too large to allow for reliable extrapolations, but the feasibility has been demonstrated and more progress can be expected for the near future. In a lattice calculation of a very different kind, the nucleon-nucleon ($NN$) potential was studied [14]. The central part of the potential shows a repulsive core plus attraction of intermediate range. This is a very promising result, but it must be noted that also in this investigation still rather large pion masses are being used. In any case, advanced lattice QCD calculations are under way and continuously improved. However, since these calculations are very time-consuming and expensive, they can only be used to check a few representative key-issues. For everyday nuclear structure physics, a more efficient approach is needed.

The efficient approach is an effective field theory. For the development of an EFT, it is crucial to identify a separation of scales. In the hadron spectrum, a large gap between the masses of the pions and the masses of the vector mesons, like $\rho(770)$ and $\omega(782)$, can clearly be identified. Thus, it is natural to assume that the pion mass sets the soft scale, $Q \sim m_\pi$, and the rho mass the hard scale, $\Lambda_\chi \sim m_\rho$, also known as the chiral-symmetry breaking scale. This is suggestive of considering an expansion in terms of the soft scale over the hard scale, $Q/\Lambda_\chi$. Concerning the relevant degrees of freedom, we noticed already that, for the ground state and the low-energy excitation spectrum of an atomic nucleus as well as for conventional nuclear reactions, quarks and gluons are ineffective degrees of freedom, while nucleons and pions are the appropriate ones. To make sure that this EFT is not just another phenomenology, it must have a firm link with QCD. The link is established by having the EFT observe all relevant symmetries of the underlying theory. This requirement is based upon a 'folk theorem' by Weinberg [9]:

> If one writes down the most general possible Lagrangian, including *all* terms consistent with assumed symmetry principles, and then calculates matrix elements with this Lagrangian to any given order of perturbation theory, the result will simply be the most general possible S-matrix consistent with analyticity, perturbative unitarity, cluster decomposition, and the assumed symmetry principles.

In summary, the EFT program consists of the following steps:

1. Identify the soft and hard scales, and the degrees of freedom appropriate for (low-energy) nuclear physics.

2. Identify the relevant symmetries of low-energy QCD and investigate if and how they are broken.

3. Construct the most general Lagrangian consistent with those symmetries and symmetry breakings.

4. Design an organizational scheme that can distinguish between more and less important contributions: a low-momentum expansion.

5. Guided by the expansion, calculate Feynman diagrams for the problem under consideration to the desired accuracy.

In the following (sub)sections, we will elaborate on these steps, one by one. Since we discussed the first step already, we will address now step two.

## 2.1. Symmetries of Low-Energy QCD

In this section, we will give a brief introduction into (low-energy) QCD, its symmetries and symmetry breakings. More detailed presentations of this topic are provided in Refs. [12, 15].

### 2.1.1. Chiral Symmetry

The QCD Lagrangian reads

$$\mathcal{L}_{\rm QCD} = \bar{q}(i\gamma^\mu \mathcal{D}_\mu - \mathcal{M})q - \frac{1}{4}\mathcal{G}_{\mu\nu,a}\mathcal{G}^{\mu\nu}_a \tag{1}$$

with the gauge-covariant derivative

$$\mathcal{D}_\mu = \partial_\mu - ig\frac{\lambda_a}{2}\mathcal{A}_{\mu,a} \tag{2}$$

and the gluon field strength tensor[1]

$$\mathcal{G}_{\mu\nu,a} = \partial_\mu \mathcal{A}_{\nu,a} - \partial_\nu \mathcal{A}_{\mu,a} + g f_{abc}\mathcal{A}_{\mu,b}\mathcal{A}_{\nu,c}\,. \tag{3}$$

In the above, $q$ denotes the quark fields and $\mathcal{M}$ the quark mass matrix. Further, $g$ is the strong coupling constant and $\mathcal{A}_{\mu,a}$ are the gluon fields. The $\lambda_a$ are the Gell-Mann matrices and the $f_{abc}$ the structure constants of the $SU(3)_{\rm color}$ Lie algebra $(a, b, c = 1, \ldots, 8)$; summation over repeated indices is always implied. The gluon-gluon term in the last equation arises from the non-Abelian nature of the gauge theory and is the reason for the peculiar features of the color force.

The masses of the up $(u)$, down $(d)$, and strange (s) quarks are [16]:

$$m_u = 2.5 \pm 0.8\ \text{MeV}, \tag{4}$$
$$m_d = 5 \pm 0.9\ \text{MeV}, \tag{5}$$
$$m_s = 101 \pm 25\ \text{MeV}. \tag{6}$$

These masses are small as compared to a typical hadronic scale, i.e., a scale of low-mass hadrons which are not Goldstone bosons, e.g., $m_\rho = 0.78$ GeV $\approx 1$ GeV.

It is therefore of interest to discuss the QCD Lagrangian in the limit of vanishing quark masses:

$$\mathcal{L}^0_{\rm QCD} = \bar{q}i\gamma^\mu \mathcal{D}_\mu q - \frac{1}{4}\mathcal{G}_{\mu\nu,a}\mathcal{G}^{\mu\nu}_a\,. \tag{7}$$

Defining right- and left-handed quark fields,

$$q_R = P_R q\,, \quad q_L = P_L q\,, \tag{8}$$

[1]For $SU(N)$ group indices, we use Latin letters, $\ldots, a, b, c, \ldots, i, j, k, \ldots$, and, in general, do not distinguish between subscripts and superscripts.

with

$$P_R = \frac{1}{2}(1+\gamma_5)\,, \quad P_L = \frac{1}{2}(1-\gamma_5)\,, \tag{9}$$

we can rewrite the Lagrangian as follows:

$$\mathcal{L}^0_{\text{QCD}} = \bar{q}_R i\gamma^\mu \mathcal{D}_\mu q_R + \bar{q}_L i\gamma^\mu \mathcal{D}_\mu q_L - \frac{1}{4}\mathcal{G}_{\mu\nu,a}\mathcal{G}^{\mu\nu}_a\,. \tag{10}$$

This equation revels that *the right- and left-handed components of massless quarks do not mix* in the QCD Lagrangian. For the two-flavor case, this is $SU(2)_R \times SU(2)_L$ symmetry, also known as *chiral symmetry*. However, this symmetry is broken in two ways: explicitly and spontaneously.

### 2.1.2. Explicit Symmetry Breaking

The mass term $-\bar{q}\mathcal{M}q$ in the QCD Lagrangian Eq. (1) breaks chiral symmetry explicitly. To better see this, let's rewrite $\mathcal{M}$ for the two-flavor case,

$$\begin{aligned}
\mathcal{M} &= \begin{pmatrix} m_u & 0 \\ 0 & m_d \end{pmatrix} \\
&= \frac{1}{2}(m_u+m_d)\begin{pmatrix} 1 & 0 \\ 0 & 1 \end{pmatrix} + \frac{1}{2}(m_u-m_d)\begin{pmatrix} 1 & 0 \\ 0 & -1 \end{pmatrix} \\
&= \frac{1}{2}(m_u+m_d)\, I + \frac{1}{2}(m_u-m_d)\, \tau_3\,.
\end{aligned} \tag{11}$$

The first term in the last equation in invariant under $SU(2)_V$ (isospin symmetry) and the second term vanishes for $m_u = m_d$. Thus, isospin is an exact symmetry if $m_u = m_d$. However, both terms in Eq. (11) break chiral symmetry. Since the up and down quark masses [Eqs. (4) and (5)] are small as compared to the typical hadronic mass scale of $\sim$ 1 GeV, the explicit chiral symmetry breaking due to non-vanishing quark masses is very small.

### 2.1.3. Spontaneous Symmetry Breaking

A (continuous) symmetry is said to be *spontaneously broken* if a symmetry of the Lagrangian is not realized in the ground state of the system. There is evidence that the (approximate) chiral symmetry of the QCD Lagrangian is spontaneously broken—for dynamical reasons of nonperturbative origin which are not fully understood at this time. The most plausible evidence comes from the hadron spectrum.

From chiral symmetry, one naively expects the existence of degenerate hadron multiplets of opposite parity, i.e., for any hadron of positive parity one would expect a degenerate hadron state of negative parity and vice versa. However, these "parity doublets" are not observed in nature. For example, take the $\rho$-meson which is a vector meson of negative parity ($J^P = 1^-$) and mass 776 MeV. There does exist a $1^+$ meson, the $a_1$, but it has a mass of 1230 MeV and, therefore, cannot be perceived as degenerate with the $\rho$. On the other hand, the $\rho$ meson comes in three charge states (equivalent to three isospin states), the $\rho^\pm$ and the $\rho^0$, with masses that differ by at most a few MeV. Thus, in the hadron spectrum, $SU(2)_V$

(isospin) symmetry is well observed, while axial symmetry is broken: $SU(2)_R \times SU(2)_L$ is broken down to $SU(2)_V$.

A spontaneously broken global symmetry implies the existence of (massless) Goldstone bosons. The Goldstone bosons are identified with the isospin triplet of the (pseudoscalar) pions, which explains why pions are so light. The pion masses are not exactly zero because the up and down quark masses are not exactly zero either (explicit symmetry breaking). Thus, pions are a truly remarkable species: they reflect spontaneous as well as explicit symmetry breaking. Goldstone bosons interact weakly at low energy. They are degenerate with the vacuum and, therefore, interactions between them must vanish at zero momentum and in the chiral limit ($m_\pi \to 0$).

## 2.2. Chiral Effective Lagrangians

The next step in our EFT program is to build the most general Lagrangian consistent with the (broken) symmetries discussed above. An elegant formalism for the construction of such Lagrangians was developed by Callan, Coleman, Wess, and Zumino (CCWZ) [17] who worked out the group-theoretical foundations of non-linear realizations of chiral symmetry.[2] It is characteristic for these non-linear realizations that, whenever functions of the Goldstone bosons appear in the Langrangian, they are always accompanied with at least one space-time derivative. The Lagrangians given below are built upon the CCWZ formalism.

As discussed, the relevant degrees of freedom are pions (Goldstone bosons) and nucleons. Since the interactions of Goldstone bosons must vanish at zero momentum transfer and in the chiral limit ($m_\pi \to 0$), the low-energy expansion of the Lagrangian is arranged in powers of derivatives and pion masses. The hard scale is the chiral-symmetry breaking scale, $\Lambda_\chi \approx 1$ GeV. Thus, the expansion is in terms of powers of $Q/\Lambda_\chi$ where $Q$ is a (small) momentum or pion mass. This is chiral perturbation theory (ChPT).

The effective Lagrangian can formally be written as,

$$\mathcal{L}_{\text{eff}} = \mathcal{L}_{\pi\pi} + \mathcal{L}_{\pi N} + \mathcal{L}_{NN} + \ \dots\,, \tag{12}$$

where $\mathcal{L}_{\pi\pi}$ deals with the dynamics among pions, $\mathcal{L}_{\pi N}$ describes the interaction between pions and a nucleon, and $\mathcal{L}_{NN}$ contains two-nucleon contact interactions which consist of four nucleon-fields (four nucleon legs) and no meson fields. The ellipsis stands for terms that involve two nucleons plus pions and three or more nucleons with or without pions, relevant for nuclear many-body forces (cf. last two terms of Eq. (18), below, and Section 5). The individual Lagrangians are organized as follows:

$$\mathcal{L}_{\pi\pi} = \mathcal{L}^{(2)}_{\pi\pi} + \mathcal{L}^{(4)}_{\pi\pi} + \dots\,, \tag{13}$$

$$\mathcal{L}_{\pi N} = \mathcal{L}^{(1)}_{\pi N} + \mathcal{L}^{(2)}_{\pi N} + \mathcal{L}^{(3)}_{\pi N} + \dots, \tag{14}$$

and

$$\mathcal{L}_{NN} = \mathcal{L}^{(0)}_{NN} + \mathcal{L}^{(2)}_{NN} + \mathcal{L}^{(4)}_{NN} + \dots\,, \tag{15}$$

where the superscript refers to the number of derivatives or pion mass insertions (chiral dimension) and the ellipsis stands for terms of higher dimensions.

[2] An accessible introduction into the rather involved CCWZ formalism can be found in Ref. [15].

Above, we have organized the Lagrangians by the number of derivatives or pion-mass insertions. This is the standard way, appropriate particularly for considerations of $\pi$-$\pi$ and $\pi$-$N$ scattering. As it turns out (cf. Section 3.1), for interactions among nucleons, it is sometimes useful to also consider the so-called index of the interaction,

$$\Delta \equiv d + \frac{n}{2} - 2\,, \tag{16}$$

where $d$ is the number of derivatives or pion-mass insertions and $n$ the number of nucleon field operators (nucleon legs). We will now write down the Lagrangian in terms of increasing values of the parameter $\Delta$ and we will do so using the so-called heavy-baryon formalism which we indicate by a "hat" [18].

The leading-order Lagrangian reads,

$$\begin{aligned}
\widehat{\mathcal{L}}^{\Delta=0} &= \frac{1}{2}\partial_\mu \boldsymbol{\pi}\cdot\partial^\mu\boldsymbol{\pi} - \frac{1}{2}m_\pi^2\boldsymbol{\pi}^2 \\
&+ \frac{1-4\alpha}{2f_\pi^2}(\boldsymbol{\pi}\cdot\partial_\mu\boldsymbol{\pi})(\boldsymbol{\pi}\cdot\partial^\mu\boldsymbol{\pi}) - \frac{\alpha}{f_\pi^2}\boldsymbol{\pi}^2\partial_\mu\boldsymbol{\pi}\cdot\partial^\mu\boldsymbol{\pi} + \frac{8\alpha-1}{8f_\pi^2}m_\pi^2\boldsymbol{\pi}^4 \\
&+ \bar{N}\left[i\partial_0 - \frac{g_A}{2f_\pi}\,\boldsymbol{\tau}\cdot(\vec{\sigma}\cdot\vec{\nabla})\boldsymbol{\pi} - \frac{1}{4f_\pi^2}\,\boldsymbol{\tau}\cdot(\boldsymbol{\pi}\times\partial_0\boldsymbol{\pi})\right]N \\
&+ \bar{N}\left\{\frac{g_A(4\alpha-1)}{4f_\pi^3}\,(\boldsymbol{\tau}\cdot\boldsymbol{\pi})\left[\boldsymbol{\pi}\cdot(\vec{\sigma}\cdot\vec{\nabla})\boldsymbol{\pi}\right] + \frac{g_A\alpha}{2f_\pi^3}\,\boldsymbol{\pi}^2\left[\boldsymbol{\tau}\cdot(\vec{\sigma}\cdot\vec{\nabla})\boldsymbol{\pi}\right]\right\}N \\
&- \frac{1}{2}C_S\bar{N}N\bar{N}N - \frac{1}{2}C_T(\bar{N}\vec{\sigma}N)\cdot(\bar{N}\vec{\sigma}N) + \ldots\,,
\end{aligned} \tag{17}$$

and subleading Lagrangians are,

$$\begin{aligned}
\widehat{\mathcal{L}}^{\Delta=1} &= \bar{N}\left\{\frac{\vec{\nabla}^2}{2M_N} - \frac{ig_A}{4M_N f_\pi}\boldsymbol{\tau}\cdot\left[\vec{\sigma}\cdot\left(\overleftarrow{\nabla}\,\partial_0\boldsymbol{\pi} - \partial_0\boldsymbol{\pi}\,\vec{\nabla}\right)\right]\right. \\
&\left. - \frac{i}{8M_N f_\pi^2}\boldsymbol{\tau}\cdot\left[\overleftarrow{\nabla}\cdot(\boldsymbol{\pi}\times\vec{\nabla}\boldsymbol{\pi}) - (\boldsymbol{\pi}\times\vec{\nabla}\boldsymbol{\pi})\cdot\vec{\nabla}\right]\right\}N \\
&+ \bar{N}\left[4c_1 m_\pi^2 - \frac{2c_1}{f_\pi^2}\,m_\pi^2\,\boldsymbol{\pi}^2 + \left(c_2 - \frac{g_A^2}{8M_N}\right)\frac{1}{f_\pi^2}(\partial_0\boldsymbol{\pi}\cdot\partial_0\boldsymbol{\pi})\right. \\
&\left. + \frac{c_3}{f_\pi^2}\,(\partial_\mu\boldsymbol{\pi}\cdot\partial^\mu\boldsymbol{\pi}) - \left(c_4 + \frac{1}{4M_N}\right)\frac{1}{2f_\pi^2}\epsilon^{ijk}\epsilon^{abc}\sigma^i\tau^a(\partial^j\pi^b)(\partial^k\pi^c)\right]N \\
&- \frac{D}{4f_\pi}(\bar{N}N)\bar{N}\left[\boldsymbol{\tau}\cdot(\vec{\sigma}\cdot\vec{\nabla})\boldsymbol{\pi}\right]N - \frac{1}{2}E(\bar{N}N)(\bar{N}\boldsymbol{\tau}N)\cdot(\bar{N}\boldsymbol{\tau}N) + \ldots
\end{aligned} \tag{18}$$

$$\widehat{\mathcal{L}}^{\Delta=2} = \mathcal{L}_{\pi\pi}^{(4)} + \widehat{\mathcal{L}}_{\pi N}^{(3)} + \widehat{\mathcal{L}}_{NN}^{(2)} + \ldots\,, \tag{19}$$

$$\widehat{\mathcal{L}}^{\Delta=4} = \widehat{\mathcal{L}}_{NN}^{(4)} + \ldots\,, \tag{20}$$

where the ellipses represent terms that are irrelevant for the derivation of nuclear forces up to fourth order. The Lagrangian $\widehat{\mathcal{L}}_{\pi N}^{(3)}$ can be found in Ref. [18] and the $NN$ contact Lagrangians are given below. The pion fields are denoted by $\boldsymbol{\pi}$ and the heavy baryon nucleon field by $N$ ($\bar{N} = N^\dagger$). Furthermore, $g_A$, $f_\pi$, $m_\pi$, and $M_N$ are the axial-vector coupling constant, pion decay constant, pion mass, and nucleon mass, respectively. Numerical values for these quantities will be given later. The $c_i$ are low-energy constants (LECs) from the dimension two $\pi N$ Lagrangian and $\alpha$ is a parameter that appears in the expansion of the pion fields, see Ref. [12] for more details. Results are independent of $\alpha$.

The lowest order (or leading order) $NN$ Lagrangian has no derivatives and reads [10]

$$\widehat{\mathcal{L}}^{(0)}_{NN} = -\frac{1}{2}C_S\bar{N}N\bar{N}N - \frac{1}{2}C_T(\bar{N}\vec{\sigma}N)\cdot(\bar{N}\vec{\sigma}N)\,, \tag{21}$$

where $C_S$ and $C_T$ are unknown constants which are determined by a fit to the $NN$ data.

The second order $NN$ Lagrangian can be stated as follows [19]

$$\begin{aligned}
\widehat{\mathcal{L}}^{(2)}_{NN} \;=\;& -C_1'\left[(\bar{N}\vec{\nabla}N)^2+(\overline{\vec{\nabla}N}N)^2\right] - C_2'(\bar{N}\vec{\nabla}N)\cdot(\overline{\vec{\nabla}N}N) - C_3'\bar{N}N\left[\bar{N}\vec{\nabla}^2N+\overline{\vec{\nabla}^2N}N\right] \\
& -iC_4'\left[\bar{N}\vec{\nabla}N\cdot(\overline{\vec{\nabla}N}\times\vec{\sigma}N)+\overline{(\vec{\nabla}N)}N\cdot(\bar{N}\vec{\sigma}\times\vec{\nabla}N)\right] \\
& -iC_5'\bar{N}N(\overline{\vec{\nabla}N}\cdot\vec{\sigma}\times\vec{\nabla}N) - iC_6'(\bar{N}\vec{\sigma}N)\cdot(\overline{\vec{\nabla}N}\times\vec{\nabla}N) \\
& -(C_7'\delta_{ik}\delta_{jl}+C_8'\delta_{il}\delta_{kj}+C_9'\delta_{ij}\delta_{kl})\left[\bar{N}\sigma_k\partial_iN\bar{N}\sigma_l\partial_jN+\overline{\partial_iN}\sigma_kN\overline{\partial_jN}\sigma_lN\right] \\
& -(C_{10}'\delta_{ik}\delta_{jl}+C_{11}'\delta_{il}\delta_{kj}+C_{12}'\delta_{ij}\delta_{kl})\,\bar{N}\sigma_k\partial_iN\overline{\partial_jN}\sigma_lN \\
& -\left(\frac{1}{2}C_{13}'(\delta_{ik}\delta_{jl}+\delta_{il}\delta_{kj})+C_{14}'\delta_{ij}\delta_{kl}\right)\left[\overline{\partial_iN}\sigma_k\partial_jN+\overline{\partial_jN}\sigma_k\partial_iN\right]\bar{N}\sigma_lN\,.
\end{aligned} \tag{22}$$

Similar to $C_S$ and $C_T$ of Eq. (21), the $C_i'$ of Eq. (22) are unknown constants which are fixed in a fit to the $NN$ data. Obviously, these contact Lagrangians blow up quite a bit with increasing order, which is why we do not give $\widehat{\mathcal{L}}^{(4)}_{NN}$ explicitly here. The $NN$ contact potentials that emerge from these Lagrangians are given in Section 4.2.

## 3. Nuclear Forces from EFT: Overview

In the beginning of Section 2, we listed the steps we have to take for carrying out the EFT program of a derivation of nuclear forces. So far, we discussed steps one to three. What is left are steps four (low-momentum expansion) and five (Feynman diagrams). In this section, we will say more about the expansion we are using and give an overview of the Feynman diagrams that arise order by order.

### 3.1. Chiral Perturbation Theory and Power Counting

Effective Langrangians have infinitely many terms, and an unlimited number of Feynman graphs can be calculated from them. Therefore, we need a scheme that makes the theory manageable and calculable. This scheme which tells us how to distinguish between large (important) and small (unimportant) contributions is chiral perturbation theory (ChPT).

In ChPT, graphs are analyzed in terms of powers of small external momenta over the large scale: $(Q/\Lambda_\chi)^\nu$, where $Q$ is generic for a momentum (nucleon three-momentum or pion four-momentum) or a pion mass and $\Lambda_\chi \sim 1$ GeV is the chiral symmetry breaking scale (hadronic scale, hard scale). Determining the power $\nu$ has become known as power counting.

For the moment, we will consider only so-called irreducible graphs; the problem of reducible or iterative diagrams and their relevance for the $NN$ system will be discussed later (cf. Section 4.3). By definition, an irreducible graph is a diagram that cannot be separated into two by cutting only nucleon lines. Following the Feynman rules of covariant perturbation theory, a nucleon propagator is $Q^{-1}$, a pion propagator $Q^{-2}$, each derivative in any interaction is $Q$, and each four-momentum integration $Q^4$. This is also known as naive dimensional analysis. Applying then some topological identities, one obtains for the power of an irreducible diagram involving $A$ nucleons [12]

$$\nu = -2 + 2A - 2C + 2L + \sum_i \Delta_i \,, \tag{23}$$

with

$$\Delta_i \equiv d_i + \frac{n_i}{2} - 2 \,, \tag{24}$$

where $C$ denotes the number of separately connected pieces and $L$ the number of loops in the diagram; $d_i$ is the number of derivatives or pion-mass insertions and $n_i$ the number of nucleon fields (nucleon legs) involved in vertex $i$; the sum runs over all vertices $i$ contained in the diagram under consideration. Note that $\Delta_i \geq 0$ for all interactions allowed by chiral symmetry. Purely pionic interactions have at least two derivatives ($d_i \geq 2, n_i = 0$); interactions of pions with a nucleon have at least one derivative ($d_i \geq 1, n_i = 2$); and nucleon-nucleon contact terms ($n_i = 4$) have $d_i \geq 0$. This demonstrates how chiral symmetry guarantees a low-energy expansion.

An important observation from power counting is that the powers are bounded from below and, specifically, $\nu \geq 0$. This fact is crucial for the convergence of the low-momentum expansion.

Moreover, the power formula Eq. (23) allows to predict the leading orders of connected multi-nucleon forces. Consider a $m$-nucleon irreducibly connected diagram ($m$-nucleon force) in an $A$-nucleon system ($m \leq A$). The number of separately connected pieces is $C = A - m + 1$. Inserting this into Eq. (23) together with $L = 0$ and $\sum_i \Delta_i = 0$ yields $\nu = 2m - 4$. Thus, two-nucleon forces ($m = 2$) start at $\nu = 0$, three-nucleon forces ($m = 3$) at $\nu = 2$ (but they happen to cancel at that order), and four-nucleon forces at $\nu = 4$ (they don't cancel). More about this in the next sub-section and Section 5.

For later purposes, we note that for an irreducible $NN$ diagram ($A = 2$, $C = 1$), the power formula collapses to the very simple expression

$$\nu = 2L + \sum_i \Delta_i \,. \tag{25}$$

In summary, the chief point of the ChPT expansion is that, at a given order $\nu$, there exists only a finite number of graphs. This is what makes the theory calculable. The expression $(Q/\Lambda_\chi)^{\nu+1}$ provides a rough estimate of the relative size of the contributions left out and, thus, of the accuracy at order $\nu$. In this sense, the theory can be calculated to any desired accuracy and has predictive power.

## 3.2. The Hierarchy of Nuclear Forces

Chiral perturbation theory and power counting imply that nuclear forces emerge as a hierarchy controlled by the power $\nu$, Fig. 1.

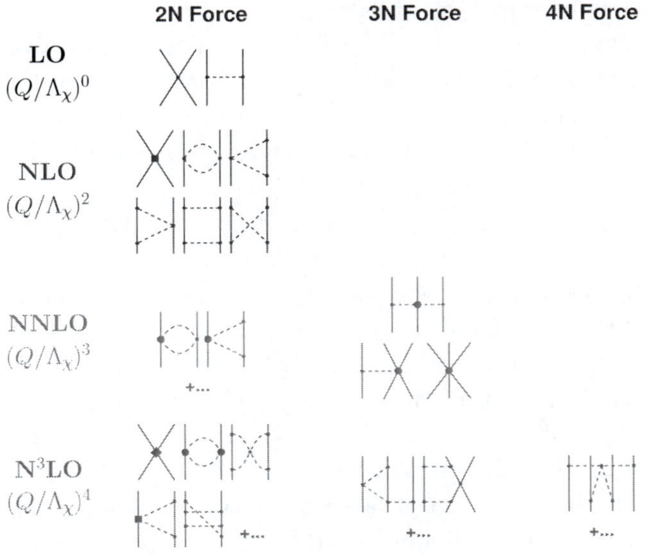


Figure 1. Hierarchy of nuclear forces in ChPT. Solid lines represent nucleons and dashed lines pions. Small dots, large solid dots, solid squares, and solid diamonds denote vertices of index $\Delta =$ 0, 1, 2, and 4, respectively. Further explanations are given in the text.

In lowest order, better known as leading order (LO, $\nu = 0$), the $NN$ amplitude is made up by two momentum-independent contact terms ($\sim Q^0$), represented by the four-nucleon-leg graph with a small-dot vertex shown in the first row of Fig. 1, and static one-pion exchange (1PE), second diagram in the first row of the figure. This is, of course, a rather rough approximation to the two-nucleon force (2NF), but accounts already for some important features. The 1PE provides the tensor force, necessary to describe the deuteron, and it explains $NN$ scattering in peripheral partial waves of very high orbital angular momentum. At this order, the two contacts which contribute only in $S$-waves provide the short- and intermediate-range interaction which is somewhat crude.

In the next order, $\nu = 1$, all contributions vanish due to parity and time-reversal invariance.

Therefore, the next-to-leading order (NLO) is $\nu = 2$. Two-pion exchange (2PE) occurs for the first time ("leading 2PE") and, thus, the creation of a more sophisticated description of the intermediate-range interaction is starting here. Since the loop involved in each pion-diagram implies already $\nu = 2$ [cf. Eq. (25)], the vertices must have $\Delta_i = 0$. Therefore, at this order, only the lowest order $\pi NN$ and $\pi\pi NN$ vertices are allowed which is why the leading 2PE is rather weak. Furthermore, there are seven contact terms of $\mathcal{O}(Q^2)$, shown by the four-nucleon-leg graph with a solid square, which contribute in $S$ and $P$ waves. The operator structure of these contacts include a spin-orbit term besides central, spin-spin, and tensor terms. Thus, essentially all spin-isospin structures necessary to describe the two-nucleon force phenomenologically have been generated at this order. The main deficiency at this stage of development is an insufficient intermediate-range attraction.

This problem is finally fixed at order three ($\nu = 3$), next-to-next-to-leading order

(NNLO). The 2PE involves now the two-derivative $\pi\pi NN$ seagull vertices (proportional to the $c_i$ LECs) denoted by a large solid dot in Fig. 1. These vertices represent correlated 2PE as well as intermediate $\Delta(1232)$-isobar contributions. It is well-known from the meson phenomenology of nuclear forces [8, 7] that these two contributions are crucial for a realistic and quantitative 2PE model. Consequently, the 2PE now assumes a realistic size and describes the intermediate-range attraction of the nuclear force about right. Moreover, first relativistic corrections come into play at this order. There are no new contacts.

The reason why we talk of a hierarchy of nuclear forces is that two- and many-nucleon forces are created on an equal footing and emerge in increasing number as we go to higher and higher orders. At NNLO, the first set of nonvanishing three-nucleon forces (3NF) occur [20, 21], cf. column '3N Force' of Fig. 1. In fact, at the previous order, NLO, irreducible 3N graphs appear already, however, it has been shown by Weinberg [11] that these diagrams all cancel. Since nonvanishing 3NF contributions happen first at order $(Q/\Lambda_\chi)^3$, they are very weak as compared to the 2NF which starts at $(Q/\Lambda_\chi)^0$.

More 2PE is produced at $\nu = 4$, next-to-next-to-next-to-leading order (N$^3$LO), of which we show only a few symbolic diagrams in Fig. 1. Two-loop 2PE graphs show up for the first time and so does three-pion exchange (3PE) which necessarily involves two loops. 3PE was found to be negligible at this order [22, 23]. Most importantly, 15 new contact terms $\sim Q^4$ arise and are represented by the four-nucleon-leg graph with a solid diamond. They include a quadratic spin-orbit term and contribute up to $D$-waves. Mainly due to the increased number of contact terms, a quantitative description of the two-nucleon interaction up to about 300 MeV lab. energy is possible, at N$^3$LO (for details, see below). Besides further 3NF, four-nucleon forces (4NF) start at this order. Since the leading 4NF come into existence one order higher than the leading 3NF, 4NF are weaker than 3NF. Thus, ChPT provides a straightforward explanation for the empirically known fact that 2NF $\gg$ 3NF $\gg$ 4NF ....

# 4. Two-Nucleon Interactions

The last section was just an overview. In this section, we will fill in all the details involved in the ChPT development of the $NN$ interaction; and 3NF and 4NF will be discussed in Section 5. We start by talking about the various pion-exchange contributions.

## 4.1. Pion-Exchange Contributions in ChPT

Based upon the effective pion Lagrangians of Section 2.2, we will now derive the pion-exchange contributions to the $NN$ interaction order by order.

As noted before, there are infinitely many pion-exchange contributions to the $NN$ interaction and, thus, we need to get organized. First, we arrange the various pion-exchange contributions according to the number of pions being exchanged between the two nucleons:

$$V_\pi = V_{1\pi} + V_{2\pi} + V_{3\pi} + \ldots , \qquad (26)$$

where the meaning of the subscripts is obvious and the ellipsis represents $4\pi$ and higher pion exchanges. Second, for each of the above terms, we assume a low-momentum expansion:

$$V_{1\pi} = V_{1\pi}^{(0)} + V_{1\pi}^{(2)} + V_{1\pi}^{(3)} + V_{1\pi}^{(4)} + \ldots \tag{27}$$

$$V_{2\pi} = V_{2\pi}^{(2)} + V_{2\pi}^{(3)} + V_{2\pi}^{(4)} + \ldots \tag{28}$$

$$V_{3\pi} = V_{3\pi}^{(4)} + \ldots , \tag{29}$$

where the superscript denotes the order $\nu$ and the ellipses stand for contributions of fifth and higher orders. Due to parity and time-reversal, there are no first order contributions. Moreover, since $n$ pions create $L = n - 1$ loops, the leading order for $n$-pion exchange ocurrs at $\nu = 2n - 2$ [cf. Eq. (25)].

In the following subsections, we will discuss $V_{1\pi}$, $V_{2\pi}$, and $V_{3\pi}$, one by one and order by order.

#### 4.1.1. One-Pion Exchange

At leading order (LO, $\nu = 0$), we have the well-known static one-pion exchange (1PE), which is given by [for notation, see Eq. (32) below]:

$$V_{1\pi}(\vec{p}\,', \vec{p}) = -\frac{g_A^2}{4f_\pi^2}\, \boldsymbol{\tau}_1 \cdot \boldsymbol{\tau}_2\, \frac{\vec{\sigma}_1 \cdot \vec{q}\, \vec{\sigma}_2 \cdot \vec{q}}{q^2 + m_\pi^2} . \tag{30}$$

On-shell and in the center-of-mass system (CMS), there are no relativistic corrections at any order. Off-shell corrections come into play when the relativistic 1PE is iterated, i.e., in four-dimensional planar box diagrams. These corrections will be taken into account in the evaluation of those box diagrams up to the given order (see below).

We use $g_A = 1.290$ (instead of $g_A = 1.276$ [24]) to account for the so-called Goldberger-Treiman discrepancy. Via the Goldberger-Treiman relation, $g_{\pi NN} = g_A M_N / f_\pi$, our value for $g_A$ together with $f_\pi = 92.4$ MeV and $M_N = 938.918$ MeV implies $g_{\pi NN}^2/4\pi = 13.67$ which is consistent with the empirical values obtained from $\pi N$ and $NN$ data analysis [25, 26]. The renormalizations of $f_\pi$, $m_\pi$, and $M_N$ are taken care of by working with their physical values.

#### 4.1.2. Two-Pion Exchange

The exchange of two or more pions always involves loop diagrams, which implies that we are faced with a non-trivial problem. To conduct such calculations, a scheme of field-theoretic perturbation theory as well as regularization and renormalization methods for dealing with divergent loop integrals must be adopted. Ordóñez, Ray, and van Kolck [19], who were the first to calculate chiral $2\pi$-exchange to NNLO, used time-ordered perturbation theory and a Gaussian cutoff function for regularization. The Juelich group [27] developed a method involving unitary transformations (to get rid of the energy-dependence of time-ordered perturbation theory) and applied so-called spectral function regularization. The Munich group [28] used covariant perturbation theory and dimensional regularization. We will follow here the method chosen by the Munich group since we believe it to be the most efficient and elegant one. In this approach, one starts with the relativistic versions of the $\pi N$ Lagrangians and sets up four-dimensional (covariant) loop integrals. Relativistic vertices

NLO

$(Q/\Lambda_\chi)^2$

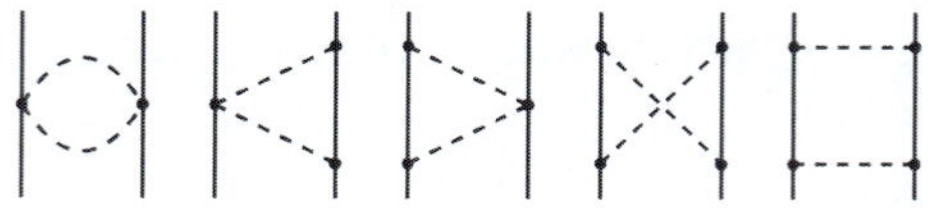

Figure 2. Leading two-pion exchange contributions to the $NN$ interaction. Notation as in Fig. 1.

and nucleon propagators are then expanded in powers of $1/M_N$. The divergences that occur in conjunction with the four-dimensional loop integrals are treated by means of dimensional regularization, a prescription which is consistent with chiral symmetry and power counting. The results derived in this way are the same obtained when starting right away with the heavy-baryon versions of the $\pi N$ Lagrangians [18]. However, as it turns out, the method used by the Munich group is more efficient in dealing with the rather tedious calculations and particularly useful in conjunction with the planar box diagram (see Appendix B of Ref. [12] for details).

The results will be stated in terms of contributions to the momentum-space $NN$ amplitude in the CMS, which takes the general form

$$\begin{aligned} V(\vec{p}\,',\vec{p}) &= V_C + \boldsymbol{\tau}_1\cdot\boldsymbol{\tau}_2 W_C \\ &+ [V_S + \boldsymbol{\tau}_1\cdot\boldsymbol{\tau}_2 W_S]\,\vec{\sigma}_1\cdot\vec{\sigma}_2 \\ &+ [V_{LS} + \boldsymbol{\tau}_1\cdot\boldsymbol{\tau}_2 W_{LS}]\left(-i\vec{S}\cdot(\vec{q}\times\vec{k})\right) \\ &+ [V_T + \boldsymbol{\tau}_1\cdot\boldsymbol{\tau}_2 W_T]\,\vec{\sigma}_1\cdot\vec{q}\,\vec{\sigma}_2\cdot\vec{q} \\ &+ [V_{\sigma L} + \boldsymbol{\tau}_1\cdot\boldsymbol{\tau}_2 W_{\sigma L}]\,\vec{\sigma}_1\cdot(\vec{q}\times\vec{k})\,\vec{\sigma}_2\cdot(\vec{q}\times\vec{k})\,, \end{aligned} \tag{31}$$

where $\vec{p}\,'$ and $\vec{p}$ denote the final and initial nucleon momenta in the CMS, respectively; moreover,

$$\begin{aligned} \vec{q} &\equiv \vec{p}\,' - \vec{p} && \text{is the momentum transfer,} \\ \vec{k} &\equiv \tfrac{1}{2}(\vec{p}\,' + \vec{p}) && \text{the average momentum,} \\ \vec{S} &\equiv \tfrac{1}{2}(\vec{\sigma}_1 + \vec{\sigma}_2) && \text{the total spin,} \end{aligned} \tag{32}$$

and $\vec{\sigma}_{1,2}$ and $\boldsymbol{\tau}_{1,2}$ are the spin and isospin operators, respectively, of nucleon 1 and 2. For on-energy-shell scattering, $V_\alpha$ and $W_\alpha$ ($\alpha = C, S, LS, T, \sigma L$) can be expressed as functions of $q$ and $k$ (with $q \equiv |\vec{q}|$ and $k \equiv |\vec{k}|$), only.

In all expressions given below, we will state only the *nonpolynomial* contributions to the $NN$ amplitude. Note, however, that dimensional regularization typically generates also polynomial terms which are, in part, infinite or scale dependent. These polynomials are absorbed by the contact interactions to be discussed in a later section.

**Next-to-leading order (NLO)** The leading two-pion exchange appears at second order ($\nu = 2$, next-to-leading order, NLO) and is shown in Fig. 2. Since a loop creates already $\nu = 2$ [cf. Eq. (25)], the vertices involved at this order have index $\Delta_i = 0$, i.e., they are from the leading order Lagrangian $\widehat{\mathcal{L}}^{\Delta=0}$, Eq. (17), where the $\pi N$ vertices carry only one derivative. These vertices are denoted by small dots in the figures. The rather complicated evaluation of these diagrams can be found in Appendix B of Ref. [12].

**NNLO**
$(Q/\Lambda_\chi)^3$

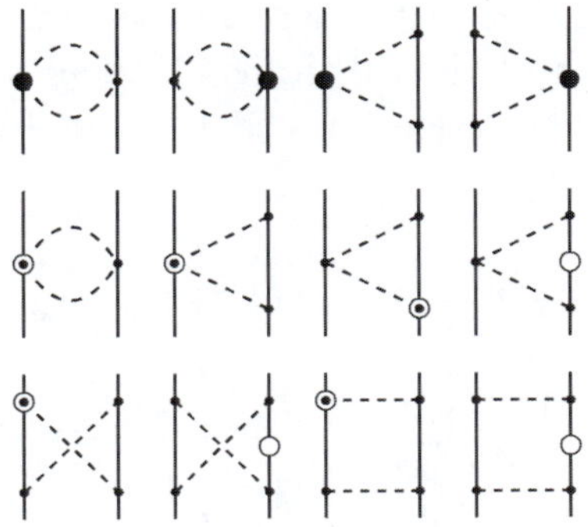

Figure 3. Two-pion exchange contributions to the $NN$ interaction at order three in small momenta (NNLO). Basic notation as in Fig. 1. Large solid dots denote vertices from the Lagrangian $\widehat{\mathcal{L}}^{\Delta=1}$, Eq. (18), proportional to the LECs $c_i$. Symbols with an open circles are relativistic $1/M_N$ corrections which are also part of $\widehat{\mathcal{L}}^{\Delta=1}$. Only a few representative examples of $1/M_N$ corrections are shown. Note that all football diagrams shown in this figure vanish (i.e., the first and second diagram in the first row, and the first diagram in the second row).

Concerning the planar box diagram in Fig. 2, we should note that we include only the non-iterative part of this diagram which is obtained by subtracting the iterated 1PE contribution. Summarizing all contributions from irreducible two-pion exchange at second order, one obtains:

$$W_C = -\frac{L(q)}{384\pi^2 f_\pi^4}\left[4m_\pi^2(5g_A^4 - 4g_A^2 - 1) + q^2(23g_A^4 - 10g_A^2 - 1) + \frac{48g_A^4 m_\pi^4}{w^2}\right] \quad (33)$$

$$V_T = -\frac{1}{q^2}V_S = -\frac{3g_A^4 L(q)}{64\pi^2 f_\pi^4}, \quad (34)$$

where

$$L(q) \equiv \frac{w}{q}\ln\frac{w+q}{2m_\pi} \quad (35)$$

and

$$w \equiv \sqrt{4m_\pi^2 + q^2}\,. \quad (36)$$

This part of the 2PE is rather weak and insufficient to properly describe the $NN$ interaction at intermediate range.

**Next-to-next-to-leading order (NNLO)** The two-pion exchange diagrams of order three ($\nu = 3$, next-to-next-to-leading order, NNLO) are very similar to the ones of order two, except that they contain one insertion from $\widehat{\mathcal{L}}^{\Delta=1}$, Eq. (18). The resulting contributions are typically either proportional to one of the low-energy constants $c_i$ or they contain a factor $1/M_N$. Notice that relativistic $1/M_N$ corrections derive from vertices and nucleon propagators. In Fig. 3, we show in row one the diagrams with one vertex proportional to $c_i$ (large solid dot), and in row two and three a few representative graphs with a $1/M_N$ correction (symbols with an open circle). The number of $1/M_N$ correction graphs is large and not all are shown. Note that all football diagrams vanish at this order, because the loop integrals involve odd powers of the time-component of the loop momentum. Again, the planar box diagram is corrected for a contribution from the iterated 1PE. One obtains for the full third order [28]:

$$V_C = \frac{3g_A^2}{16\pi f_\pi^4}\left\{\frac{g_A^2 m_\pi^5}{16M_N w^2} - \left[2m_\pi^2(2c_1 - c_3) - q^2\left(c_3 + \frac{3g_A^2}{16M_N}\right)\right]\widetilde{w}^2 A(q)\right\} \quad (37)$$

$$W_C = \frac{g_A^2}{128\pi M_N f_\pi^4}\left\{3g_A^2 m_\pi^5 w^{-2} - \left[4m_\pi^2 + 2q^2 - g_A^2(4m_\pi^2 + 3q^2)\right]\widetilde{w}^2 A(q)\right\} \quad (38)$$

$$V_T = -\frac{1}{q^2}V_S = \frac{9g_A^4 \widetilde{w}^2 A(q)}{512\pi M_N f_\pi^4}, \quad (39)$$

$$W_T = -\frac{1}{q^2}W_S = -\frac{g_A^2 A(q)}{32\pi f_\pi^4}\left[\left(c_4 + \frac{1}{4M_N}\right)w^2 - \frac{g_A^2}{8M_N}(10m_\pi^2 + 3q^2)\right], \quad (40)$$

$$V_{LS} = \frac{3g_A^4 \widetilde{w}^2 A(q)}{32\pi M_N f_\pi^4}, \quad (41)$$

$$W_{LS} = \frac{g_A^2(1 - g_A^2)}{32\pi M_N f_\pi^4} w^2 A(q), \quad (42)$$

with

$$A(q) \equiv \frac{1}{2q}\arctan\frac{q}{2m_\pi} \quad (43)$$

and

$$\widetilde{w} \equiv \sqrt{2m_\pi^2 + q^2}. \quad (44)$$

This contribution to the 2PE is the crucial one, because it provides an intermediate-range attraction of proper strength. The iso-scalar central potential, $V_C$, is strong and attractive due to the LEC $c_3$, which is negative and of large magnitude (cf. Table 5, below). Via resonance saturation, $c_3$ is associated with $\pi$-$\pi$ correlations ('$\sigma$ meson') and virtual $\Delta$-isobar excitations, which create the most crucial contributions to 2PE in the frame work of conventional meson theory [5, 8]. The configuration-space expressions, which correspond to the above momentum-space potentials, are given in Ref. [28], where also a detailed comparison with meson-exchange potentials is conducted.

If the relativistic Blackenbecler-Sugar (BbS) equation [12] is used as the scattering equation, the $1/M_N$ terms are slightly different. The changes are taken care of by adding to Eqs. (37)-(40) the following terms:

$$V_C = -\frac{3g_A^4}{256\pi f_\pi^4 M_N}(m_\pi w^2 + \widetilde{w}^4 A(q)), \quad (45)$$

$$W_C = \frac{g_A^4}{128\pi f_\pi^4 M_N}(m_\pi w^2 + \widetilde{w}^4 A(q)), \quad (46)$$

$$V_T = -\frac{1}{q^2}V_S = \frac{3g_A^4}{512\pi f_\pi^4 M_N}(m_\pi + w^2 A(q)), \quad (47)$$

$$W_T = -\frac{1}{q^2}W_S = -\frac{g_A^4}{256\pi f_\pi^4 M_N}(m_\pi + w^2 A(q)). \quad (48)$$

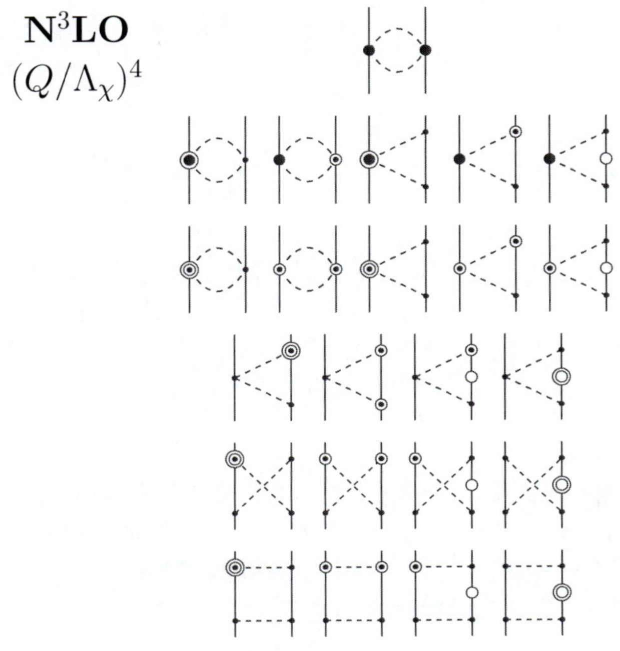


Figure 4. One-loop $2\pi$-exchange contributions to the $NN$ interaction at order four. Notation as in Fig. 3. Moreover, symbols with a large solid dot and an open circle denote $1/M_N$ corrections of vertices proportional to $c_i$. Symbols with two open circles mark relativistic $1/M_N^2$ corrections. Both corrections are part of the Lagrangian $\widehat{\mathcal{L}}^{\Delta=2}$, Eq. (19). Representative examples for all types of one-loop graphs that occur at this order are shown.

**Next-to-next-to-next-to-leading order (N$^3$LO)** This order ($\nu = 4$) is very involved. The contributions can be subdivided into two groups, one-loop graphs, Fig. 4, and two-loop diagrams, Fig. 5. Applying Eq. (25), it is easy to verify that all contributions result in $\nu = 4$. The comprehensive mathematical expressions of this order can be found in Appendix D of Ref. [12]. The net effect of this contribution is small indicating a trend towards convergence.

#### 4.1.3. Three-Pion Exchange

Since the exchange of three pions involves two loops, three-pion exchange (3PE) starts at order four (N$^3$LO). Two loops generate $\nu = 4$ and, therefore, at this order, all vertices have to be from the leading order Lagrangian $\widehat{\mathcal{L}}^{\Delta=0}$, Eq. (17). The 3PE contributions at N$^3$LO have been calculated by the Munich group and found to be negligible [22, 23]. This is not surprising since only the leading vertices are involved which are known to be weak. As discussed, for similar reasons, the leading 2PE also turned out to be rather small (even though not negligible). We will ignore the 3PE at N$^3$LO because of its negligible strength.

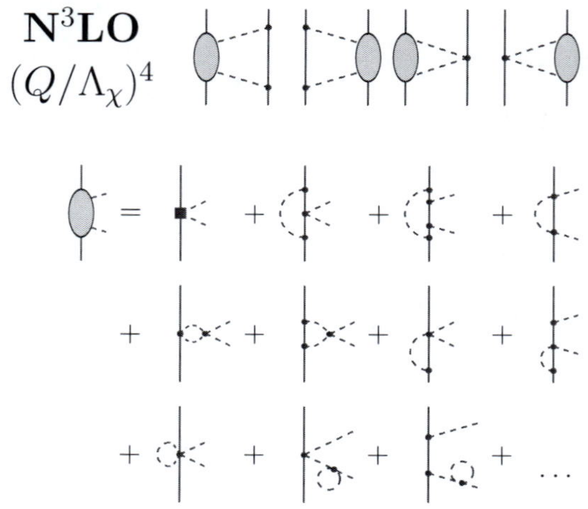


Figure 5. Two-loop $2\pi$-exchange contributions at order four. Basic notation as in Fig. 1. The oval stands for all one-loop $\pi N$ graphs some of which are shown in the lower part of the figure. The solid square represents vertices proportional to the LECs $d_i$ introduced in $\widehat{\mathcal{L}}^{(3)}_{\pi N}$ which is part of $\widehat{\mathcal{L}}^{\Delta=2}$, Eq. (19).

### 4.2. $NN$ Contact Terms

Pion-exchange contributions describe the long- and intermediate-range parts of the nuclear force, which rule the peripheral partial waves. However, for a "complete" nuclear force, we have to describe correctly all partial waves, including the lower ones. In fact, in calculations of $NN$ observables at low energies (cross sections, analyzing powers, etc.), the lower partial waves with $L \leq 2$ are the most important ones, generating the largest contributions. The same is true for microscopic nuclear structure calculations. These lower partial waves are dominated by the dynamics at short distances. Therefore, we need to look now into the short-range part of the $NN$ potential.

In conventional meson theory, the short-range nuclear force is described by the exchange of heavy mesons, notably the $\omega(782)$. Qualitatively, the short-distance behavior of the $NN$ potential is obtained by Fourier transform of the propagator of a heavy meson,

$$\int d^3q \frac{e^{i\vec{q}\cdot\vec{r}}}{m_\omega^2 + \vec{q}^2} \sim \frac{e^{-m_\omega r}}{r} . \tag{49}$$

ChPT is an expansion in small momenta $Q$, too small to resolve structures like a $\rho(770)$ or $\omega(782)$ meson, because $Q \ll \Lambda_\chi \approx m_{\rho,\omega}$. But the latter relation allows us to expand the propagator of a heavy meson into a power series,

$$\frac{1}{m_\omega^2 + Q^2} \approx \frac{1}{m_\omega^2}\left(1 - \frac{Q^2}{m_\omega^2} + \frac{Q^4}{m_\omega^4} - + \dots\right), \tag{50}$$

where the $\omega$ is representative for any heavy meson of interest. The above expansion suggests that it should be possible to describe the short distance part of the nuclear force simply in terms of powers of $Q/m_\omega$, which fits in well with our over-all power expansion since $Q/m_\omega \approx Q/\Lambda_\chi$.

A second purpose of contact terms is renormalization. Dimensional regularization of the loop integrals that occur in multi-pion exchange diagrams typically generates polynomial terms with coefficients that are, in part, infinite or scale dependent (cf. Appendix B of Ref. [12]). Contact terms pick up infinities and remove scale dependences, which is why they are also known as counter terms.

The partial-wave decomposition of a power $Q^\nu$ has an interesting property. First note that $Q$ can only be either the momentum transfer between the two interacting nucleons $q$ or the average momentum $k$ [cf. Eq. (32) for their definitions]. In any case, for even $\nu$,

$$Q^\nu = f_{\frac{\nu}{2}}(\cos\theta)\,, \tag{51}$$

where $f_m$ stands for a polynomial of degree $m$ and $\theta$ is the CMS scattering angle. The partial-wave decomposition of $Q^\nu$ for a state of orbital-angular momentum $L$ involves the integral

$$I_L^{(\nu)} = \int_{-1}^{+1} Q^\nu P_L(\cos\theta) d\cos\theta = \int_{-1}^{+1} f_{\frac{\nu}{2}}(\cos\theta) P_L(\cos\theta) d\cos\theta\,, \tag{52}$$

where $P_L$ is a Legendre polynomial. Due to the orthogonality of the $P_L$,

$$I_L^{(\nu)} = 0 \quad \text{for} \quad L > \frac{\nu}{2}\,. \tag{53}$$

Consequently, contact terms of order zero contribute only in $S$-waves, while order-two terms contribute up to $P$-waves, order-four terms up to $D$-waves, etc..

Due to parity, only even powers of $Q$ are allowed. Thus, the expansion of the contact potential is formally given by

$$V_{\text{ct}} = V_{\text{ct}}^{(0)} + V_{\text{ct}}^{(2)} + V_{\text{ct}}^{(4)} + \ldots\,, \tag{54}$$

where the supersript denotes the power or order.

We will now present, one by one, the various orders of $NN$ contact terms which result from the contact Lagrangians shown at the end of Section 2.2.

### 4.2.1. Zeroth Order (LO)

The contact Lagrangian $\widehat{\mathcal{L}}_{NN}^{(0)}$, Eq. (21), which is part of $\widehat{\mathcal{L}}^{\Delta=0}$, Eq. (17), leads to the following $NN$ contact potential,

$$V_{\text{ct}}^{(0)}(\vec{p}\,', \vec{p}) = C_S + C_T\,\vec{\sigma}_1 \cdot \vec{\sigma}_2\,, \tag{55}$$

and, in terms of partial waves, we have

$$\begin{aligned} V_{\text{ct}}^{(0)}(^1S_0) &= \widetilde{C}_{^1S_0} = 4\pi\,(C_S - 3\,C_T) \\ V_{\text{ct}}^{(0)}(^3S_1) &= \widetilde{C}_{^3S_1} = 4\pi\,(C_S + C_T)\,. \end{aligned} \tag{56}$$

### 4.2.2. Second Order (NLO)

The contact Lagrangian $\widehat{\mathcal{L}}^{(2)}_{NN}$, Eq. (22), which is part of $\widehat{\mathcal{L}}^{\Delta=2}$, Eq. (19), generates the following $NN$ contact potential

$$\begin{aligned} V_{\rm ct}^{(2)}(\vec{p}',\vec{p}) &= C_1\, q^2 + C_2\, k^2 \\ &+ \left(C_3\, q^2 + C_4\, k^2\right) \vec{\sigma}_1 \cdot \vec{\sigma}_2 \\ &+ C_5 \left(-i\vec{S}\cdot(\vec{q}\times\vec{k})\right) \\ &+ C_6\, (\vec{\sigma}_1\cdot\vec{q})\,(\vec{\sigma}_2\cdot\vec{q}) \\ &+ C_7\, (\vec{\sigma}_1\cdot\vec{k})\,(\vec{\sigma}_2\cdot\vec{k})\,. \end{aligned} \tag{57}$$

The coefficients $C_i$ used here in the contact potential are, of course, related to the coefficients $C'_i$ that occur in the Lagrangian $\widehat{\mathcal{L}}^{(2)}_{NN}$, Eq. (22). The relation, which is unimportant for our purposes, can be found in Refs. [19, 27].

There are many ways to perform the partial-wave decomposition of the above potential. We perceive the method presented by Erkelenz, Alzetta, and Holinde [29] as the most elegant one. Thus, one obtains

$$\begin{aligned} V_{\rm ct}^{(2)}(^1S_0) &= C_{^1S_0}(p^2+p'^2) \\ V_{\rm ct}^{(2)}(^3P_0) &= C_{^3P_0}\, pp' \\ V_{\rm ct}^{(2)}(^1P_1) &= C_{^1P_1}\, pp' \\ V_{\rm ct}^{(2)}(^3P_1) &= C_{^3P_1}\, pp' \\ V_{\rm ct}^{(2)}(^3S_1) &= C_{^3S_1}(p^2+p'^2) \\ V_{\rm ct}^{(2)}(^3S_1-{}^3D_1) &= C_{^3S_1-{}^3D_1}p^2 \\ V_{\rm ct}^{(2)}(^3D_1-{}^3S_1) &= C_{^3S_1-{}^3D_1}p'^2 \\ V_{\rm ct}^{(2)}(^3P_2) &= C_{^3P_2}\, pp' \end{aligned} \tag{58}$$

with

$$\begin{aligned} C_{^1S_0} &= 4\pi\left(C_1+\frac{1}{4}C_2-3C_3-\frac{3}{4}C_4-C_6-\frac{1}{4}C_7\right) \\ C_{^3P_0} &= 4\pi\left(-\frac{2}{3}C_1+\frac{1}{6}C_2-\frac{2}{3}C_3+\frac{1}{6}C_4-\frac{2}{3}C_5+2C_6-\frac{1}{2}C_7\right) \\ C_{^1P_1} &= 4\pi\left(-\frac{2}{3}C_1+\frac{1}{6}C_2+2C_3-\frac{1}{2}C_4+\frac{2}{3}C_6-\frac{1}{6}C_7\right) \end{aligned}$$

$$
\begin{aligned}
C_{^3P_1} &= 4\pi\left(-\frac{2}{3}C_1+\frac{1}{6}C_2-\frac{2}{3}C_3+\frac{1}{6}C_4-\frac{1}{3}C_5-\frac{4}{3}C_6+\frac{1}{3}C_7\right)\\
C_{^3S_1} &= 4\pi\left(C_1+\frac{1}{4}C_2+C_3+\frac{1}{4}C_4+\frac{1}{3}C_6+\frac{1}{12}C_7\right)\\
C_{^3S_1-^3D_1} &= 4\pi\left(-\frac{2\sqrt{2}}{3}C_6-\frac{\sqrt{2}}{6}C_7\right)\\
C_{^3P_2} &= 4\pi\left(-\frac{2}{3}C_1+\frac{1}{6}C_2-\frac{2}{3}C_3+\frac{1}{6}C_4+\frac{1}{3}C_5\right)\,. \qquad (59)
\end{aligned}
$$

#### 4.2.3. Fourth Order (N$^3$LO)

The contact potential of order four reads

$$
\begin{aligned}
V_{\rm ct}^{(4)}(\vec{p}\,',\vec{p}) &= D_1\,q^4+D_2\,k^4+D_3\,q^2k^2+D_4\,(\vec{q}\times\vec{k})^2\\
&+\left(D_5\,q^4+D_6\,k^4+D_7\,q^2k^2+D_8\,(\vec{q}\times\vec{k})^2\right)\vec{\sigma}_1\cdot\vec{\sigma}_2\\
&+(D_9\,q^2+D_{10}\,k^2)\left(-i\vec{S}\cdot(\vec{q}\times\vec{k})\right)\\
&+(D_{11}\,q^2+D_{12}\,k^2)\,(\vec{\sigma}_1\cdot\vec{q})\,(\vec{\sigma}_2\cdot\vec{q})\\
&+(D_{13}\,q^2+D_{14}\,k^2)\,(\vec{\sigma}_1\cdot\vec{k})\,(\vec{\sigma}_2\cdot\vec{k})\\
&+D_{15}\left(\vec{\sigma}_1\cdot(\vec{q}\times\vec{k})\;\vec{\sigma}_2\cdot(\vec{q}\times\vec{k})\right). \qquad (60)
\end{aligned}
$$

The rather lengthy partial-wave expressions of this order are given in Appendix E of Ref. [12].

### 4.3. Definition of $NN$ Potential

We have now rounded up everything needed for a realistic nuclear force—long, intermediate, and short ranged components—and so we can finally proceed to the lower partial waves. However, here we encounter another problem. The two-nucleon system at low angular momentum, particularly, in $S$ waves, is characterized by the presence of a shallow bound state (the deuteron) and large scattering lengths. Thus, perturbation theory does not apply. In contrast to $\pi$-$\pi$ and $\pi$-$N$, the interaction between nucleons is not suppressed in the chiral limit ($Q \to 0$). Weinberg [10] showed that the strong enhancement of the scattering amplitude arises from purely nucleonic intermediate states ("infrared enhancement"). He therefore suggested to use perturbation theory to calculate the $NN$ potential (i.e., the irreducible graphs) and to apply this potential in a scattering equation to obtain the $NN$ amplitude. We will follow this prescription and discuss potential problems in the next subsection.

The full $NN$ potential is given by,

$$
V = V_{1\pi} + V_{2\pi} + V_{\rm ct}\,. \qquad (61)
$$

with

$$V_{2\pi} = V_{2\pi}^{(2)} + V_{2\pi}^{(3)} + V_{2\pi}^{(4)} + \ldots , \tag{62}$$

which was calculated in Section 4.1.2: $V_{2\pi}^{(2)}$ is given by the contributions of Eqs. (33) and (34), $V_{2\pi}^{(3)}$ is made up from Eqs. (37)-(48), and $V_{2\pi}^{(4)}$ is given in Appendix D of Ref. [12].

The contact potentials come in even orders,

$$V_{\rm ct} = V_{\rm ct}^{(0)} + V_{\rm ct}^{(2)} + V_{\rm ct}^{(4)} + \ldots , \tag{63}$$

and were presented in Section 4.2.

In summary, the $NN$ potential $V$, calculated to certain orders, is given by:

$$V_{\rm LO} = V_{1\pi} + V_{\rm ct}^{(0)} \tag{64}$$

$$V_{\rm NLO} = V_{\rm LO} + V_{2\pi}^{(2)} + V_{\rm ct}^{(2)} \tag{65}$$

$$V_{\rm NNLO} = V_{\rm NLO} + V_{2\pi}^{(3)} \tag{66}$$

$$V_{\rm N^3LO} = V_{\rm NNLO} + V_{2\pi}^{(4)} + V_{\rm ct}^{(4)} \tag{67}$$

The potential $V$ satisfies the relativistic BbS equation [12], which reads explicitly,

$$T(\vec{p}\,',\vec{p}) = V(\vec{p}\,',\vec{p}) + \int \frac{d^3p''}{(2\pi)^3}\, V(\vec{p}\,',\vec{p}\,'')\, \frac{M_N^2}{E_{p''}}\, \frac{1}{p^2 - p''^2 + i\epsilon}\, T(\vec{p}\,'',\vec{p}) \tag{68}$$

with $E_{p''} \equiv \sqrt{M_N^2 + p''^2}$. The advantage of using a relativistic scattering equation is that it automatically includes relativistic corrections to all orders. Thus, in the scattering equation, no propagator modifications are necessary when raising the order to which the calculation is conducted.

Defining

$$\widehat{V}(\vec{p}\,',\vec{p}) \equiv \frac{1}{(2\pi)^3} \sqrt{\frac{M_N}{E_{p'}}}\; V(\vec{p}\,',\vec{p})\; \sqrt{\frac{M_N}{E_p}} \tag{69}$$

and

$$\widehat{T}(\vec{p}\,',\vec{p}) \equiv \frac{1}{(2\pi)^3} \sqrt{\frac{M_N}{E_{p'}}}\; T(\vec{p}\,',\vec{p})\; \sqrt{\frac{M_N}{E_p}} , \tag{70}$$

where the factor $1/(2\pi)^3$ is added for convenience, the BbS equation collapses into the usual, nonrelativistic Lippmann-Schwinger (LS) equation,

$$\widehat{T}(\vec{p}\,',\vec{p}) = \widehat{V}(\vec{p}\,',\vec{p}) + \int d^3p''\, \widehat{V}(\vec{p}\,',\vec{p}\,'')\, \frac{M_N}{p^2 - p''^2 + i\epsilon}\, \widehat{T}(\vec{p}\,'',\vec{p}) . \tag{71}$$

Since $\widehat{V}$ satisfies Eq. (71), it can be used like a usual nonrelativistic potential, and $\widehat{T}$ may be perceived as the conventional nonrelativistic T-matrix. In applications, it is more convenient to use the K-matrix instead of the T-matrix and to have the LS equation decomposed into partial waves: all these technical issues are explained in detail in Appendix A of Ref. [6] where also the formulas for the calculation of $np$ and $pp$ (the latter with Coulomb) phase shifts are provided. The partial wave decomposition of the operators by which the potential is represented can be found in section 4 of Ref. [29], and numerical methods for solving the LS equation are explained in Ref. [30].

### 4.4. Regularization and Non-perturbative Renormalization

Iteration of $\widehat{V}$ in the LS equation, Eq. (71), requires cutting $\widehat{V}$ off for high momenta to avoid infinities. This is consistent with the fact that ChPT is a low-momentum expansion which is valid only for momenta $Q \ll \Lambda_\chi \approx 1$ GeV. Therefore, the potential $\widehat{V}$ is multiplied with the regulator function $f(p', p)$,

$$\widehat{V}(\vec{p}\,', \vec{p}) \longmapsto \widehat{V}(\vec{p}\,', \vec{p})\, f(p', p) \tag{72}$$

with

$$f(p', p) = \exp[-(p'/\Lambda)^{2n} - (p/\Lambda)^{2n}]\,, \tag{73}$$

such that

$$\widehat{V}(\vec{p}\,', \vec{p})\, f(p', p) \approx \widehat{V}(\vec{p}\,', \vec{p}) \left\{ 1 - \left[ \left(\frac{p'}{\Lambda}\right)^{2n} + \left(\frac{p}{\Lambda}\right)^{2n} \right] + \ldots \right\} . \tag{74}$$

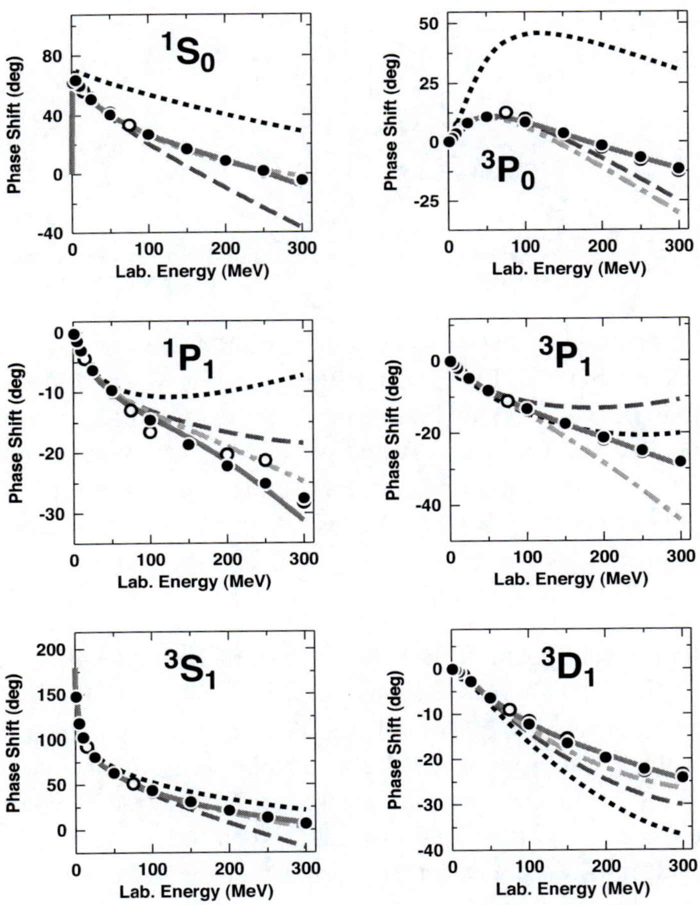


Figure 6. Continued on next page.

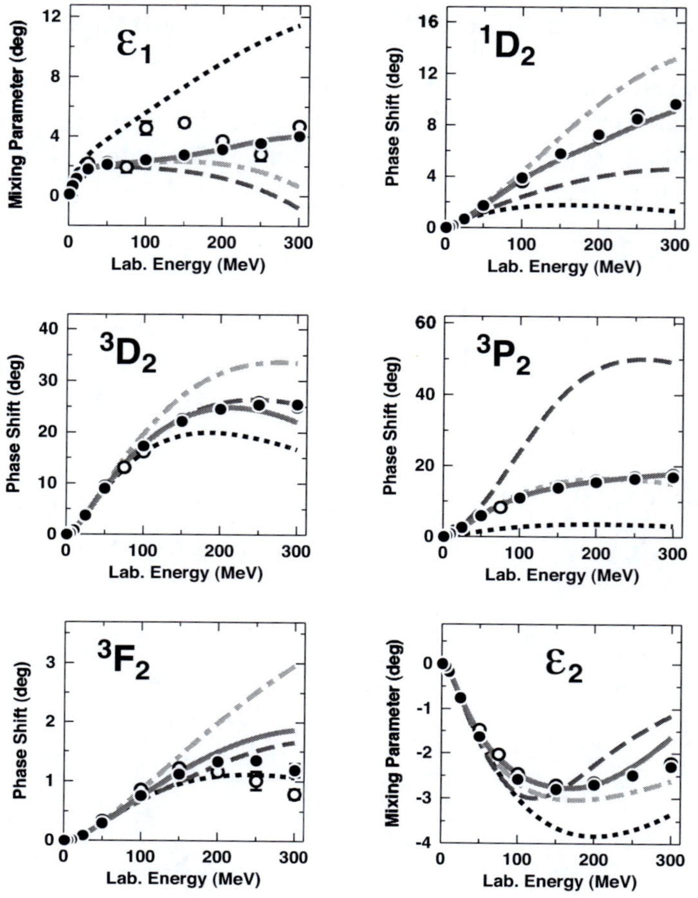


Figure 6. Phase shifts of $np$ scattering as calculated from $NN$ potentials at different orders of ChPT. The black dotted line is LO(500), the blue dashed is NLO(550/700) [31], the green dash-dotted NNLO(600/700) [31], and the red solid N$^3$LO(500) [32], where the numbers in parentheses denote the cutoffs in MeV. Partial waves with total angular momentum $J \leq 2$ are displayed. The solid dots and open circles are the results from the Nijmegen multi-energy $np$ phase shift analysis [33] and the VPI/GWU single-energy $np$ analysis SM99 [34], respectively.

Typical choices for the cutoff parameter $\Lambda$ that appears in the regulator are $\Lambda \approx 0.5$ GeV $< \Lambda_\chi \approx 1$ GeV. At N$^3$LO, an appropriate choice for $n$ is $n = 3$.

Equation (74) provides an indication of the fact that the exponential cutoff does not necessarily affect the given order at which the calculation is conducted. For sufficiently large $n$, the regulator introduces contributions that are beyond the given order. Assuming a good rate of convergence of the chiral expansion, such orders are small as compared to the given order and, thus, do not affect the accuracy at the given order. In calculations, one uses, of course, the exponential form, Eq. (73), and not the expansion Eq. (74). On a similar note, we also do not expand the square-root factors in Eqs. (69-70) because they are kinematical factors which guarantee relativistic elastic unitarity.

**Table 1. Columns three and four show the $\chi^2$/datum for the reproduction of the 1999 $np$ database [35] (subdivided into energy intervals) by families of $np$ potentials at NLO and NNLO constructed by the Juelich group [31]. The $\chi^2$/datum is stated in terms of ranges which result from a variation of the cutoff parameters used in the regulator functions. The values of these cutoff parameters in units of MeV are given in parentheses. $T_{\rm lab}$ denotes the kinetic energy of the incident neutron in the laboratory system**

| $T_{\rm lab}$ bin (MeV) | # of $np$ data | — *Juelich $np$ potentials* — NLO (550/700–400/500) | NNLO (600/700–450/500) |
|---|---|---|---|
| 0–100 | 1058 | 4–5 | 1.4–1.9 |
| 100–190 | 501 | 77–121 | 12–32 |
| 190–290 | 843 | 140–220 | 25–69 |
| 0–290 | 2402 | 67–105 | 12–27 |

It is pretty obvious that results for the $T$-matrix may depend sensitively on the regulator and its cutoff parameter. The removal of such regulator dependence is known as renormalization. Because of lack of space we cannot discuss this issue here and refer the interested reader to Section 4.5 of Ref. [12].

## 4.5. Constructing Quantitative Chiral $NN$ Potentials

### 4.5.1. What Order?

As discussed, the $NN$ potential can be calculated up to various orders, cf. Eqs. (64)-(67), and the accuracy increases as the order increases. That triggers the obvious question: To what order do we have to go for an accuracy that we would perceive as necessary and sufficient for, e. g., reliable microscopic nuclear structure calculations? How well ChPT converges in the important lower partial waves is demostrated in Fig. 6, where we show the $J \leq 2$ phase parameters for potentials constructed at LO, NLO, NNLO, and N$^3$LO. These figures clearly reveal substantial improvements in the reproduction of the empirical phase shifts as the orders go up.

There is an even better way to confront theory with experiment. We can calculate observables of $NN$ scattering and compare to the experimental data. It is customary to state the result of such a comparison in terms of the $\chi^2$/datum where a value around unity would signify a perfect fit.

Let's start with potentials developed to NLO and NNLO. In Table 1, we show the $\chi^2$/datum for the fit of the world $np$ data below 290 MeV for families of $np$ potentials at NLO and NNLO constructed by the Juelich group [31]. The NLO potentials produce the very large $\chi^2$/datum between 67 and 105, and the NNLO are between 12 and 27. The rate of improvement from one order to the other is very encouraging, but the quality of the reproduction of the $np$ data at NLO and NNLO is obviously insufficient for reliable predictions.

Based upon these facts, it has been pointed out in 2002 by Entem and Machleidt [36, 37]

**Table 2. Number of parameters needed for fitting the $np$ data in phase-shift analysis and by a high-precision $NN$ potential *versus* the total number of $NN$ contact terms of EFT based potentials to different orders**

| | Nijmegen partial-wave analysis [33] | CD-Bonn high-precision potential [6] | — *Contact Potentials* — $Q^0$ (LO) | $Q^2$ (NLO, NNLO) | $Q^4$ ($N^3$LO) |
|---|---|---|---|---|---|
| $^1S_0$ | 3 | 4 | 1 | 2 | 4 |
| $^3S_1$ | 3 | 4 | 1 | 2 | 4 |
| $^3S_1$-$^3D_1$ | 2 | 2 | 0 | 1 | 3 |
| $^1P_1$ | 3 | 3 | 0 | 1 | 2 |
| $^3P_0$ | 3 | 2 | 0 | 1 | 2 |
| $^3P_1$ | 2 | 2 | 0 | 1 | 2 |
| $^3P_2$ | 3 | 3 | 0 | 1 | 2 |
| $^3P_2$-$^3F_2$ | 2 | 1 | 0 | 0 | 1 |
| $^1D_2$ | 2 | 3 | 0 | 0 | 1 |
| $^3D_1$ | 2 | 1 | 0 | 0 | 1 |
| $^3D_2$ | 2 | 2 | 0 | 0 | 1 |
| $^3D_3$ | 1 | 2 | 0 | 0 | 1 |
| $^3D_3$-$^3G_3$ | 1 | 0 | 0 | 0 | 0 |
| $^1F_3$ | 1 | 1 | 0 | 0 | 0 |
| $^3F_2$ | 1 | 2 | 0 | 0 | 0 |
| $^3F_3$ | 1 | 2 | 0 | 0 | 0 |
| $^3F_4$ | 2 | 1 | 0 | 0 | 0 |
| $^3F_4$-$^3H_4$ | 0 | 0 | 0 | 0 | 0 |
| $^1G_4$ | 1 | 0 | 0 | 0 | 0 |
| $^3G_3$ | 0 | 1 | 0 | 0 | 0 |
| $^3G_4$ | 0 | 1 | 0 | 0 | 0 |
| $^3G_5$ | 0 | 1 | 0 | 0 | 0 |
| Total | 35 | 38 | 2 | 9 | 24 |

that one has to proceed to $N^3$LO. Consequently, the first $N^3$LO potential was published in 2003 [32].

At $N^3$LO, there are a total of 24 contact terms (24 parameters) which contribute to the partial waves with $L \leq 2$ (cf. Section 4.2). These 24 LECs are essentially free constants which parametrize the short-ranged phenomenological part of the interaction. In Table 2, column '$Q^4$ ($N^3$LO)', we show how these terms are distributed over the partial waves. Most important for the improved reproduction of the $NN$ phase shifts (and $NN$ observables) at $N^3$LO is the fact that contacts appear for the first time in $D$-waves. $D$-waves are not truely peripheral and, therefore, 1PE plus 2PE alone do not describe them well (Fig. 6). The $D$-wave contacts provide the necessary short-range corrections to get the $D$-phases right.

**Table 3. Columns three to five display the $\chi^2$/datum for the reproduction of the 1999 $np$ database [35] (subdivided into energy intervals) by various $np$ potentials. For the chiral potentials, the $\chi^2$/datum is stated in terms of ranges which result from a variation of the cutoff parameters used in the regulator functions. The values of these cutoff parameters in units of MeV are given in parentheses. $T_{\rm lab}$ denotes the kinetic energy of the incident nucleon in the laboratory system**

| $T_{\rm lab}$ bin (MeV) | # of $np$ data | *Idaho* N$^3$LO (500–600) [32] | *Juelich* N$^3$LO (600/700–450/500) [40] | Argonne $V_{18}$ [39] |
|---|---|---|---|---|
| 0–100 | 1058 | 1.0–1.1 | 1.0–1.1 | 0.95 |
| 100–190 | 501 | 1.1–1.2 | 1.3–1.8 | 1.10 |
| 190–290 | 843 | 1.2–1.4 | 2.8–20.0 | 1.11 |
| 0–290 | 2402 | 1.1–1.3 | 1.7–7.9 | 1.04 |

Besides this, at N$^3$LO, another contact is added to each $P$-wave, which leads to substantial improvements, particularly, in $^3P_0$ and $^3P_1$ above 100 MeV (cf. Fig. 6).

In Table 2, we also show the number of parameters used in the Nijmegen partial wave analysis (PWA93) [33] and in the high-precision CD-Bonn potential [6]. The table reveals that, for $S$ and $P$ waves, the number of parameters used in high-precision phenomenology and in EFT at N$^3$LO are about the same. **Thus, the EFT approach provides retroactively a justification for the phenomenology used in the 1990's to obtain high-precision fits.**

At NLO and NNLO, the number of contact parameters is substantially smaller than for PWA93 and CD-Bonn, which explains why these orders are insufficient for a quantitative potential. The 24 parameters of N$^3$LO are close to the 30+ used in PWA93 and high precision potentials. Consequently (see following sections for details), at N$^3$LO, a fit of the $NN$ data is possible that is of about the same quality as the one by the high-precision $NN$ potentials [6, 38, 39]. Thus, one may perceive N$^3$LO as the order of ChPT that is necessary and sufficient for a reliable $NN$ potential.

### 4.5.2. A Quantitative $NN$ Potential at N$^3$LO

After previous sections have thoroughly prepared the terrain, we will now present a quantitative $NN$ potential at N$^3$LO, including details of construction and results. We choose

**Table 4. Same as Table 3 but for $pp$**

| $T_{\rm lab}$ bin (MeV) | # of $pp$ data | *Idaho* N$^3$LO (500–600) [32] | *Juelich* N$^3$LO (600/700–450/500) [40] | Argonne $V_{18}$ [39] |
|---|---|---|---|---|
| 0–100 | 795 | 1.0–1.7 | 1.0–3.8 | 1.0 |
| 100–190 | 411 | 1.5–1.9 | 3.5–11.6 | 1.3 |
| 190–290 | 851 | 1.9–2.7 | 4.3–44.4 | 1.8 |
| 0–290 | 2057 | 1.5–2.1 | 2.9–22.3 | 1.4 |

the Idaho N$^3$LO potential [32] as example. The isospin symmetric part of the chiral $NN$ potential at N$^3$LO is defined in Eq. (67). Numerous parameters are involved which can be subdivided into three groups: $\pi N$ LECs (cf. Table 5, below), $NN$ contact parameters, and the cutoff parameter $\Lambda$ of the regulator Eq. (73). For $\Lambda$ we choose initially 500 MeV. Within a certain reasonable range, results should not depend sensitively on $\Lambda$ (cf. discussion in Section 4.4). Therefore, we have also made a second fit for $\Lambda = 600$ MeV.

The $\chi^2$/datum for the fit of the $np$ data below 290 MeV are shown in Table 3, and the corresponding ones for $pp$ are given in Table 4. These tables reveal that at N$^3$LO a $\chi^2$/datum comparable to the high-precision Argonne $V_{18}$ [39] potential can, indeed, be achieved. The Idaho N$^3$LO potential [32] with $\Lambda = 500$ MeV produces a $\chi^2$/datum = 1.1 for the world $np$ data below 290 MeV which compares well with the $\chi^2$/datum = 1.04 by the Argonne potential. In 2005, also the Juelich group produced several N$^3$LO $NN$ potentials [40], the best of which fits the $np$ data with a $\chi^2$/datum = 1.7 and the worse with 7.9 (Table 3).

Turning to $pp$, the $\chi^2$ for $pp$ data are typically larger than for $np$ because of the higher precision of $pp$ data. Thus, the Argonne $V_{18}$ produces a $\chi^2$/datum = 1.4 for the world $pp$ data below 290 MeV and the best Idaho N$^3$LO $pp$ potential obtains 1.5. The fit by the best

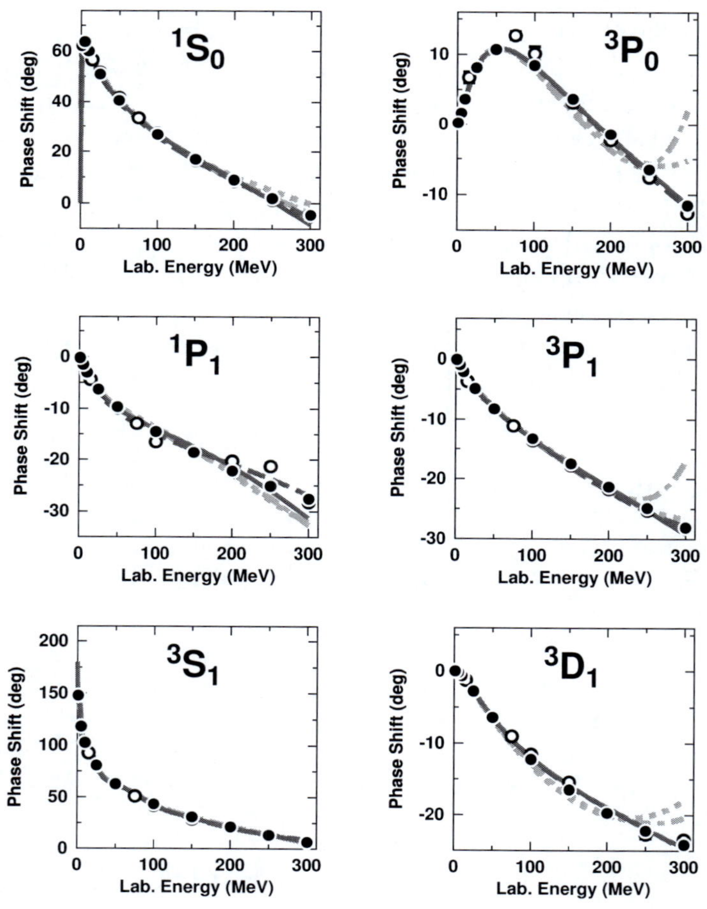


Figure 7. Continued on next page.

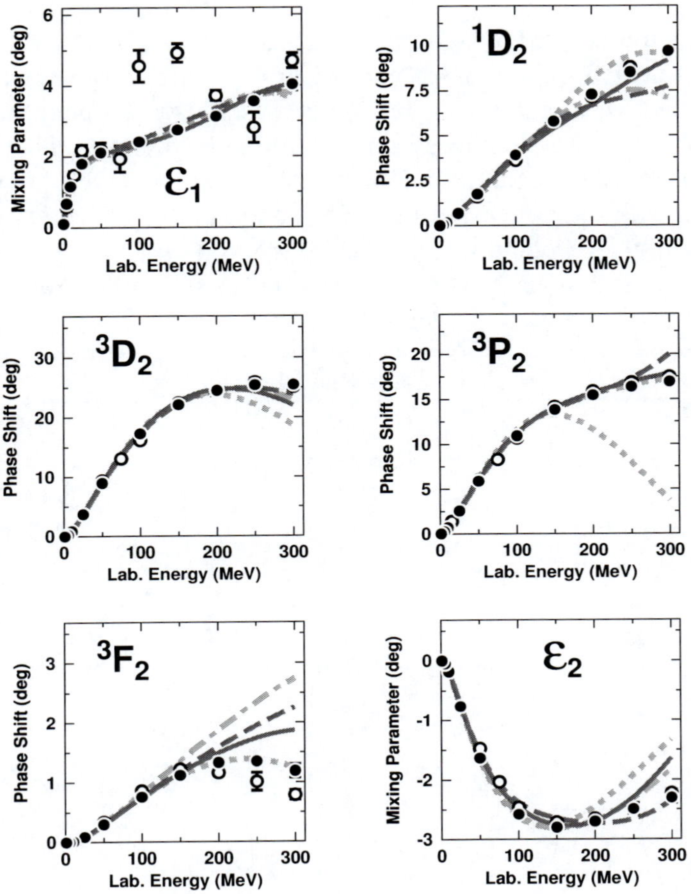


Figure 7. Neutron-proton phase parameters as described by various chiral potentials at $N^3LO$. The (red) solid and the dashed curves are calculated from Idaho $N^3LO$ potentials [32] with $\Lambda = 500$ and 600 MeV, respectively; while the (green) dash-dotted and the dotted curves are based upon Juelich $N^3LO$ potentials [40] with cutoff combinations 600/700 and 450/500 MeV, respectively. Partial waves with total angular momentum $J \leq 2$ are displayed. Empirical phase shifts (solid dots and open circles) as in Fig. 6

Juelich $N^3LO$ $pp$ potential results in a $\chi^2$/datum = 2.9 and the worst produces 22.3.

Phase shifts of $np$ scattering from two Idaho (solid and dashed lines) and two Juelich (dash-dotted and dotted lines) $N^3LO$ $np$ potentials are shown in Fig. 7. The phase shifts confirm what the corresponding $\chi^2$ have already revealed.

## 5. Nuclear Many-Body Forces

The chiral 2NF discussed in the previous section has been applied in microscopic calculations of nuclear structure with, in general, a great deal of success. However, from high-precision studies conducted in the 1990s, it is well-known that certain few-nucleon reactions and nuclear structure issues require 3NFs for their microscopic explanation. Outstanding

**Table 5. Hadron masses and low-energy constants (LECs). $m_{\pi^\pm}$ and $m_{\pi^0}$ denote the charged- and neutral-pion masses, and $M_p$ and $M_n$ are the proton and neutron masses, respectively, in units of MeV. $g_A$ is the axial-vector coupling constant (dimensionless) and $f_\pi$ the pion decay constant (in units of MeV). The $c_i$ belong to the dimension-two $\pi N$ Lagrangian, Eq. (18), and are in units of GeV$^{-1}$, whereas the $\bar{d}_i$ are associated with the dimension-three Lagrangian, Eq. (19), and are in units of GeV$^{-2}$. Column "$NN$ Potential" lists the values used for a $NN$ potential at N$^3$LO presented in Section 4.5.2. The last column displays values from empirical determinations**

| | $NN$ Potential (Sec. 4.5.2) | Empirical |
|---|---|---|
| $m_{\pi^\pm}$ | 139.5702 | 139.57018(35) [16] |
| $m_{\pi^0}$ | 134.9766 | 134.9766(6) [16] |
| $M_p$ | 938.2720 | 938.272013(23) [16] |
| $M_n$ | 939.5653 | 939.565346(23) [16] |
| $g_A$ | 1.29 | 1.2759(45) [24] |
| $f_\pi$ | 92.4 | $92.2 \pm 0.2$ [16] |
| $c_1$ | –0.81 | $-0.81 \pm 0.15$ [43] |
| $c_2$ | 2.80 | $3.28 \pm 0.23$ [44] |
| $c_3$ | –3.20 | $-4.69 \pm 1.34$ [43] |
| $c_4$ | 5.40 | $3.40 \pm 0.04$ [43] |
| $\bar{d}_1 + \bar{d}_2$ | 3.06 | $3.06 \pm 0.21$ [44] |
| $\bar{d}_3$ | –3.27 | $-3.27 \pm 0.73$ [44] |
| $\bar{d}_5$ | 0.45 | $0.45 \pm 0.42$ [44] |
| $\bar{d}_{14} - \bar{d}_{15}$ | –5.65 | $-5.65 \pm 0.41$ [44] |

examples are the $A_y$ puzzle of $N$-$d$ scattering [41] and the ground state of $^{10}$B [42]. As noted before, an important advantage of the EFT approach to nuclear forces is that it creates two- and many-nucleon forces on an equal footing (cf. the overview given in Fig. 1). In this section, we will now explain in some detail chiral three- and four-nucleon forces.

## 5.1. Three-Nucleon Forces

Nuclear three-body forces in ChPT were initially discussed by Weinberg [11] and the 3NF at NNLO was first derived by van Kolck [20].

For a 3NF, we have $A = 3$ and $C = 1$ and, thus, Eq. (23) implies

$$\nu = 2 + 2L + \sum_i \Delta_i \,. \tag{75}$$

We will use this equation to analyze 3NF contributions order by order.

### 5.1.1. Next-to-Leading Order

The lowest possible power is obviously $\nu = 2$ (NLO), which is obtained for no loops ($L = 0$) and only leading vertices ($\sum_i \Delta_i = 0$). As discussed by Weinberg [11] and van

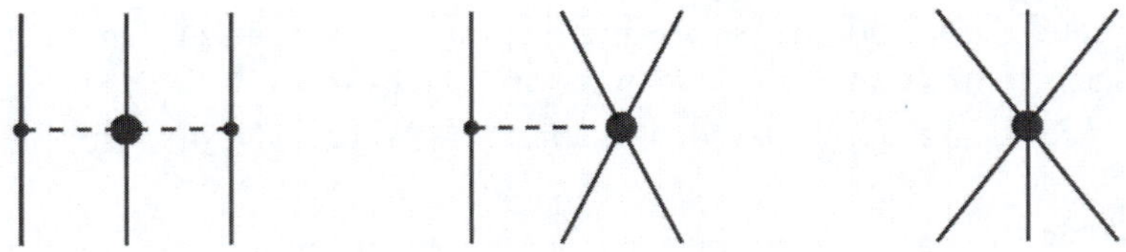

Figure 8. The three-nucleon force at NNLO. From left to right: 2PE, 1PE, and contact diagrams. Notation as in Fig. 1.

Kolck [20], the contributions from these diagrams vanish at NLO. So, the bottom line is that there is no genuine 3NF contribution at NLO. The first non-vanishing 3NF appears at NNLO.

### 5.1.2. Next-to-Next-to-Leading Order

The power $\nu = 3$ (NNLO) is obtained when there are no loops ($L = 0$) and $\sum_i \Delta_i = 1$, i.e., $\Delta_i = 1$ for one vertex while $\Delta_i = 0$ for all other vertices. There are three topologies which fulfill this condition, known as the two-pion exchange (2PE), 1PE, and contact graphs [20, 21] (Fig. 8).

The 2PE 3N-potential is derived to be

$$V^{\rm 3NF}_{\rm 2PE} = \left(\frac{g_A}{2f_\pi}\right)^2 \frac{1}{2} \sum_{i\neq j\neq k} \frac{(\vec\sigma_i \cdot \vec q_i)(\vec\sigma_j \cdot \vec q_j)}{(q_i^2+m_\pi^2)(q_j^2+m_\pi^2)} F^{ab}_{ijk}\, \tau^a_i \tau^b_j \tag{76}$$

with $\vec{q_i} \equiv \vec{p_i}' - \vec{p_i}$, where $\vec{p_i}$ and $\vec{p_i}'$ are the initial and final momenta of nucleon $i$, respectively, and

$$F^{ab}_{ijk} = \delta^{ab}\left[-\frac{4c_1 m_\pi^2}{f_\pi^2} + \frac{2c_3}{f_\pi^2}\, \vec q_i \cdot \vec q_j\right] + \frac{c_4}{f_\pi^2} \sum_c \epsilon^{abc}\, \tau^c_k\, \vec\sigma_k \cdot [\vec q_i \times \vec q_j]\ . \tag{77}$$

There are great similarities between this force and earlier derivations of 2PE 3NFs, notably the 50-year old Fujita-Miyazawa [45] and the Tucson-Melbourne (TM) [46] forces.

Note that the 2PE 3NF does not contain any new parameters, because the LECs $c_1$, $c_3$, and $c_4$ appear already in the 2PE 2NF (Section 4.1.2) and are fixed by $\pi N$ and/or $NN$ data.

The other two 3NF contributions shown in Fig. 8 are easily derived by taking the last two terms of the $\Delta = 1$ Langrangian, Eq. (18), into account. The 1PE contribution is

$$V^{\rm 3NF}_{\rm 1PE} = -D\, \frac{g_A}{8f_\pi^2} \sum_{i\neq j\neq k} \frac{\vec\sigma_j \cdot \vec q_j}{q_j^2+m_\pi^2} (\boldsymbol{\tau}_i \cdot \boldsymbol{\tau}_j)(\vec\sigma_i \cdot \vec q_j) \tag{78}$$

and the 3N contact potential reads

$$V^{\rm 3NF}_{\rm ct} = E\, \frac{1}{2} \sum_{i\neq j\neq k} \boldsymbol{\tau}_i \cdot \boldsymbol{\tau}_j\ . \tag{79}$$

These 3NF terms involve the two new parameters $D$ and $E$, which do not appear in the 2N problem. There are many ways to pin these two parameters down. In Ref. [21], the triton binding energy and the $nd$ doublet scattering length $^2a_{nd}$ were used. One may also choose the binding energies of $^3$H and $^4$He [47] or an optimal over-all fit of the properties of light nuclei [48]. Once $D$ and $E$ are fixed, the results for other 3N, 4N, etc. observables are predictions.

The 3NF at NNLO has been applied in calculations of few-nucleon reactions, structure of light- and medium-mass nuclei [49, 50], and nuclear and neutron matter [51, 52] with a good deal of success. Yet, the famous '$A_y$ puzzle' of nucleon-deuteron scattering is not resolved [21]. When only 2NFs are applied, the analyzing power in $p$-$^3$He scattering is even more underpredicted than in $p$-$d$. However, when the NNLO 3NF is added, the $p$-$^3$He $A_y$ substantially improves (more than in $p$-$d$) [53]—but a discrepancy remains. Furthermore, the spectra of light nuclei leave room for improvement [48].

To summarize, the 3NF at NNLO is a remarkable contribution: It represents the leading many-body force within the scheme of ChPT; it includes terms that were advocated already some 50 years ago; and it produces noticeable improvements in few-nucleon reactions and the structure of light nuclei. But unresolved problems remain. Moreover, in the case of the 2NF, we have seen that one has to proceed to N$^3$LO to achieve sufficient accuracy. Therefore, the 3NF at N$^3$LO is needed for at least two reasons: for consistency with the 2NF and to hopefully resolve outstanding problems in microscopic structure and reactions.

### 5.1.3. Next-to-Next-to-Next-to-Leading Oorder

At N$^3$LO, there are loop and tree diagrams. For the loops (Fig. 9), we have $L = 1$ and, therefore, all $\Delta_i$ have to be zero to ensure $\nu = 4$. Thus, these one-loop 3NF diagrams can include only leading order vertices, the parameters of which are fixed from $\pi N$ and $NN$ analysis. One sub-group of these diagrams (the 2PE graphs, cf. Fig. 9) has been calculated by Ishikawa and Robilotta [54], and the other topologies have been evaluated by the Bochum-Bonn group [55, 56]. The N$^3$LO 2PE 3NF has been applied in the calculation of nucleon-deuteron observables in Ref. [54] causing little impact. Very recently, the long-range part of the chiral N$^3$LO 3NF has been tested in the triton [57] and in three-nucleon scattering [58] yielding only moderate effects. The long- and short-range parts of this force have been used in neutron matter calculations (together with the N$^3$LO 4NF) producing relatively large contributions from the 3NF [59]. Thus, the ultimate assessment of the N$^3$LO 3NF is still outstanding and will require more few- and many-body applications.

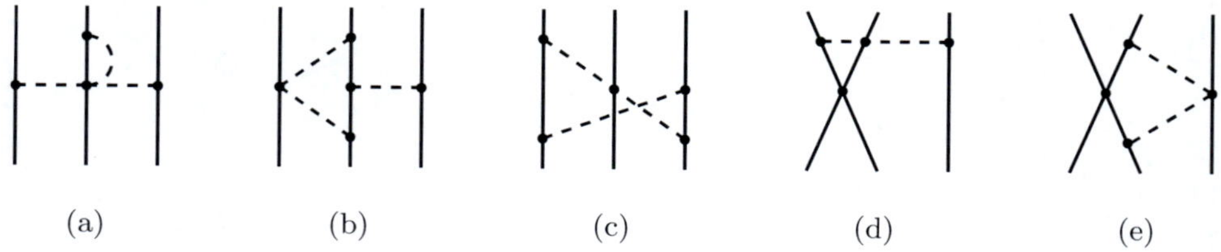


Figure 9. Leading one-loop 3NF diagrams at N$^3$LO. We show one representative example for each of five topologies, which are: (a) 2PE, (b) 1PE-2PE, (c) ring, (d) contact-1PE, (e) contact-2PE. Notation as in Fig. 1.

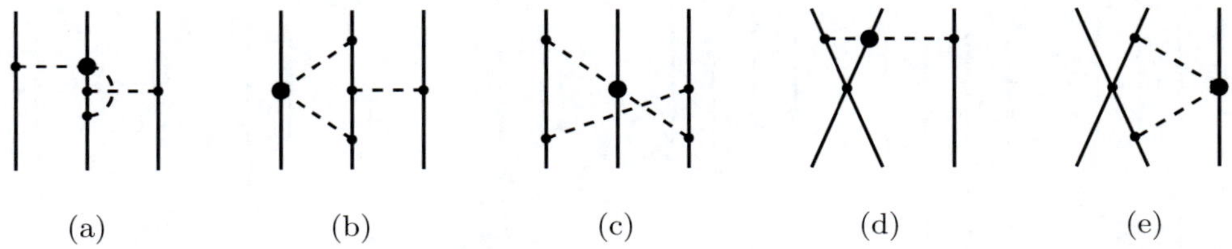


Figure 10. Sub-leading one-loop 3NF diagrams which appear at N$^4$LO with topologies similar to Fig. 9. Notation as in Fig. 1.

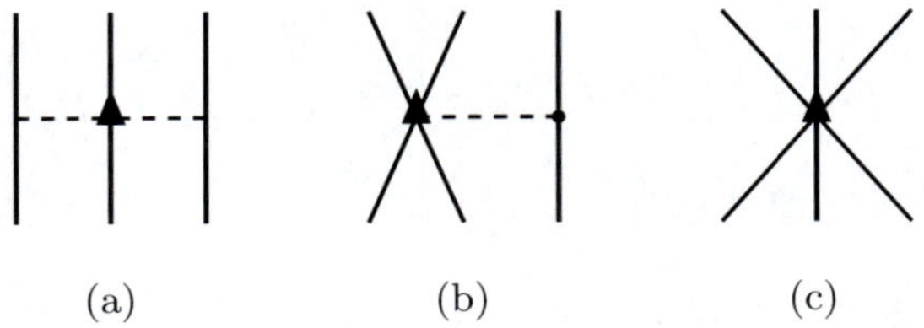


Figure 11. 3NF tree graphs at N$^4$LO ($\nu = 5$) denoted by: (a) 2PE, (b) 1PE-contact, and (c) contact. Solid triangles represent vertices of index $\Delta_i = 3$. Other notation as in Fig. 1.

### 5.1.4. The 3NF at N$^4$LO

In the meantime, it is of interest to take already a look at the next order of 3NFs, which is N$^4$LO or $\nu = 5$ (of the $\Delta$-less theory which we have silently assumed so far). The loop contributions that occur at this order are obtained by replacing in the N$^3$LO loops one vertex by a $\Delta_i = 1$ vertex (with LEC $c_i$), Fig. 10, which is why these loops may be more sizable than the N$^3$LO loops. The 2PE and 1PE-2PE topologies have been evaluated [60] so far. In addition, we have three 'tree' topologies (Fig. 11), which include a new set of 3N contact interactions, which have recently been derived by the Pisa group [61]. Contact terms are typically simple (as compared to loop diagrams) and their coefficients are essentially free. Therefore, it would be an attractive project to test some terms (in particular, the spin-orbit terms) of the N$^4$LO contact 3NF [61] in calculations of few-body reactions (specifically, the p-d and p-$^3$He $A_y$) and spectra of light nuclei.

## 5.2. Four-Nucleon Forces

For connected ($C = 1$) $A = 4$ diagrams, Eq. (23) yields

$$\nu = 4 + 2L + \sum_i \Delta_i \,. \tag{80}$$

Therefore, a connected 4NF appears for the first time at $\nu = 4$ (N$^3$LO), with no loops and only leading vertices, Fig. 12. This 4NF includes no new parameters and does not vanish [62]. Some graphs in Fig. 12 appear to be reducible (iterative). Note, however, that these are Feynman diagrams, which are best analyzed in terms of time-ordered perturbation theory. The various time-orderings include also some irreducible topologies (which are, by

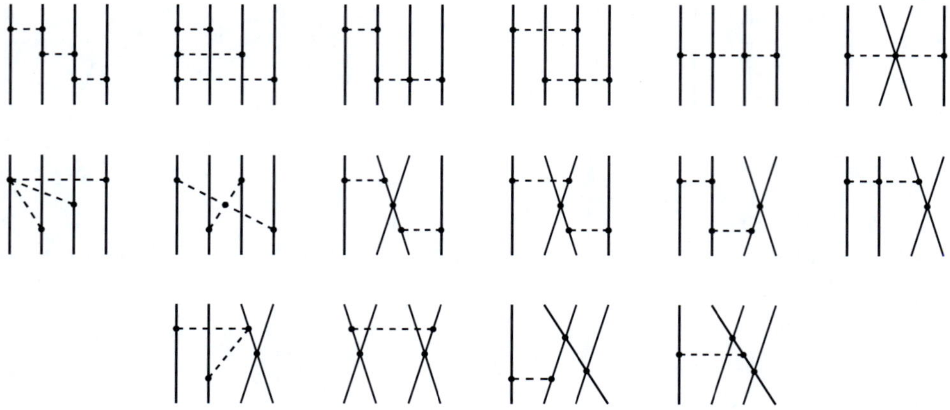

Figure 12. Leading four-nucleon force at $N^3LO$.

definition, 4NFs). Or, in other words, the Feynman diagram minus the reducible part of it yields the (irreducible) contribution to the 4NF. The reducible part depends on the two-, three-, and four-body scattering equations used.

Assuming a good rate of convergence, a contribution of order $(Q/\Lambda_\chi)^4$ is expected to be rather small. Thus, ChPT predicts 4NF to be essentially insignificant, consistent with experience. Still, nothing is fully proven in physics unless we have performed explicit calculations. Recently, the leading 4NF (Fig. 12) has been applied in a calculation of the $^4$He binding energy, where it contributes a few 100 keV [63]. It should be noted that this preliminary calculation involves many approximations, but it certainly provides the right order of magnitude of the result, which is indeed very small as compared to the full $^4$He binding energy of 28.3 MeV.

## 6. Introducing Δ-Isobar Degrees of Freedom

The lowest excited state of the nucleon is the $\Delta(1232)$ resonance or isobar (a $\pi$-$N$ $P$-wave resonance with both spin and isospin 3/2) with an excitation energy of $\Delta M = M_\Delta - M_N =$ 293 MeV. Because of its strong coupling to the $\pi$-$N$ system and low excitation energy, it is an important ingredient for models of pion-nucleon scattering in the $\Delta$-region and pion production from the two-nucleon system at intermediate energies, where the particle production proceeds prevailingly through the formation of $\Delta$ isobars [5]. At low energies, the more sophisticated conventional models for the $2\pi$-exchange contribution to the $NN$ interaction include the virtual excitation of $\Delta$'s, which in these models accounts for about 50% of the intermediate-range attraction of the nuclear force—as demonstrated by the Bonn potential [8, 64].

Because of its relatively small excitation energy, it is not clear from the outset if, in an EFT, the $\Delta$ should be taken into account explicitly or integrated out as a "heavy" degree of freedom. If it is included, then $\Delta M \sim m_\pi$ is considered as another small expansion parameter, besides the pion mass and small external momenta. This scheme has become known as the small scale expansion (SSE). Note, however, that this extension is of phenomenological

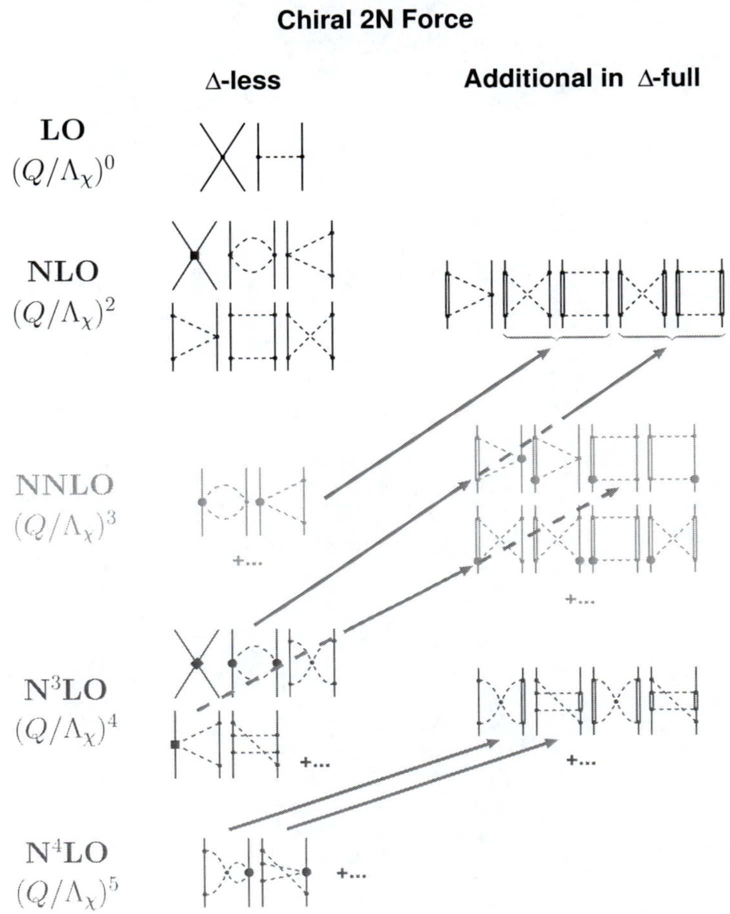


Figure 13. Chiral 2NF without and with Δ-isobar degrees of freedom. Arrows indicate the shift of strength when explicit Δ's are added to the theory. Note that the Δ-full theory consists of the diagrams involving Δ's *plus* the Δ-less ones. Double lines represent Δ-isobars; remaining notation as in Fig. 1.

character, since $\Delta M$ does not vanish in the chiral limit.

In the chiral EFT discussed so far in this article (also known as the "Δ-less" theory), the effects due to Δ isobars are taken into account implicitly. Note that the dimension-two LECs, the $c_i$, have unnaturally large values (cf. Table 5). The reason for this is that the Δ-isobar (and some meson resonances) contribute considerably to the $c_i$—a mechanism that has become known as resonance saturation. Therefore, the explicit inclusion of the Δ ("Δ-full" theory) will take strength out of these LECs and move this strength to a lower order. As a consequence, the convergence of the expansion improves, which is another motivation for introducing explicit Δ-degrees of freedom. We observed that, in the Δ-less theory, the subleading 2PE and 3PE contributions to the 2NF are larger than the leading ones. The promotion of large contributions by one order in the Δ-full theory fixes this problem.

In the heavy baryon formalism, the leading Lagrangian involving Δ's reads [19, 20]

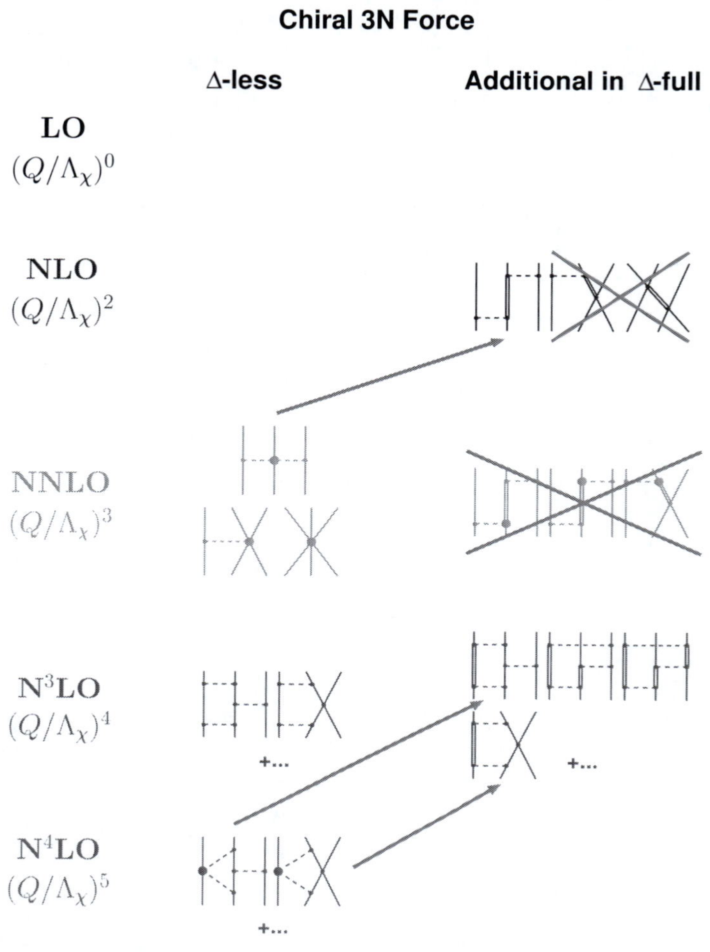


Figure 14. The 3NF without and with Δ-isobar degrees of freedom. Arrows indicate the shift of strength when explicit Δ's are added to the theory. Note that the Δ-full theory consists of the diagrams involving Δ's *plus* the Δ-less ones. Double lines represent Δ-isobars; remaining notation as in Fig. 1.

(listing only terms relevant to our present discussion)

$$\widehat{\mathcal{L}}_{\Delta}^{\Delta_i=0} = \bar{\Delta}(i\partial_0 - \Delta M)\Delta - \frac{h_A}{2f_\pi}\left(\bar{N}\mathbf{T}\vec{S}\Delta + \text{h.c.}\right)\cdot\nabla\boldsymbol{\pi} - D_T\,\bar{N}\boldsymbol{\tau}\vec{\sigma}N\cdot\left(\bar{N}\mathbf{T}\vec{S}\Delta + \text{h.c.}\right), \tag{81}$$

where $\Delta$ is a four-component spinor in both spin and isospin space representing the Δ-isobar and $h_A$ and $D_T$ are LECs. Moreover, $S^i$ are $2 \times 4$ spin transition matrices which satisfy $S^i S^{j\dagger} = (2\delta^{ij} - i\epsilon^{ijk}\sigma^k)/3$ and $T^a$ are similar isospin matrices with $T^a T^{b\dagger} = (2\delta^{ab} - i\epsilon^{abc}\tau^c)/3$. Notice that, due to the heavy baryon expansion, the mass of the Δ-isobar, $M_\Delta$, has disappeared and only the small mass difference $\Delta M$ enters.

The LECs of the $\pi N$ Lagrangian are usually extracted in the analysis of $\pi$-$N$ scattering data and clearly come out differently in the $\Delta$-full theory as compared to the $\Delta$-less one. While in the $\Delta$-less theory, the magnitude of the LECs $c_3$ and $c_4$ is about 3-5 GeV$^{-1}$ (cf. Table 5), they turn out to be around 1 GeV$^{-1}$ in the $\Delta$-full theory [65].

In the 2NF, the virtual excitation of $\Delta$-isobars requires at least one loop and, thus, the contribution occurs first at $\nu = 2$ (NLO), see Fig. 13. The $\Delta$ contributions to the 2PE were first evaluated in Ref. [19] using time-ordered perturbation theory and later by Kaiser et al. [66] in covariant perturbation theory. Recently, also the NNLO contributions have been worked out [65]. Krebs et al. [65] verified the consistency between the $\Delta$-full and $\Delta$-less theories by showing that the contributions due to intermediate $\Delta$-excitations, expanded in powers of $1/\Delta M$, can be absorbed into a redefinition of the LECs of the $\Delta$-less theory. The corresponding shift of the LECs $c_3$, $c_4$ is given by

$$c_3 = -2c_4 = -\frac{h_A^2}{9\Delta M} . \tag{82}$$

Using $h_A = 3g_A/\sqrt{2}$ (large $N_c$ value), almost all of $c_3$ and an appreciable part of $c_4$ is explained by the $\Delta$ resonance.

The studies of Refs. [66, 65] confirm that a large amount of the intermediate-range attraction of the 2NF is shifted from NNLO to NLO with the explicit introduction of the $\Delta$-isobar. However, it is also found that the NNLO 2PE potential of the $\Delta$-less theory provides a very good approximation to the NNLO potential in the $\Delta$-full theory.

The $\Delta$ isobar also changes the 3NF scenario, see Fig. 14. The leading 2PE 3NF is promoted to NLO. In the $\Delta$-full theory, this term has the same mathematical form as the corresponding term in the $\Delta$-less theory, Eqs. (76) and (77), provided one chooses $c_1 = 0$ and $c_3$, $c_4$ according to Eq. (82). Note that the other two NLO 3NF terms involving $\Delta$'s vanish [67] as a consequence of the antisymmetrisation of the 3N states. The $\Delta$ contributions to the 3NF at NNLO [67] vanish at this order, because the subleading $N\Delta\pi$ vertex contains a time-derivative, which demotes the contributions by one order. However, substantial 3NF contributions are expected at N$^3$LO from one-loop diagrams with one, two, or three intermediate $\Delta$-excitations, which correspond to diagrams of order N$^4$LO, N$^5$LO, and N$^6$LO, respectively, in the $\Delta$-less theory.

To summarize, the inclusion of explicit $\Delta$ degrees of freedom does certainly improve the convergence of the chiral expansion by shifting sizable contributions from NNLO to NLO. On the other hand, at NNLO the results for the $\Delta$-full and $\Delta$-less theory are essentially the same. Note that the $\Delta$-full theory consists of the diagrams involving $\Delta$'s plus all diagrams of the $\Delta$-less theory. Thus, the $\Delta$-full theory is much more involved. Moreover, in the $\Delta$-full theory, $1/M_N$ 2NF corrections appear at NNLO (not shown in Fig. 13), which were found to be uncomfortably large by Kaiser et al. [66]. Thus, it appears that up to NNLO, the $\Delta$-less theory is more manageable.

The situation could, however, change at N$^3$LO where potentially large contributions enter the picture. It may be more efficient to calculate these terms in the $\Delta$-full theory, because in the $\Delta$-less theory they are spread out over N$^3$LO, N$^4$LO and, in part, N$^5$LO. These higher order contributions are a crucial test for the convergence of the chiral expansion of nuclear forces and represent a challenging topic for the future.

# 7. Conclusion

The past 15 years have seen great progress in our understanding of nuclear forces in terms of low-energy QCD. Key to this development was the realization that low-energy QCD is equivalent to an effective field theory (EFT) which allows for a perturbative expansion that has become known as chiral perturbation theory (ChPT). In this framework, two- and many-body forces emerge on an equal footing and the empirical fact that nuclear many-body forces are substantially weaker than the two-nucleon force is explained naturally.

In this review, I have shown in detail how the two-nucleon force is derived from ChPT and demonstrated that, at $N^3LO$, the accuracy can be achieved that is necessary and sufficient for reliable microscopic nuclear structure predictions. First calculations applying the $N^3LO$ $NN$ potential [32] in the conventional shell model, the *ab initio* no-core shell model, and the coupled cluster formalism have produced promising results.

We also discussed nuclear many-body forces based upon chiral EFT. The 3NF at NNLO has been known for a while [20, 21] and applied in few-nucleon reactions, structure of light- and medium-mass nuclei [49, 50], and nuclear and neutron matter [51, 52] with some success. However, the famous '$A_y$ puzzle' of nucleon-deuteron scattering is not resolved by the 3NF at NNLO. Thus, one important open issue are the few-nucleon forces beyond NNLO ("sub-leading few-nucleon forces") which, besides the $A_y$ puzzle, may also resolve some important outstanding nuclear structure problems. As explained, this may require going even beyond $N^3LO$.

Another open question is the convergence of the chiral expansion (of the two- as well as the three-nucleon potentials) at orders beyond $N^3LO$, for which the inclusion of $\Delta$-isobar degrees of freedom may be useful. Furthermore, the non-perturbative renormalizations of the chiral potentials require more work and a better understanding.

Having identified some of the open issues, I hope that this review will be helpful for motivating future research.

If the outstanding problems are resolved within the next few years, then, after 80 years of desperate struggle, we may finally claim that the nuclear force problem is essentially under control. The greatest beneficiary of such progress will be the field of *ab initio* nuclear structure physics.

## Acknowledgments

I gratefully acknowledge the instrumental role that my collaborator D. R. Entem has played in the development of the field reviewed in this article. The research by the author is supported in part by the US Department of Energy under Grant No. DE-FG02-03ER41270.

# References

[1] Yukawa, H. *Proc. Phys. Math. Soc. Japan* 1935, 17, 48.

[2] Bryan, R. A.; Scott, B. L. *Phys. Rev.* 1964, 135, B434; *ibid.* 1969, 177, 1435.

[3] Erkelenz, K. *Phys. Rep.* 1974, 13C, 191.

[4] Holinde, K.; Machleidt, R. *Nucl. Phys.* 1975, A247, 495; *ibid.* 1976, A256, 479, 497.

[5] For a thorough and pedagogical introduction into the meson theory of nuclear forces, see:

Machleidt, R. *Adv. Nucl. Phys.* 1989, 19, 189.

[6] Machleidt, R. *Phys. Rev. C* 2001, 63, 024001.

[7] Lacombe, M.; Loiseau, B.; Richard, J. M.; Vinh Mau, R.; Côté, J.; Pires, P.; de Tourreil, R. *Phys. Rev. C* 1980, 21, 861.

[8] Machleidt, R.; Holinde, K.; Elster, Ch. *Phys. Rep.* 1987, 149, 1.

[9] Weinberg, S. *Physica* 1979, 96A, 327.

[10] Weinberg, S. *Nucl. Phys.* 1991, B363, 3.

[11] Weinberg, S. *Phys. Lett. B* 1992, 295, 114.

[12] A comprehensive and accessible presentation of the chiral effective field theory approach to nuclear forces can be found in:

Machleidt, R.; Entem, D. R. *Phys. Rep.* 2011, 503, 1.

[13] Beane, S. R.; Bedaque, P. F.; Orginos, K.; Savage, M. J. *Phys. Rev. Lett.* 2006, 97, 012001.

[14] Ishii, N.; Aoki, S.; Hatsuda, T. *Phys. Rev. Lett.* 2007, 99, 022001.

[15] Scherer, S. *Adv. Nucl. Phys.* 2003, 27, 277.

[16] Nakamura. K. et al. (Particle Data Group) *J. Phys. G: Nucl. Part. Phys.* 2010, 37, 075021.

[17] Coleman, S.; Wess, J.; Zumino, B. *Phys. Rev.* 1969, 177, 2239; Callan, C. G.; Coleman, S.; Wess, J.; Zumino, B. *ibid.* 1969, 177, 2247.

[18] Fettes, N.; Meißner, U.-G.; Mojžiš, M.; Steininger, S. *Ann. Phys. (N.Y.)* 2000, 283, 273; *ibid.* 2001, 288, 249.

[19] Ordóñez, C.; Ray, L.; van Kolck, U. *Phys. Rev. C* 1996, 53, 2086.

[20] van Kolck, U. *Phys. Rev. C* 1994, 49, 2932.

[21] Epelbaum, E.; Nogga, A.; Glöckle, W.; Kamada, H.; Meißner, U.-G.; Witala, H. *Phys. Rev. C* 2002, 66, 064001.

[22] Kaiser, N. *Phys. Rev. C* 2000, 61, 014003.

[23] Kaiser, N. *Phys. Rev. C* 2000, 62, 024001.

[24] Liu J. et al. *Phys. Rev. Lett.* 2010, 105, 181803.

[25] Pavan, M. M.; Arndt, R. A.; Strakovsky, I. I.; Workman, R. L. *Physica Scripta* 2000, T87, 65.

[26] Arndt, R. A.; Strakovsky, I. I.; Workman, R. L.; Pavan, M. M. *Physica Scripta* 2000, T87, 62.

[27] Epelbaum, E.; Glöckle, W.; Meißner, U.-G. *Nucl. Phys.* 1998, A637, 107.

[28] Kaiser, Brockmann, R.; Weise, W. *Nucl. Phys.* 1997, A625, 758.

[29] Erkelenz, K.; Alzetta, R.; Holinde, K. *Nucl. Phys.* 1971, A176, 413.[3]

[30] Machleidt, R. In*Computational Nuclear Physics 2 – Nuclear Reactions*; Langanke, K.; Maruhn, J. A.; Koonin, S. E; Eds.; Springer: New York, NY, 1993; pp. 1-29.

[31] Epelbaum, E.; Glöckle, W.; Meißner, U.-G. *Eur. Phys. J.* 2004, A19, 401.

[32] Entem, D. R.; Machleidt, R. *Phys. Rev. C* 2003 68, 041001.

[33] Stoks, V. G. J.; Klomp, R. A. M.; Rentmeester, M. C. M.; de Swart, J. J. *Phys. Rev. C* 1993, 48, 792.

[34] Arndt, R. A.; Strakovsky, I. I.; Workman, R. L. SAID, Scattering Analysis Interactive Dial-in computer facility, George Washington University (formerly Virginia Polytechnic Institute), solution SM99 (Summer 1999); for more information see, e. g., Arndt, R. A.; Strakovsky, I. I.; Workman, R. L. *Phys. Rev. C* 1994, 50, 2731.

[35] The 1999 $NN$ data base is defined in Ref. [6].

[36] Entem, D. R.; Machleidt, R. *Phys. Lett. B* 2002, 524, 93.

[37] Entem, D. R.; Machleidt, R. *Phys. Rev. C* 2002, 66, 014002.

[38] Stoks, V. G. J.; Klomp, R. A. M.; Terheggen, C. P. F.; de Swart, J. J. *Phys. Rev. C* 1994, 49, 2950.

[39] Wiringa, R. B.; Stoks, V. G. J.; Schiavilla, R. *Phys. Rev. C* 1995, 51, 38.

[40] Epelbaum, E.; Glöckle, W.; Meißner, U.-G. *Nucl. Phys.* 2005, A747, 362.

[41] Entem, D. R.; Machleidt, R.; Witala, H. *Phys. Rev. C* 2002, 65, 064005, and references to earlier work therein.

[42] Caurier, E.; Navratil, P.; Ormand, W. E.; Vary, J. P. *Phys. Rev. C* 2002, 66, 024314.

[3]note that there is an error in equation (4.22) of this paper [ref. 29] where it should read

$$^{-}W^{J}_{LS} = 2qq' \, \frac{J-1}{2J-1} \left[ A^{J-2,(0)}_{LS} - A^{J(0)}_{LS} \right]$$

and

$$^{+}W^{J}_{LS} = 2qq' \, \frac{J+2}{2J+3} \left[ A^{J+2,(0)}_{LS} - A^{J(0)}_{LS} \right] .$$

[43] Büttiker, P.; Meißner, U.-G. *Nucl. Phys.* 2000, A668, 97.

[44] Fettes, N.; Meißner, U.-G.; Steiniger, S. *Nucl. Phys.* 1998, A640, 199.

[45] Fujita, J.-I.; Miyazawa, H. *Prog. Theor. Phys.* 1957, 17, 360.

[46] Coon, S. A.; Scadron, M. D.; McNamee, P. C.; Barrett, B. R.; Blatt, D. W. E.; McKeller, B. H. J. *Nucl. Phys.* 1979, A317, 242.

[47] Nogga, A,; Navratil, P.; Barrett, B. R.; Vary, J. P. *Phys. Rev. C* 2006, 73, 064002.

[48] Navratil, P.; Gueorguiev, V. G.; Vary,J. P.; Ormand, W. E.; Nogga, A. *Phys. Rev. Lett.* 2007, 99, 042501.

[49] Hagen, G.; Hjorth-Jensen, M.; Jansen, G. R.; Machleidt, R.; Papenbrock, T. *Phys. Rev. Lett.* 2012, 108, 242501.

[50] Hagen, G.; Hjorth-Jensen, M.; Jansen, G. R.; Machleidt, R.; Papenbrock, T. *Phys. Rev. Lett.* 2012, 109, 032502.

[51] Sammarruca, F.; Chen, F. B.; Coraggio, L.; Itaco, N.; Machleidt, R. *Phys. Rev. C* 86, 054317 (2012).

[52] Coraggio, L.; Holt, J. W.; Itaco, N.; Machleidt, R.; Sammarruca, F. *Phys. Rev. C* 2013, 87, 014322.

[53] Viviani, M.; Giarlanda, L.; Kievsky, A.; Marcucci, L. E. *Phys. Rev. Lett.* 2013, 111, 172302.

[54] Ishikawa S.; Robilotta, M. R. *Phys. Rev. C* 2007, 76, 014006.

[55] Bernard, V.; Epelbaum, E.; Krebs, H.; Meißner, U.-G. *Phys. Rev. C* 2008, 77, 064004.

[56] Bernard, V.; Epelbaum, E.; Krebs, H.; Meißner, U.-G. *Phys. Rev. C* 2011, 84, 054001.

[57] Skibinski, R. et al., *Phys. Rev. C* 2011, 84, 054005.

[58] Witala, H. et al. *Few-Body Syst.* 2013, 54, 897.

[59] Tews, I.; Krüger, T.; Hebeler, K.; Schwenk, A. *Phys. Rev. Lett.* 2013, 110, 032504.

[60] Krebs, H.; Gasparyan, A.; Epelbaum, E. *Phys. Rev. C* 2012, 85, 054006; *ibid.* 2013, 87, 054007.

[61] Girlanda, L.; Kievsky, A.; Viviani, M. *Phys. Rev. C* 2011, 84, 014001.

[62] Epelbaum, E. *Eur. Phys. J.* 2007, A34, 197.

[63] Rozpedzik, D.; Golak, J.; Skibinski, R.; Witala, H.; Glöckle, W.; Epelbaum, E.; Nogga, A.; Kamada, H. *Acta Phys. Polon.* 2006, B37, 2889; arXiv:nucl-th/0606017.

[64] Holinde, K.; Machleidt, R. *Nucl. Phys.* 1977, A280, 429.

[65] Krebs, H.; Epelbaum, E.; Meißner, U.-G. *Eur. Phys. J.* 2007, A32, 127.

[66] Kaiser, N.; Gerstendörfer, S.; Weise, W. *Nucl. Phys.* 1998, A637, 395.

[67] Epelbaum, E.; Krebs, H.; Meißner, U.-G. *Nucl. Phys.* 2008, A806, 65.

In: NN and 3N Interactions
Editors: Leonid Blokhintsev and Igor Strakovsky
ISBN: 978-1-63321-053-0


*Chapter 2*

# THE SPIN DEPENDENCE OF NEUTRON-PROTON ELASTIC SCATTERING STUDIED IN THE DEUTERON CHARGE-EXCHANGE REACTION

***Colin Wilkin****
Physics and Astronomy Department,
UCL, London, UK

**Abstract**

The existing data on the deuteron charge exchange on hydrogen, $dp \to ppn$, that lead to a final $pp$ system at low excitation energy are reviewed. In particular it is shown that the proton analyzing power, the deuteron tensor analyzing powers, and the deuteron-proton spin correlations can contribute useful information on the spin dependence of near-backward neutron-proton elastic scattering in the GeV region. At energies where the $np$ database is rich, and the partial wave solutions are reliable, the theoretical description works very well. However at higher energies the data suggest that the current $np$ solutions might need to be refined in order to reproduce the spin dependence of deuteron-proton charge exchange.

**PACS** 25.45.De, 25.45.Kk, 24.70.+s

**Keywords:** Neutron-proton amplitudes, Deuteron charge exchange, Polarization effects

## 1. Introduction

A thorough understanding of the nucleon-nucleon interaction remains one of the principal goals of nuclear and hadronic physics. Apart from their intrinsic importance for the study of nuclear forces, elastic *NN* scattering data are also necessary ingredients in the modeling of meson production and other nuclear reactions at intermediate energies. It therefore goes without saying that all facilities should try to fill in any residual gaps in our knowledge in this field. The diverse sources of information are important elements in partial wave

*E-mail address: c.wilkin@ucl.ac.uk

analyses (PWA), most notably that of the SAID *NN* program [1, 2], hosted by the George Washington University.

The large amount of data on proton-proton elastic scattering, sometimes involving the measurement of up to three of the spins, has led to isospin $I = 1$ solutions that have few uncertainties for beam energies $T_p$ below about 2 GeV and have useful predictive power up to around 2.5 GeV [2]. The situation is less promising in the $I = 0$ case because the equivalent data set of neutron-proton measurements is far less populated. The quality of the neutron beams is generally much poorer and a deuterium target is often used as a substitute for a free neutron target. Although the current SAID *np* solution can be applied up to 1.3 GeV [2], the data on which it is based are certainly far more sparse above about 800 MeV. One therefore has to find ingenious methods to try to expand this database in useful ways.

It was pointed out many years ago by Pomeranchuk [3] that at very low excitation energy $E_{pp}$ in a $pp$ system the Pauli principle requires the diproton to be in the $^1S_0$ configuration with antiparallel spins, here denoted by $\{pp\}_s$. As a consequence the deuteron charge exchange $dp \to \{pp\}_s n$ involves a spin-flip from the spin-triplet deuteron to the spin-singlet diproton. These ideas were made much more concrete by Bugg and Wilkin [4], who showed that measurements with polarized deuterons could yield information on the spin structure of large angle neutron-proton elastic scattering provided that $E_{pp}$ and the momentum transfer $q$ between the deuteron and diproton were both small. This led to a series of experimental programs each of which had slightly different priorities. The purpose of this chapter is to discuss what we have learned about neutron-proton elastic scattering amplitudes from measurements of deuteron charge exchange on hydrogen.

The impulse approximation model is described in section 2 in terms of an appropriate form of the neutron-proton amplitude basis. Here it is assumed that there is a quasi-free $np$ charge exchange followed by some final state interaction (FSI) between the two protons. The effects of relaxing the severe cut on $E_{pp}$, which allows a greater contribution from final spin-triplet $P$-waves, are discussed quantitatively. The probability that the final $pp$ pair will be in the $^1S_0$ configuration after the interaction with just one of the nucleons in the deuteron falls rapidly with $q$. For momentum transfers above about 160 MeV/$c$ there are significant corrections to the predictions of the spin observables that arise from the interaction with both nucleons in the deuteron and these are treated within an eikonal approach.

The existing experimental results are discussed in section 3. The data show that the SAID partial wave solution gives a very good description of deuteron charge exchange at 600 MeV per nucleon but that this needs some refinement at 1.135 GeV in order to reproduce the measurements of the deuteron tensor analyzing powers and the deuteron-proton spin correlation parameters. The $d\vec{p} \to \{pp\}_s n$ proton analyzing power is particularly poorly described at this higher energy. How valuable the deuteron charge-exchange approach might be is touched upon in our conclusions of section 4.

## 2. Theoretical Description

It is convenient for our purposes to write the elementary $np \to pn$ (charge-exchange) amplitude in terms of five scalar amplitudes in the center-of-mass (cm) system as

$$\begin{aligned} f_{np\to pn} &= \alpha(q) - i\gamma(q)(\vec{\sigma}_1+\vec{\sigma}_2)\cdot\hat{n} + \beta(q)(\vec{\sigma}_1\cdot\hat{n})(\vec{\sigma}_2\cdot\hat{n}) \\ &\quad +\delta(q)(\vec{\sigma}_1\cdot\hat{m})(\vec{\sigma}_2\cdot\hat{m}) + \varepsilon(q)(\vec{\sigma}_1\cdot\hat{l})(\vec{\sigma}_2\cdot\hat{l}), \end{aligned} \tag{1}$$

where $q = \sqrt{-t}$ is the three-momentum transfer between the initial neutron and final proton and the Pauli matrices $\vec{\sigma}$ are sandwiched between neutron and proton spinors. Here $\alpha$ represents the spin-independent amplitude, $\gamma$ is a spin-orbit contribution, and $\beta$, $\delta$, and $\varepsilon$ are three spin-spin terms. It must be stressed that, although $\alpha(q)$ is generally quite small at low $q$, this would certainly not be true if $f_{np\to pn}$ were written in terms of the spin dependence between the initial and final protons.

The orthogonal unit vectors used in Eq. (1) are defined in terms of the initial neutron ($\vec{K}$) and final proton ($\vec{K}'$) cm momenta;

$$\hat{l} = \frac{\vec{K}'+\vec{K}}{|\vec{K}'+\vec{K}|}, \quad \hat{m} = \frac{\vec{K}'-\vec{K}}{|\vec{K}'-\vec{K}|}, \quad \hat{n} = \hat{l}\times\hat{m} = \frac{\vec{K}\times\vec{K}'}{|\vec{K}\times\vec{K}'|}. \tag{2}$$

The amplitudes are normalized such that the unpolarized $np \to pn$ differential cross section has the form

$$\left(\frac{d\sigma}{dt}\right)_{np\to pn} = |\alpha(q)|^2 + |\beta(q)|^2 + 2|\gamma(q)|^2 + |\delta(q)|^2 + |\varepsilon(q)|^2. \tag{3}$$

### 2.1. Impulse Approximation Formalism

We now turn to the deuteron charge-exchange reaction in impulse approximation, as illustrated in Fig. 1. Providing that the excitation energy in the diproton system is neglected, then the corresponding amplitude can be evaluated in the Breit frame, where the initial deuteron has momentum $-\frac{1}{2}\vec{q}$ and the final diproton $\frac{1}{2}\vec{q}$. There is no energy transfer between the $d$ and $pp$ system and $q^2 = -t$, where $\vec{q}$ is the three-momentum transfer. The initial and final nuclear states are handled symmetrically and, at low $q$, can be described by non-relativistic wave functions. The price to be paid is that the predictions have then to be transformed to the frame where the experiments are carried out and this necessarily involves a Wigner rotation of the spins.

In the Breit frame, the $dp \to \{pp\}n$ amplitude is given by [4, 5]

$$F_{dp\to\{pp\}n}(\vec{K},\vec{q},\vec{k},S,\nu_f,M,m_n,m_p) = \langle \Psi_{pp}^{(-)S,\nu_f}, m_n | f_{np\to pn}(q) e^{i\vec{q}\cdot\vec{r}/2} | \Phi_d^M, m_p \rangle, \tag{4}$$

where $\Psi_{pp,\vec{k}}^{(-)S,\nu_f}(\vec{r})$ describes the outgoing $pp$ wave function with relative momentum $2\vec{k}$, total spin $S$ (= 0 or 1) and projection $\nu_f$. $\Phi_d^M(\vec{r})$ is the initial deuteron wave function of spin projection $M$. These wave functions have argument $\vec{r} = \vec{r}_1 - \vec{r}_2$, where in the initial state particles 1 and 2 are the neutron and proton in the deuteron and 3 the colliding proton. In the final state it is the two protons that are labeled 1 and 2. The magnetic quantum numbers of the struck proton and recoil neutron are designated by $m_p$ and $m_n$, respectively.

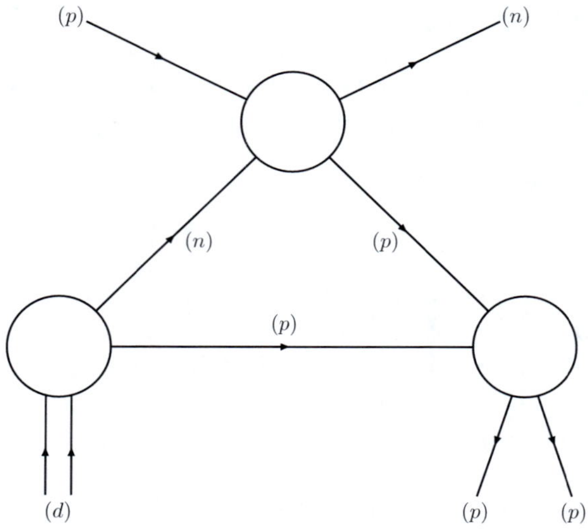


Figure 1. Impulse approximation diagram for the proton charge exchange on the deuteron.

In the full impulse approximation the nucleon-nucleon amplitude $f_{np\to pn}$ should be integrated over the deuteron Fermi momentum. Providing this amplitude varies slowly with momentum (on the internal deuteron momentum scale), one can take it out of the integration at some optimal value of the kinematics to leave an amplitude proportional to $f_{np\to pn}$ times a form factor that reflects the overlap between the initial deuteron wave function and that of the outgoing diproton. In the Breit frame the spectator proton in Fig. 1 is assumed to have zero momentum. It is then argued [5] that, by putting the struck nucleon on its mass shell, the laboratory kinetic energy in *pn* collisions in the laboratory frame should be evaluated at $(1+3q^2/32m^2)T_p+7q^2/32m$, where $T_p$ is the laboratory proton energy in the *pd* reaction and $m$ is the nucleon mass. Fortunately this corresponds to only a few MeV change for $q \lesssim m_\pi$.

The results of the model are particularly transparent at very low values of $E_{pp}=k^2/m_p$, when the *pp* system must be in the $^1S_0$ state, denoted by $\{pp\}_s$. The $dp\to\{pp\}_s n$ reaction then necessarily involves a spin flip from the $S=1$ of the deuteron to the $S=0$ of the diproton. In this case the contribution from the spin-independent amplitude $\alpha$ drops out and one is left with only the spin-flip cross sections. The Cartesian observables corresponding to the deuteron ($A_y^d$) and proton ($A_y^p$) vector analyzing powers, the deuteron tensor analyzing powers ($A_{xx}$ and $A_{yy}$), and the deuteron-proton spin correlations ($C_{x,x}$, $C_{y,y}$, and $C_{yy,y}$) are linked to the amplitudes through:

$$\begin{aligned}
\frac{d^4\sigma}{dtd^3k} &= \tfrac{1}{3}I\left\{S^-(k,\tfrac{1}{2}q)\right\}^2, \\
IA_y^d &= 0, \\
IA_y^p &= -2Im(\beta^*\gamma), \\
IA_{xx} &= |\beta|^2+|\gamma|^2+|\varepsilon|^2-2|\delta|^2R^2(k,\tfrac{1}{2}q), \\
IA_{yy} &= |\delta|^2R^2(k,\tfrac{1}{2}q)+|\varepsilon|^2-2|\beta|^2-2|\gamma|^2, \\
IC_{y,y} &= -2Re(\varepsilon^*\delta)R(k,\tfrac{1}{2}q), \\
IC_{x,x} &= -2Re(\varepsilon^*\beta), \\
C_{yy,y} &= -2A_y^p,
\end{aligned} \tag{5}$$

where the spin-flip intensity

$$I=|\beta|^2+|\gamma|^2+|\varepsilon|^2+|\delta|^2R^2(k,\tfrac{1}{2}q). \tag{6}$$

There are two transition form factors that depend on the $S$- and $D$-states of the deuteron:

$$S^+(k,q) = \langle\{pp\}_s|j_0(qr)|S\rangle+\sqrt{2}\langle\{pp\}_s|j_2(qr)|D\rangle, \tag{7}$$

$$S^-(k,q) = \langle\{pp\}_s|j_0(qr)|S\rangle-\langle\{pp\}_s|j_2(qr)|D\rangle/\sqrt{2}, \tag{8}$$

Their ratio,

$$R(k,q)=S^+(k,q)/S^-(k,q), \tag{9}$$

is equal to 1 in the forward direction.

It must however be stressed again that the predictions given in Eq. (5) correspond to the Breit frame and the observables are subject to a (generally small) Wigner rotation before being compared to experiment.

In the forward direction the value of the tensor analyzing powers in the $^1S_0$ limit is fixed by the longitudinal and transverse spin-transfer parameters between an initial proton and final neutron in elastic $pn$ scattering [6, 7]:

$$A_{xx}(0)=A_{yy}(0)=\frac{2[K_{0ll0}(\pi)-K_{0nn0}(\pi)]}{3-K_{0ll0}(\pi)-2K_{0nn0}(\pi)}, \tag{10}$$

or linked to the value of the spin-transfer parameter $K_{x,x}$ in the $d(\vec{p},\vec{n})\{pp\}_s$ reaction:

$$A_{xx}(0)=-\tfrac{1}{2}[1+3K_{x,x}(0)]. \tag{11}$$

Similarly the deuteron-proton spin-correlation parameter is determined by a combination of neutron-proton spin-correlation and spin-transfer parameters through

$$C_{x,x}(0)=C_{y,y}(0)=\frac{2[A_{00nn}(\pi)-D_{n0n0}(\pi)]}{3-K_{0ll0}(\pi)-2K_{0nn0}(\pi)}. \tag{12}$$

The tensor analyzing powers for the $P$-wave spin-triplet $pp$ states are generally of opposite sign to the $S$-wave results of Eq. (5). Although the strong $pp$ final state interaction in the $^1S_0$ state enhances the $S$-wave components in the final state, at finite $E_{pp}$ there are significant dilutions of the analyzing powers from the $P$-waves. This can be taken into account by evaluating the impulse approximation with $pp$ final wave functions using realistic nuclear potentials that include the Coulomb potential. This was done in the work of Ref. [5], where plane waves were assumed for the $F$ and higher waves, which means that the $P$–$F$ coupling through the tensor force was neglected. Though the initial estimates were made using the de Tourreil and Sprung soft core nucleon-nucleon potential [8], no measurable differences were found when this is replaced by any of a wide variety of more modern $NN$ potentials. Even with a cut as low as 3 MeV in the $pp$ excitation energy, this program shows that there are significant deviations from the $^1S_0$ predictions of Eq. (5), as can be seen in the 1.6 GeV case (800 MeV per nucleon) for $A_{xx}$ and $A_{yy}$ shown in Fig. 2.

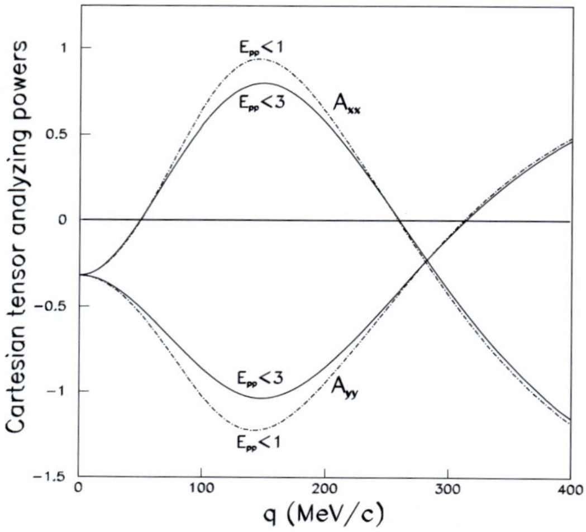


Figure 2. Predictions of the impulse approximation model [5] for the tensor analyzing powers $A_{xx}$ and $A_{yy}$ for the $\vec{d}p \to \{pp\}n$ reaction as a function of the momentum transfer $q$ between the deuteron and diproton at 800 MeV per nucleon. The effects are shown of putting cuts of 1 MeV and 3 MeV on the diproton excitation energy.

The effects of the dilution of the signals through the excitation of final spin-triplet $pp$ waves depends critically upon the value of the momentum transfer. In the forward direction any effect is very small and the signal should be largely independent of the $E_{pp}$ cut. In fact it is possible to evaluate the total spin-flip contribution to forward $np$ elastic scattering by invoking a sum rule and summing the unpolarized cross section over all values of $E_{pp}$ [9]. Experimental data of this type are available up to 2 GeV [10].

The contributions from the higher partial waves can also studied by investigating the dependence on the angle between the momentum transfer $\vec{q}$ and the diproton relative momentum $\vec{k}$ since $P$-waves cannot be excited when these two vectors are perpendicular [4].

## 2.2. Multiple Scattering Corrections

Since the reactions discussed in this chapter involve a deuteron rather than a free neutron, it is natural to wonder whether this might introduce significant corrections to the simple picture so far outlined. In the energy region of several hundred MeV per nucleon it is not expected that the 2.2 MeV binding in the deuteron is significant. However, the associated Fermi motion of the neutron in the deuteron would broaden any narrow structure that might be present in the $np$ system.

The most important correction though is the multiple scattering of both nucleons in the deuteron on the target proton, which is most easily handled using Glauber theory [11]. In this approach the single-scattering amplitude of Eq. (4) is modified to become

$$\begin{aligned} F_{dp\to\{pp\}n} &= \langle\Psi_{pp}^{(-)S,\nu_f}, m_n|f_{np\to pn}(q)e^{i\vec{q}\cdot\vec{r}/2} \\ &+ \frac{i}{2\pi^{3/2}}\int e^{i(\vec{q}\,'-\vec{q}/2)\cdot\vec{r}} f_{np\to pn}(\vec{q}\,')f_{NN}(\vec{q}-\vec{q}\,')\,d^2q'|\Phi_d^M, m_p\rangle, \end{aligned} \tag{13}$$

where the integration over $\vec{q}\,'$ is carried out over the plane perpendicular to the mean momentum $\frac{1}{2}(\vec{K}+\vec{K}')$.

The small-angle charge-averaged nucleon-nucleon amplitude $f_{NN}(q)$ is dominated by the spin-independent contribution so that it is a good approximation to make the replacement $f_{NN}(q) = \bar{\alpha}(q) = \frac{1}{2}(\alpha_{pp}(q)+\alpha_{pn}(q))$. If further one only considers the $^1S_0$ final state, it is possible to derive the explicit expressions for the modified scalar amplitudes $\beta$, $\delta$, $\gamma$, and $\varepsilon$, which become respectively [4][1]:

$$\begin{aligned} &\beta(q)S^-(k,\tfrac{1}{2}q)+\frac{i}{2\pi^{3/2}}\int d^2q'\,\bar{\alpha}(\vec{q}-\vec{q}\,')\Big[\beta(q')\Big\{S^-(k,Q)\cos^2\phi+\frac{3qq'}{2\sqrt{2}Q^2}\cos\phi\sin^2\phi\,S_D(k,Q)\Big\} \\ &\qquad +\delta(q')\Big\{S^+(k,Q)\sin^2\phi+\frac{3}{2\sqrt{2}Q^2}(qq'\cos\phi-\tfrac{1}{2}q^2)S_D(k,Q)\sin^2\phi\Big\}\Big], \\ &\delta(q)S^+(k,\tfrac{1}{2}q)+\frac{i}{2\pi^{3/2}}\int d^2q'\,\bar{\alpha}(\vec{q}-\vec{q}\,')\Big[\delta(q')\Big\{S^+(k,Q)\cos^2\phi-\frac{3qq'}{2\sqrt{2}Q^2}\cos\phi\sin^2\phi\,S_D(k,Q)\Big\} \\ &\qquad +\beta(q')\Big\{S^-(k,Q)\sin^2\phi+\frac{3}{2\sqrt{2}Q^2}(qq'\cos\phi+\tfrac{1}{2}q^2)S_D(k,Q)\sin^2\phi\Big\}\Big], \\ &\gamma(q)S^-(k,\tfrac{1}{2}q)+\frac{i}{2\pi^{3/2}}\int d^2q'\,\bar{\alpha}(\vec{q}-\vec{q}\,')\gamma(q')\Big\{S^-(k,Q)\cos\phi+\frac{3qq'}{2\sqrt{2}Q^2}\sin^2\phi\,S_D(k,Q)\Big\}, \\ &\varepsilon(q)S^-(k,\tfrac{1}{2}q)+\frac{i}{2\pi^{3/2}}\int d^2q'\,\bar{\alpha}(\vec{q}-\vec{q}\,')\varepsilon(q')S^-(k,Q), \end{aligned} \tag{14}$$

where $\phi$ is the azimuthal angle of $\vec{q}\,'$ and $\vec{Q} = \vec{q}\,' - \frac{1}{2}\vec{q}$.

It is seen that the double scattering corrections for $\delta(q)$ and $\beta(q)$ are coupled through the integration over $\phi$ but that, as $q \to 0$, the double-scattering corrections to $\beta(q)-\delta(q)$ and $\gamma(q)$ tend to zero, as they must do on invariance grounds.

The moduli of the four charge-exchange spin amplitudes at 800 MeV, evaluated from the current SAID $NN$ solution [2], are shown in Fig. 3. These amplitudes are mainly real and what is striking is the profound minimum in $|\delta(q)|$ for $q \sim 140$ MeV/$c$ that arises from the strong influence of the one-pion-exchange term in this amplitude. It is this which gives rise to much of the structure observed in the deuteron charge-exchange reaction.

---

[1] Two typographical errors in Eq. (7.3) of Ref. [4] have been corrected here.

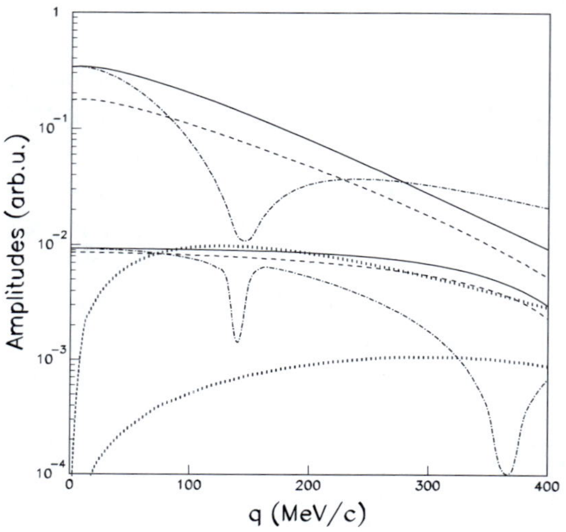


Figure 3. Magnitudes of the $\beta(q)$ (solid line), $\delta(q)$ (dot-dashed line), $\varepsilon(q)$ (dashed line), and $\gamma(q)$ (dotted line) charge-exchange amplitudes at 800 MeV derived from the current SAID *NN* solution [2] as functions of momentum transfer. Also shown below them are the predictions for the magnitudes of the interference terms for these four amplitudes.

Also shown in the figure are the magnitudes of the interference amplitudes, which are defined as the difference between the moduli squared of the single+double scattering and the single scattering terms divided by twice the modulus of the single scattering [4]. Evaluating these corrections using Eq. (14), it is seen that in the forward direction the $\varepsilon(0)$ amplitude is reduced by about 5.0% whereas the reduction in $\beta(0) = \delta(0)$ is only 2.8%. The smaller correction in this case is caused by the steep variation of $\delta(q)$ with momentum transfer.

The effects of the double scattering corrections on the tensor analyzing powers are illustrated for $A_{xx}$ and $A_{yy}$ in Fig. 4. The changes are very small for $q \lesssim 140$ MeV/*c* but become significant above about 200 MeV/*c* and this increase is probably influenced by the minimum in $|\delta(q)|$ for $q \approx m_\pi$.

The double-scattering effects are more important for the unpolarized differential cross section, whose predictions at 800 MeV per nucleon are shown in Fig. 5. Reductions of 6%, 15%, and 20% are to be expected at momentum transfers of 0, 140, and 400 MeV/*c*, respectively. One generally assumes that cross sections on the deuteron are reduced uniformly through the shadowing effect by about 5-6% [11] and this takes into account the change in the forward cross section but not the larger reductions predicted at non-zero angles.

Double scatterings are expected to be less significant for the *P* and higher *pp* waves until larger momentum transfers are reached. As a consequence, the changes shown in Figs. 4 and 5 should be understood as upper limits of what is to be expected in any experiment. The modifications of the analyzing powers are negligible for $q \lesssim 140$ MeV/*c* and remain small until at least 200 MeV/*c*. The effects are somewhat bigger for the unpolarized cross section where the value at 140 MeV/*c* is reduced by about 10% compared to the forward direction. It therefore seems that the ratios of $|\beta(q)|^2$, $|\delta(q)|^2$, and $|\varepsilon(q)|^2$ can be extracted

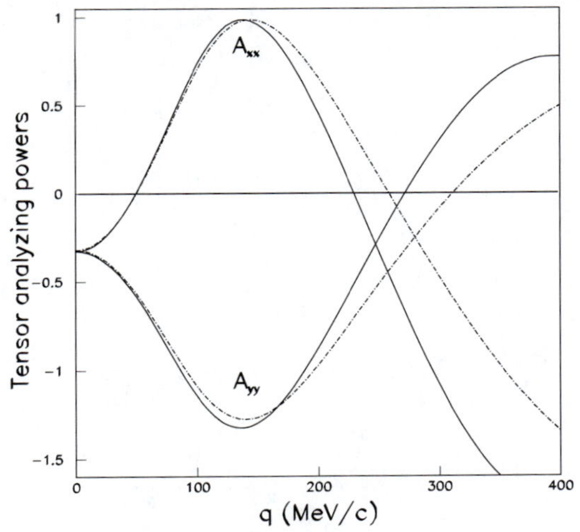


Figure 4. Predictions for the tensor analyzing powers $A_{xx}$ and $A_{yy}$ in the $\vec{d}p \to \{pp\}_s n$ reaction at 800 MeV per nucleon estimated using the current SAID $NN$ amplitudes [2]. The dot-dashed curves are single-scattering results and the solid ones include the effects of double scattering in the eikonal limit of Eq. (14).

from deuteron charge-exchange data for $q \lesssim 140$ MeV/$c$ using the impulse approximation, but that this might overestimate their sum by several per cent.

# 3. Experimental Data

The $p(\vec{d},2p)n$ reaction was studied at RCNP at 70 and 56 MeV [12], with the aim of comparing the results in the forward direction with the corresponding spin-transfer parameter in the $d(\vec{p},\vec{n})pp$ reaction [13], thus testing the relation given by Eq. (11). For small angles the magnetic spectrograph used restricted the excitation energy of the final protons $E_{pp} < 200$ keV, which easily satisfies the $^1S_0$ criterion. However, at such low beam energies many driving mechanisms can be important and the impulse approximation of Fig. 1 can at best provide a qualitative description. The results are therefore not relevant for the purpose of enhancing the $np$ database.

## 3.1. The EMRIC Experiments

The primary motivation of the EMRIC collaboration was to provide data to help in the design of a polarimeter to measure the tensor polarization of the recoil deuteron in elastic electron-deuteron scattering. With this in mind, they measured the analyzing powers of the $p(\vec{d},2p)n$ reaction at deuteron beam energies of 200 and 350 MeV [14, 15]. This eventually led to the construction of the POLDER polarimeter that was used to separate the contributions from the deuteron electromagnetic monopole and quadrupole form factors at JLab [16]. Although, as shown in Fig. 6, the EMRIC data are generally in very good agreement with impulse approximation estimates at small momentum transfers [4, 5], there

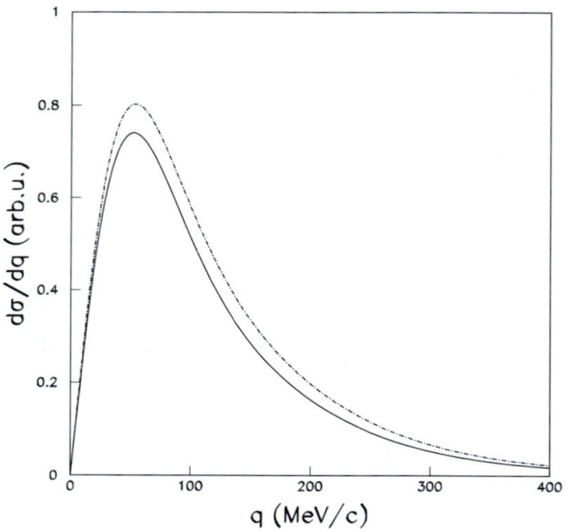


Figure 5. Predictions for the unpolarized differential cross section for the $\vec{d}p \to \{pp\}_s n$ reaction in the $^1S_0$ limit at 800 MeV per nucleon estimated using the current SAID *NN* amplitudes [2]. The dot-dashed curve is the single-scattering result and the solid one includes the effects of double scattering in the eikonal limit of Eq. (14).

is very little ambiguity in the neutron-proton amplitude analysis at such low energies. We shall therefore only discuss in detail the experiments carried out at higher energies using the SPES-IV and ANKE magnetic spectrometers.

### 3.2. The SPES-IV Experiments

The Franco-Scandinavian collaboration used the high resolution magnetic spectrometer SPES-IV at Saclay to study the $p(\vec{d},2p)n$ reaction at 1.6 and 2.0 GeV, with some data being taken also at 0.65 GeV [17, 18, 19, 20]. Although hydrogen and deuterium targets were used, the main thrust of the experiments was actually to study the spin-longitudinal and -transverse responses for the $^{12}$C target that was also employed. Since both protons were detected in the spectrometer, the small angular acceptance ($1.7^\circ \times 3.4^\circ$), combined with the momentum bite of $\Delta p/p \approx 7\%$, ensured that only events with very low *pp* excitation energies were registered. Monte Carlo simulations showed that the peak of the $E_{pp}$ distribution was around 650 keV and, under these circumstances, any contamination from *P*-waves in the *pp* system is very small. Because of the small angular acceptance, the deflection angle in the spectrometer had to be changed in order to measure the differential cross section and the tensor analyzing power at different values of the momentum transfer $q$.

In the SPES-IV experiment the quantization axis for the beam polarization was perpendicular to the horizontal plane of the accelerator. Denoting by $\phi$ the azimuthal angle of the diproton measured from the horizontal plane, the tensor analyzing power signal becomes

$$A_{\phi\phi} = [(1+\cos 2\phi)A_{yy} + (1-\cos 2\phi)A_{xx}]/2\,. \qquad (15)$$

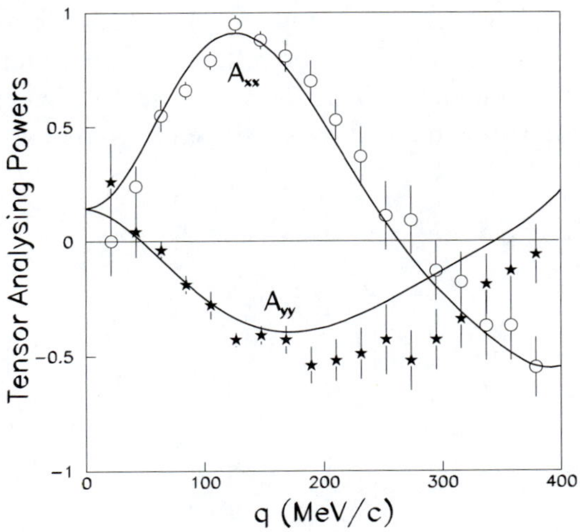


Figure 6. The deuteron tensor analyzing powers in the $\vec{d}p \to \{pp\}_s n$ reaction at 350 MeV for $E_{pp} < 1$ MeV. The experimental data from the EMRIC collaboration [15] are compared to their impulse approximation estimates that were based upon the *np* amplitudes used in Ref. [4].

This signal was then averaged over the $1.7° \times 3.4°$ angular acceptance to give

$$\langle A_{\phi\phi} \rangle = [(1 + \langle \cos 2\phi \rangle) A_{yy} + (1 - \langle \cos 2\phi \rangle) A_{xx}]/2. \tag{16}$$

In the forward direction, a Monte Carlo simulation suggests that $\langle \cos 2\phi \rangle = -0.28$ at 1.6 GeV, the negative value arising because the vertical acceptance was larger than the horizontal. At finite angle the positioning of the spectrometer arm in the horizontal plane means that the data were primarily sensitive to $A_{yy}$, with $\langle \cos 2\phi \rangle = +0.92$ at a laboratory angle of 2.7°, rising to +0.99 at 10°.

The experimental values of $\langle A_{\phi\phi} \rangle$ are shown in Fig. 7 at 1.6 and 2.0 GeV[2]. These were obtained using a deuteron beam with a (spherical) tensor polarization of $\rho_{20} = 0.61 \pm 0.01$. It should be noted that, due to the finite aperture, the average value of $q$ was non-zero even in the forward direction. Furthermore, the effects of the rms spread in $q$ were significant, especially at forward angles. Since the values of $\langle A_{\phi\phi} \rangle$ were deduced from the data without taking the $\langle \cos 2\phi \rangle$ average into account, this has to be included in the modeling. The single-scattering predictions (dashed curve) and those with the inclusion of double scattering (solid curve) taken from Ref. [20] at 1.6 GeV were evaluated on the basis of 1988 *np* amplitudes [21] that were rather similar to those used in Ref. [4]. As already noted, there is very little change to the predictions for $q < 160$ MeV/$c$ when the double scattering is included and the agreement is good for the first two points.

The situation changes radically when modern *np* amplitudes, such as those provided by the SAID program [2], are used. The predicted magnitude of $\langle A_{\phi\phi} \rangle$ at 140 MeV/$c$ is then

[2] In the SPES-IV publications this quantity was actually called $A_{yy}$, though this is only strictly true at large momentum transfers.

reduced in the single-scattering approximation by about 20% and any agreement with the experimental data is much poorer. Although no reliable $np$ amplitudes were available at the time of the publication [17], the 2.0 GeV data show a similar picture, with the 160 MeV/$c$ point in Fig. 7b lying well below the prediction derived from modern $np$ amplitudes [2].

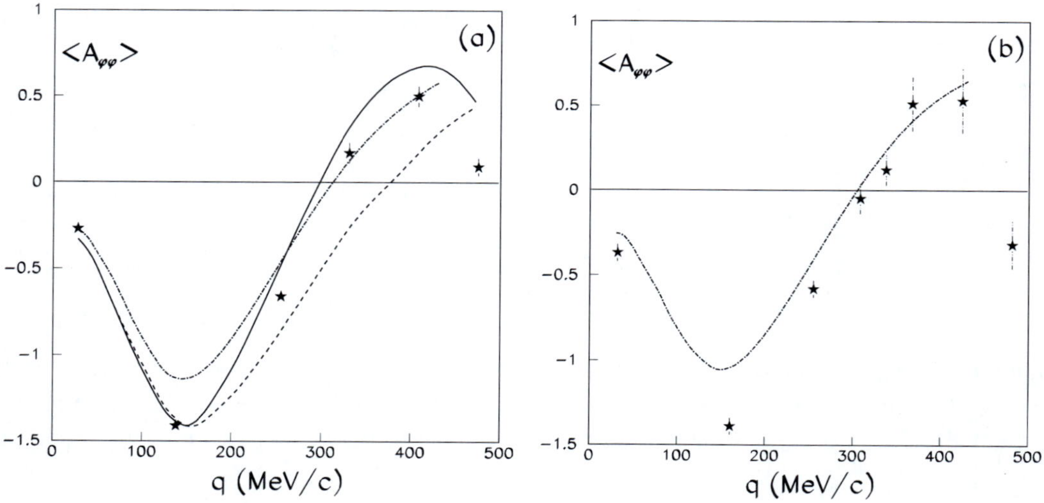


Figure 7. Values of the polarization response defined by Eq. (16) measured at beam energies of (a) 1.6 GeV and (b) 2.0 GeV with the SPES-IV spectrometer [17, 18, 19, 20]. The dashed line in (a) is the impulse approximation prediction given by the authors in Ref. [20] that takes into account the variation of $\langle\cos 2\phi\rangle$ with momentum transfer. The solid curve includes also the effects of double scattering in the Glauber approximation. The neutron-proton charge-exchange amplitudes used were similar to those employed in Ref. [4]. Taking the current SAID amplitudes [2] leads to the very different single-scattering predictions shown by the dot-dashed curve in (a). The analogous predictions are also shown at 2.0 GeV in panel (b).

Whereas it is very helpful to note the great sensitivity of the data to the $np$ input amplitudes, it remains a mystery as to why there is such a big discrepancy between the SPES-IV data and the modern single-scattering prediction for $q \approx 140$ MeV/$c$. The dip in $A_{yy}$ in this region is caused by the near vanishing of the $\delta(q)$ amplitude for $q \approx m_\pi$ [4] and its depth is then directly controlled by the size of $|\varepsilon(q)|^2$ compared to $|\beta(q)|^2$ at this momentum transfer. Taken at face value, the SPES-IV data would suggest a much lower value of $|\varepsilon(q=m_\pi)|^2$ than that delivered by the SAID partial wave analysis [2]. In complete contrast, as is shown in the next section, recent experimental data from ANKE at 1.2, 1.6, and 1.8 GeV are actually completely consistent with predictions based on the newer SAID $np$ amplitudes.

In the final run at 1.6 GeV [20], the SPES-IV group introduced a powerful superconducting solenoid that allowed the deuteron beam polarization to be rotated through an angle $\phi$ of up to 45°. This led to measurements of $A_{xx}$ as well as $A_{yy}$ but only for momentum transfers of $q \gtrsim 270$ MeV/$c$, and so this is of little help in resolving the SPES-IV/ANKE controversy.

It was not possible at SPES-IV to determine absolute cross sections because of difficulties in evaluating the overall acceptance of the spectrometer but relative values could be extracted from the data [20]. These are compared in Fig. 8 with the authors' single and single + double-scattering estimates, both normalized on the 28 MeV/$c$ point. It is important to note that a single-scattering calculation using modern $np$ amplitudes gives results that are essentially indistinguishable from those shown by the dashed curve in Fig. 8.

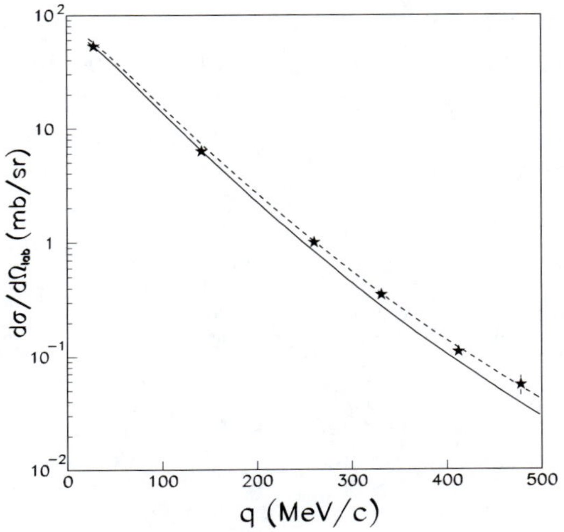


Figure 8. The laboratory differential cross section for the $dp \to \{pp\}_s n$ reaction at 1.6 GeV [20]. The overall normalization was not determined and the authors' single-scattering curve (dashed) and that with the inclusion of double scattering (solid curve) have been scaled to pass through the lowest point. The single-scattering prediction obtained using modern $np$ amplitudes is essentially identical to the dashed curve shown here.

In summary, the 800 MeV $np$ amplitudes from 1988 [21] predict a similar angular dependence for the unpolarized $dp \to \{pp\}_s n$ differential cross section to that of the current SAID amplitudes [2]. However, there are big changes in the (mainly) $A_{yy}$ signal for $q \approx m_\pi$ where the SPES-IV experimental data at neither 1.6 nor 2.0 GeV agree with the current predictions. The situation needed clarification and this is to some extent provided by the ANKE experiments, to which we now turn.

### 3.3. The ANKE Experiments

In contrast to earlier experiments, those carried out by the ANKE collaboration were the first where the basic motivation was the investigation of the spin structure of the neutron-proton elastic charge-exchange amplitudes at small angles. Although the procedures were developed over several years [22, 23, 24], all the important results are collected in Refs. [25, 26].

The experiments reported in this section were carried out using the ANKE magnetic spectrometer that is placed at an internal target position of the COoler SYnchrotron and storage ring (COSY) of the Forschungszentrum Jülich. The $\vec{d}p \to \{pp\}_s n$ reaction was measured at 1.2, 1.6, 1.8, and 2.27 GeV with an unpolarized hydrogen cluster target and, by introducing a polarized internal hydrogen target, these studies were later extended to the measurement of spin correlations in the $\vec{d}\vec{p} \to \{pp\}_s n$ reaction at 1.2 and 2.27 GeV. It should be noted here that 2.27 GeV represents the maximum deuteron beam energy that is accessible at COSY.

The characteristics of the ANKE and SPES-IV spectrometers were very different, with ANKE operating on a circulating beam in a chicane inside the COSY ring. Its energy resolution was inferior to that of SPES-IV and the intensity of the polarized deuteron beam was much less than that of the unpolarized. On the other hand, the large acceptance means that many reactions could be studied simultaneously over a wide range of angles. This is illustrated in Fig. 9, which shows the detection of single particles in the ANKE forward detector that was used in the charge-exchange studies.

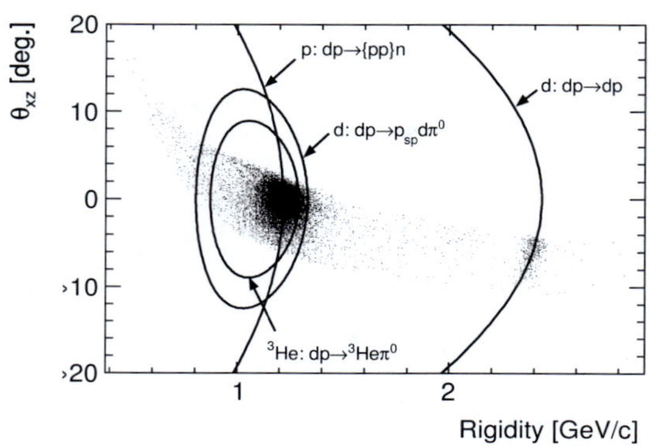


Figure 9. Scatter plot of single particles detected in ANKE from the interaction of 1.2 GeV deuterons with a hydrogen cluster-jet target in terms of the laboratory production angle in the horizontal plane and the magnetic rigidity. Loci corresponding to four common nuclear reactions are also shown. These include the $dp \to \{pp\}_s n$ reaction at zero $E_{pp}$.

The multiwire chambers in the forward detector serve for track reconstruction and the three layers of the scintillation hodoscope permit the measurements of the arrival time and energy-loss that are required for particle identification. This timing information allows a clear separation between the two protons arising from the $dp \to \{pp\}_s n$ reaction and the $dp$ pairs coming from the $dp \to p_{\text{spec}} d\pi^0$ reaction, where the proton $p_{\text{spec}}$, which has a momentum about half that of the deuteron beam, is interpreted as being a “spectator”, which only influences the $np \to d\pi^0$ pion production through the kinematics. The momentum distribution of this proton in the rest frame of the incident deuteron is completely consistent with this interpretation but, in order to enhance the spectator contribution, only events where the Fermi momentum was below 60 MeV/$c$ were included in the analysis.

The distribution in the excitation energy $E_{pp}$ of the diproton is consistent with that expected from the $pp$ final-state interaction in the $^1S_0$ channel. Though the energy resolution would have allowed a lower cut in $E_{pp}$, in order to acquire viable statistics, the ANKE data were summed over $E_{pp} < 3$ MeV [25]. As can be seen from Fig. 2, this leads to a significant dilution of the analyzing power signals through spin-triplet final $pp$ pairs, and this has to be taken into account in the modeling [5].

The normalized differential cross section is obtained using the luminosity determined by comparing the counts from the quasi-free $np \to d\pi^0$ reaction with the cross sections that have been measured for the $pp \to d\pi^+$ reaction [27], taking into account the factor of two coming from charge-independence arguments. The results shown at four energies in Fig. 10 do not include any shadowing correction since, near the forward direction, this largely cancels out between the $dp \to \{pp\}_s n$ and $dp \to p_{\text{spec}} d\pi^0$ reactions.

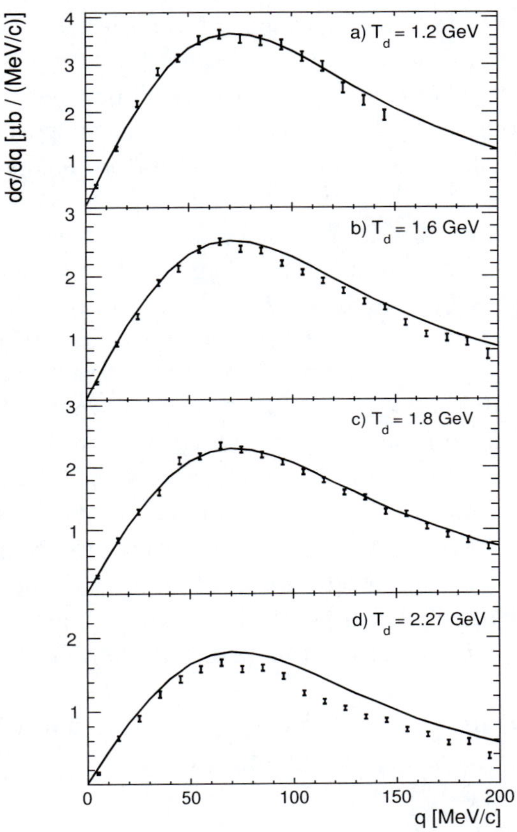


Figure 10. Differential cross sections for the $dp \to \{pp\}_s n$ reaction at four different energies compared with impulse approximation predictions [5] based upon current SAID $np \to np$ amplitudes [2]. The data are integrated over $E_{pp} < 3$ MeV. Only statistical errors are shown. A large fraction of the systematic errors of up to 6% arise from uncertainties in the $pp \to d\pi^+$ database [27].

The results for the differential cross sections are compared in Fig. 10 with the single scattering predictions integrated over the $E_{pp} < 3$ MeV interval using the current SAID $np$ amplitudes [2]. As illustrated in Fig. 5, the double scattering would be expected to reduce the predictions by only a few percent at the largest values of the momentum transfers. The agreement is very satisfactory at the three lower energies but at 2.27 GeV the predictions significantly overestimate the measured cross section. This is the first indication that the SAID amplitude analysis might need some refinement above about 1 GeV. This overprediction of the spin-spin contribution to $np$ charge exchange at high energies is consistent with the findings of inclusive $d(p,n)$ measurements [10].

The ANKE data with an unpolarized target were taken with nine different combinations of vector and tensor polarized beams. Although ideal values of the polarizations could be estimated from the settings of the ion source, these were never realized in practice and the beam polarizations had to be determined for the circulating beam in ANKE. One crucial difference with the SPES-IV measurements at an external deuteron beam at Saclay is that for an internal measurement at a storage ring one has far less control over the beam-target overlap and so the luminosity has also to be measured for each polarization mode. This is most accurately done using the inclusive production of spectator protons. The number of single-track events is enormous for all beam energies and no dependence of the rates on the tensor polarization of the deuteron beam was found for proton spectator momenta below 60 MeV/$c$.

The vector polarization of the COSY beam could be measured by comparing the results of the quasi-free $\vec{n}p \to d\pi^0$ at both small and large deuteron production angles with the amplitude analysis of $\vec{p}p \to d\pi^+$ provided by the SAID program [27]. Reactions inside the ANKE acceptance that are sensitive to the deuteron tensor polarization include deuteron-proton elastic scattering and $\vec{d}p \to {}^3\mathrm{He}\,\pi^0$. Using these reactions to calibrate the tensor polarizations of the deuteron beam at 1.17 GeV, it was shown that the analyzing powers in the deuteron charge exchange reaction $\vec{d}p \to \{pp\}_s n$ were well described in impulse approximation [5] out to momentum transfers of at least 150 MeV/$c$ [24]. As a consequence, the $\vec{d}p \to \{pp\}_s n$ reaction at 1.2 GeV was used to measure the tensor polarizations in the different source modes. This procedure, which fits in well with the philosophy of the original theoretical description [4], has the advantages of the large counting rates combined with being insensitive to the deuteron vector polarization. Due to the small anomalous magnetic moment, no depolarization is expected for deuterons in the COSY energy range and none was ever detected.

The numbers $N(q,\phi)$ of diprotons produced at momentum transfer $q$ and azimuthal angle $\phi$ depend on the beam tensor polarization $P_{zz}$ through

$$\frac{N(q,\phi)}{N_0(q)} = C_n\left\{1+\tfrac{1}{4}P_{zz}\left[A_{xx}(q)(1-\cos 2\phi)+A_{yy}(q)(1+\cos 2\phi)\right]\right\}, \tag{17}$$

where $N_0(q)$ are the numbers for an unpolarized beam and $C_n$ is the relative luminosity of the polarized beam that is determined from the inclusive spectator production. By dividing by the unpolarized counting rates one avoids any uncertainties in the evaluation of the ANKE acceptance in $\phi$. Unlike the SPES-IV case, there is useful acceptance in $\phi$ out to at least $q \approx 160$ MeV/$c$ and so the values of $A_{xx}$ and $A_{yy}$ could be obtained independently by fitting the experimental data to the form of Eq. (17).

The values of $A_{xx}$ and $A_{yy}$ deduced at 1.6, 1.8, and 2.27 GeV are compared in Fig. 11 with single-scattering predictions using as input the current SAID neutron-proton elastic amplitudes [2]. The description of the data is reasonable at the two lower energies. Though the value of $A_{yy}$ at $q \approx 140$ MeV/$c$ lies perhaps very slightly below the curve at 1.6 GeV, any disagreement is much less than that found for the SPES-IV data shown in Fig. 7a.

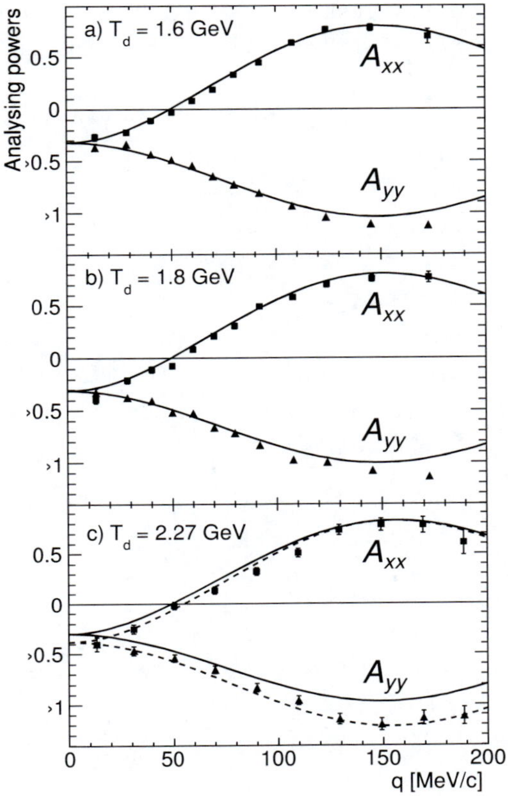


Figure 11. Tensor analysing powers $A_{xx}$ (squares) and $A_{yy}$ (triangles) of the $\vec{d}p \to \{pp\}_s n$ reaction at three beam energies for $E_{pp} < 3$ MeV compared to impulse approximation predictions based upon the current SAID $np$ amplitudes [2]. The dashed curves at 2.27 GeV correspond to a uniform reduction of the spin-longitudinal $\varepsilon(q)$ amplitude by 25%. The error bars include the uncertainties from the relative luminosity $C_n$ and the beam polarizations determined at 1.2 GeV.

At 2.27 GeV there is a clear divergence between the ANKE data and the predictions. As was stressed before, in the $q \approx 140$ MeV/$c$ region the depth of the minimum depends very sensitively on the magnitude of $\varepsilon(q)$ and reducing this globally by 25% gives the much more satisfactory agreement that is illustrated by the dashed lines in Fig. 11. The ANKE $A_{yy}$ data at 1.6 and 1.8 GeV might both be described a little better if there were a slight reduction in $\varepsilon(q)$ but the effect is marginal given the experimental uncertainties. It is an open question whether some of the discrepancy in the 2.0 GeV SPES-IV $A_{yy}$ point at $q = 160$ MeV/$c$ shown in Fig. 7b could be due to an overestimate of $|\varepsilon(q)|^2$ in the SAID partial wave analysis.

A measurement of the spin-correlation parameters $C_{x,x}$ and $C_{y,y}$ requires a vector polarized beam and target. The latter was achieved at ANKE through the use of cell target that was supplied with polarized hydrogen from an atomic beam source. The backgrounds in this case were more severe due to some of the beam halo hitting the walls of the storage cell. The polarizations of both the beam and target were established by measuring the asymmetries in the quasi-free $\vec{n}p \to d\pi^0$ and $n\vec{p} \to d\pi^0$ reactions, respectively.

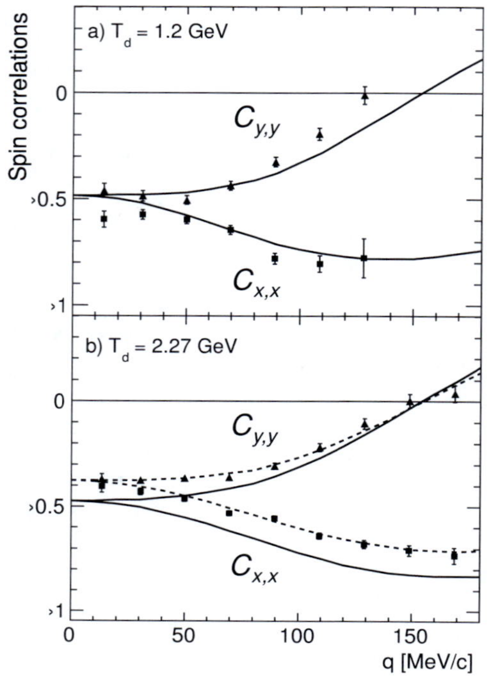


Figure 12. The spin-correlation coefficients $C_{x,x}$ and $C_{y,y}$ for the $dp \to \{pp\}_s n$ reaction at $T_d = 1.2$ and 2.27 GeV for $E_{pp} < 3$ MeV. The error bars do not include the 11% uncertainties in the combined beam and target polarizations. The curves are impulse approximation predictions; dashed curves at 2.27 GeV correspond to $\varepsilon(q)$ being reduced by 25%.

The spin-correlation studies were carried out only at 1.2 and 2.27 GeV and the results are compared in Fig. 12 with impulse approximation predictions using the current SAID $np$ amplitudes. The value at $q = 0$ can also be deduced from the predictions for the neutron-proton spin observables on the SAID WEB site [2] through the use of Eq. (12). The agreement at 1.2 GeV is broadly satisfactory, though there are some apparent discrepancies for $C_{y,y}$ at the largest momentum transfers measured. Now it is seen from Eq. (5) that, in the $^1S_0$ limit, the spin correlations are sensitive to the interference of $\varepsilon(q)$ with either $\beta(q)$ or $\delta(q)$. Reducing the value of $\varepsilon(q)$ by 25% gives the very satisfactory agreement shown by the dashed curves at 2.27 GeV. Thus both the deuteron tensor analyzing powers and the deuteron-proton spin correlations in the $dp \to \{pp\}_s n$ reaction suggest that $\varepsilon(q)$ is too large at 1.135 GeV in the current SAID partial wave analysis.

As a by-product to the spin-collaboration studies, the ANKE collaboration also measured the proton analyzing power $A_y^p$ which, in the $^1S_0$ limit, is proportional to the interference between $\beta(q)$ and $\gamma(q)$. The current SAID $np$ amplitudes describe the magnitude and angular distribution of this observable very well at 1.2 GeV shown in Fig. 13 but the corresponding predictions at 2.27 GeV are over an order of magnitude too small. It is therefore apparent that significant changes are also needed in either the magnitude or phase of the spin-orbit amplitude $\gamma(q)$ at 1.135 GeV. It was already shown in earlier measurements [24] that the deuteron analyzing power was, as expected, consistent with zero.

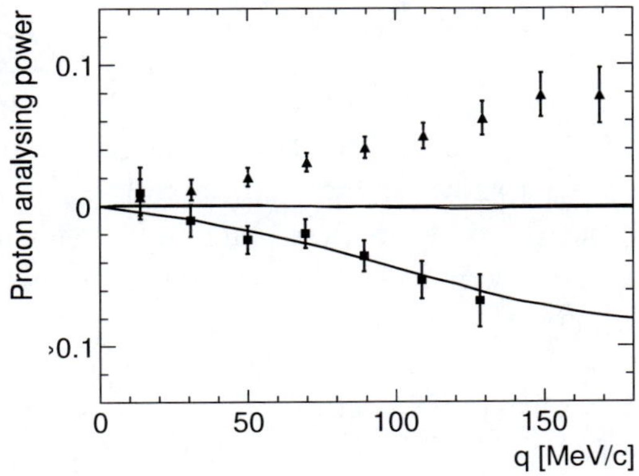


Figure 13. Proton analyzing powers $A_y^p$ for the $dp \to \{pp\}_s n$ reaction at $T_d = 1.2$ (squares) and 2.27 GeV (triangles) for $E_{pp} < 3$ MeV. The error bars do not include the 5% polarization uncertainty. The impulse approximation curve describes well the 1.2 GeV data but at 2.27 GeV the $A_y^p$ prediction is hardly visible on this scale.

## 4. Conclusion

The existing data on the deuteron charge exchange on hydrogen, $p(d,2p)n$, reaction in the GeV region have been examined with the aim of extracting more information on the spin dependence of large angle neutron-proton elastic scattering. Although even unpolarized data can pick out the total amount of $np \to pn$ spin flip at small momentum transfers $q$ between the neutron and proton, the full sensitivity arises when using tensor polarized deuterons or when measuring the spin correlation between vector polarized deuterons and protons in the initial state. Such data might help to refine the current partial wave solutions.

In the forward direction, $\delta(0) = \beta(0)$ and so the value of $A_{yy}(0)$ is fixed completely by the ratio $|\varepsilon(0)|^2/|\beta(0)|^2$. In the neutron-proton partial wave solution used by Bugg and Wilkin [4], the value of $|\varepsilon(0)|$ was quite low and this led to the rather large negative value of $A_{yy} = -0.37$ at 1.6 GeV compared to the $-0.32$ of the current SAID solution [2]. In the absence of direct neutron-proton spin-transfer measurements, good $A_{yy}$ data could easily help to choose between these two solutions.

Smaller differences are seen within the SAID solutions themselves. Equation (10) shows that the forward value of $A_{yy}$ in the $\vec{d}p \to \{pp\}_s n$ reaction is closely linked to the values of the $\vec{n}p \to n\vec{p}$ spin-longitudinal and -transverse transfer coefficients in the backward direction. In the ${}^1S_0$ limit, the impulse approximation prediction for $A_{yy}(0)$ at 1.6 GeV changes from the current $-0.32$ [2] to $-0.34$ if the SAID SP96 solution [1] is used.

However, a much bigger signal is to be found in the region around 140 MeV/$c$, where there is a strong dip in $A_{yy}$ caused by $Re\{\delta(q)\}$ changing sign for $q \approx m_\pi$. The depth of the dip is governed largely by the value of $|\varepsilon(q)|^2/|\beta(q)|^2$ in this region. The effects can be quite significant, as shown by the lowest panel in Fig. 11. The low values of $|\varepsilon(q)|$ used by Bugg and Wilkin at 1.6 GeV lead to a very deep minimum with $A_{yy} = -1.53$, which is far more negative than the minimum in Fig. 2 that has been estimated using the current SAID $np$ solution.

In contrast the maximum in $A_{xx}$ is little changed by even a 25% reduction in $\varepsilon(q = m_\pi)$. This lack of sensitivity is easily understood from Eq. (5) in the ${}^1S_0$ limit since, in this case, $A_{xx} = 1$ whenever $\delta(q)$ vanishes, independent of the size of $\varepsilon(q)$.

It is important to note that, as shown in Fig. 5, the double scattering in the deuteron has relatively little effect on either $A_{xx}$ or $A_{yy}$ until momentum transfers of at least 160 MeV/$c$. On the other hand, a precision description of the data at higher $q$ will certainly involve the consideration of multiple scattering corrections.

In summary, the values of $A_{yy}$ for the $\vec{d}p \to \{pp\}_s n$ reaction at $q \lesssim 140$ MeV/$c$ are particularly sensitive to one aspect of the neutron-proton partial wave solutions, namely the relative size of $|\varepsilon(q)|$ compared to $|\beta(q)|$. On the other hand, the values of $C_{x,x}$, $C_{y,y}$, and $A_y^p$ depend on the relative phases of the amplitudes and not merely their magnitudes.

Turning now to the experimental data, it is hard to reconcile the results presented by ANKE [25] and SPES-IV [17, 20], especially at their common energy of 1.6 GeV. Of the two SPES-IV points in the region of overlap, the kinematics of the lower one are not well defined because they are averaged over a very wide range in momentum transfer and the conditions of the spectrometer means that the polarization response at this point was more dependent on $A_{xx}$ than $A_{yy}$. These caveats do not apply to the point at $q \approx 140$ MeV/$c$, where the SPES-IV value of $A_{yy}$ is far more negative than that found at ANKE, even after taking the $E_{pp}$ cut and the value of $\langle\cos 2\phi\rangle$ into account. The fact that the SPES-IV data were well reproduced by the Bugg 1988 amplitudes [21] whereas the extensive ANKE data are described by the current SAID solution [2] must be treated as a coincidence.

Although the original SPES-IV data at 2.0 GeV [17] were not compared with predictions based upon credible $np$ amplitudes, similar features are to be found in these results, with the minimum in $A_{yy}$ of Fig. 7b being much deeper than that predicted from current $np$ amplitudes. Now from looking at Fig. 11 it is seen that the reduction in $\varepsilon(q)$ gives a better description of the 2.27 GeV ANKE data. It might then be argued that the 1.6 and 1.8 GeV data would also benefit from a similar but smaller diminution of the spin-longitudinal strength. This could alleviate marginally the deviations in the SPES-IV data in the $q \approx 140$ MeV/$c$ region.

The published ANKE data at 1.2, 1.6, 1.8, and 2.27 GeV [25] were taken in far smaller steps in $q$ than at SPES-IV and over a region where double scattering is expected to influence the spin-dependent observables in only minor ways. Furthermore, the study of the dependence on the azimuthal angle allowed $A_{xx}$ to be separated from $A_{yy}$ and $C_{x,x}$ from $C_{y,y}$. The tensor analyzing power data at the highest of these energies clearly indicates that the spin-longitudinal amplitude $\varepsilon(q)$ is overestimated in the current $np$ solution. This is not intrinsically impossible because there is a severe lack of $(\vec{n},\vec{p})$ spin-transfer measurements in this region that would constrain this amplitude. This conclusion is reinforced by the measurements of the spin-correlation coefficients and all the ANKE data seem to be better described if $\varepsilon(q)$ is reduced globally by a factor of $\approx 0.75$. It is an open question whether a similar, but possibly smaller reduction, might be appropriate also for the 1.6 and 1.8 GeV data. This would certainly help to resolve some of the differences noted with the SPES-IV work [17, 20].

In a sense the results in Fig. 13 are more surprising because the proton analyzing powers in the $d\vec{p} \to \{pp\}_s n$ reaction are very well described by the single-scattering model at 1.2 GeV using the current SAID $np$ amplitudes but the theory is over an order of magnitude too low at 2.27 GeV. The reaction model itself would be expected to get even better at the higher energies. According to Eq. (5), this observable is proportional to $Im(\beta^*\gamma)$. Since the magnitude of $\beta$ is well determined from the differential cross section and tensor analyzing power measurements, either the relative phase or, more likely, the magnitude of $\gamma$ is poorly described in the current SAID solution. Again the basic culprit here is the lack of appropriate $np$ data above 1 GeV. These would involve quite complicated experiments because the $\vec{n}p \to pn$ analyzing power is influenced strongly also by the interference of $\gamma$ with the spin-independent $\alpha$.

How could the deuteron charge-exchange approach be pushed forwards? The basic limitation at ANKE, as it was with SPES-IV, is that the maximum deuteron energy achievable at COSY is around 2.3 GeV. To go higher in energy, one has either to move to another accelerator or work in inverse kinematics, with a proton beam incident on a polarized deuterium target. The charge-exchange reaction could then be studied purely by measuring two slow protons in silicon tracking telescopes (STT) [28] placed in the target chamber. In the current design, the STT used at ANKE have three layers. In order to get a good kinematic determination, the proton has to pass through the first, which requires a minimum energy of $\approx 2.5$ MeV, i.e., a momentum of $\approx 70$ MeV/$c$. If the protons are detected in separate STT, $q$ can be small but $E_{pp} > 2 \times 2.5$ MeV. On the other hand, if the protons are detected in the same STT, then $E_{pp}$ can be small but $q > 2 \times 70$ MeV/$c$. Data taken in inverse kinematics can therefore be used to extend the $E_{pp} < 3$ MeV results obtained with a deuteron beam out to larger momentum transfers.

The placing of the STT in the horizontal plane with respect to the target means that the resulting data were insensitive to the values of $A_{xx}$. *Preliminary results* [26] for $A_{yy}$ at large momentum transfers at a proton beam energy of 0.6 GeV are shown in Fig. 14 along with older data obtained in $(\vec{d}, pp)$ kinematics at $T_d = 1.2$ GeV [22]. The smooth continuation between the old and new data is very encouraging. The single-scattering predictions are shown by the solid curves but, at these large momentum transfers, double-scattering effects have to be considered before such data might usefully contribute to the neutron-proton database.

In the $^1S_0$ limit the $(d, \{pp\}_s)$ reaction involves both a spin and isospin flip from the spin-triplet deuteron to the spin-singlet diproton. At low momentum transfers the resulting observables depend sensitively upon the spin-flip amplitudes in near-backward neutron-proton elastic scattering. The ANKE results suggest that one of these, the longitudinal spin-spin term, is overestimated in the current SAID partial wave solution at 1.135 GeV and, furthermore, that the spin-orbit term is far too small. However, it must be acknowledged that the deuteron charge-exchange reaction, though very focussed, is only a substitute for the measurement of a very limited number of free neutron-proton spin-dependent observables and the results must be used in combination with other $np \to pn$ data in a full partial wave analysis.

We have only discussed here information that could be extracted on elastic neutron-

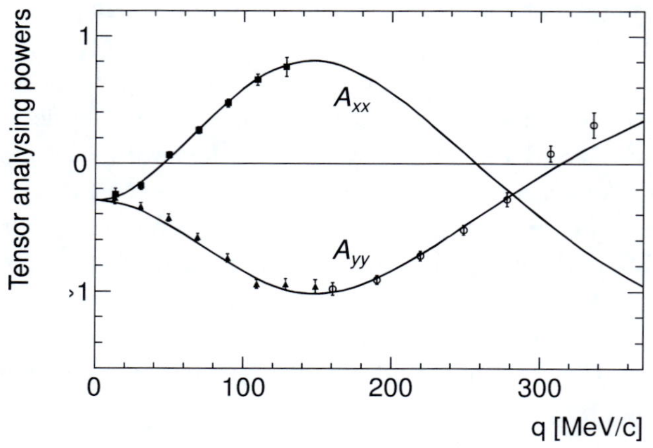


Figure 14. Tensor analysing powers $A_{xx}$ (squares) and $A_{yy}$ (triangles) of the $\vec{d}p \rightarrow \{pp\}_s n$ reaction from early ANKE measurements with a polarized deuteron beam at $T_d$ = 1.2 GeV [22]. The new *preliminary* $A_{yy}$ results for $q > 160$ MeV/$c$ (open circles) were obtained in inverse kinematics with a 600 MeV unpolarized proton beam incident on a polarized deuterium target [26]. In all cases the data are selected with a $E_{pp} < 3$ MeV cut.

proton scattering. Subjects such as the analysis of deuteron charge-exchange on heavier targets [20] or the excitation of the $\Delta(1232)$ isobar [18, 29] fall outside the remit of this book. The same is also true for the use of the deuteron charge-exchange reaction to measure beam polarizations in other processes.

## Acknowledgments

My interest in the deuteron charge exchange was kindled by the ideas of David Bugg but in recent years it has been members of the ANKE collaboration, primarily David Chiladze, Andro Kacharava, and David Mchedlishvili, who have driven the work forward. All the numerical studies were carried out in collaboration with Jaume Carbonell on the basis of the SAID neutron-proton input amplitudes that were generously provided by Igor Strakovsky. Many thanks are due to Thomas Sams, who so patiently explained the finer details of the SPES-IV experiments. Finally, special mention should be made to František Lehar, who worked in the nucleon-nucleon area all his life and who influenced my thinking in this field.

# References

[1] Arndt, R. A.; Strakovsky, I. I.; Workman, R. L. *Phys. Rev. C* 2000, 62, 034005 (1-16).

[2] Arndt, R. A.; Briscoe, W. J.; Strakovsky, I. I.; Workman, R. L. *Phys. Rev. C* 2007, 76, 025209 (1-10); `http://gwdac.phys.gwu.edu`.

[3] Pomeranchuk, I. *Dokl. Akad. Nauk. USSR* 1951, 77, 249-253.

[4] Bugg, D. V.; Wilkin, C. *Nucl. Phys. A* 1987, 467, 575-620.

[5] Carbonell, J.; Barbaro M. B.; Wilkin C. *Nucl. Phys. A* 1991, 529, 653-674.

[6] Lehar, F.; Wilkin, C. *Eur. Phys. J.* 2008, 37, 143-149.

[7] Lehar, F.; Wilkin, C. *Phys. Elementary Part. Atomic Nuclei* 2010, 7, 393-423.

[8] de Tourreil, R.; Sprung, D. W.; *Nucl. Phys. A* 1973, 201, 193-214.

[9] Dean, N. W. *Phys. Rev. D* 1972, 5, 2832-2835.

[10] Sharov, V. I. et al., *Eur. Phys. J. A* 2009, 39, 267-280.

[11] Glauber, R. J., *Lectures in Theoretical Physics*, Ed. Brittin, W. E., (Interscience, N.Y., 1959) 1, 315-414.

[12] Motobayashi, T. et al., *Nucl. Phys. A* 1988, 481 207-216.

[13] Sakai, H. et al., *Phys. Lett. B* 1986, 177, 155-158.

[14] Kox, S. et al., *Phys. Lett. B* 1991, 266, 264-268.

[15] Kox, S. et al., *Nucl. Phys. A* 1993, 556, 621-640.

[16] Abbott, D. et al., *Phys. Rev. Lett.* 2000, 84, 5053-5057.

[17] Ellegaard, C. et al., *Phys. Rev. Lett.* 1987, 59, 974-977.

[18] Ellegaard, C. et al., *Phys. Lett. B* 1989, 231, 365-369.

[19] Sams, T., PhD thesis, Nils Bohr Institute Copenhagen (1991); `www.physics.dk/Sams1990.pdf`.

[20] Sams, T. et al., *Phys. Rev. C* 1995, 51, 1945-1960.

[21] Bugg, D. V. (private communication to T. Sams in 1988).

[22] Chiladze, D. et al., *Phys. Lett. B* 2006, 637, 170-175.

[23] Chiladze, D. et al., *Phys. Rev. ST Accel. Beams* 2006, 9, 050101 (1-13).

[24] Chiladze, D. et al., *Eur. Phys. J. A* 2009, 40, 23-33.

[25] Mchedlishvili, D. et al., *Eur. Phys. J. A* 2013, 49, 49 (1-13).

[26] Mchedlishvili, D., PhD thesis, Tbilisi State University, 2013.

[27] Arndt, R. A.; Strakovsky, I. I.; Workman, R. L.; Bugg, D. V., *Phys. Rev. C* 1993, 48, 1926-1938; `http://gwdac.phys.gwu.edu`.

[28] Schleichert, R. et al., *IEEE Trans. Nucl. Sci.* 2003, 50, 301-306.

[29] Mchedlishvili, D. et al., *Phys. Lett. B* 2013, 726, 145-150.

In: NN and 3N Interactions
Editors: Leonid Blokhintsev and Igor Strakovsky
ISBN: 978-1-63321-053-0


*Chapter 3*

# EXPONENTIAL NUCLEON-NUCLEON POTENTIAL WITH FORBIDDEN STATE

***S. B. Dubovichenko***[1,2,*]
[1]V. G. Fessenkov Astrophysical Institute "NCSRT" NSA RK, Observatory 23, Kamenskoe plato, Almaty, Kazakhstan
[2]Institute of Nuclear Physics CAE MINT RK, Almaty, Kazakhstan

## ABSTRACT

The possibility of constructing the nucleon-nucleon potential of the exponential form and the OPEP with the node of a wave function only in the *S*-wave is shown, which allows one to describe the low energy np characteristics, the elastic scattering phase shifts up to 500 MeV, the energy of deuteron, form factors and the $e^2H$ polarization up to 5 – 6 $fm^{-1}$ correctly.

**PACS Number(s):** 13.75.Cs, 13.85.Dz, 21.45.Bc, 21.60.Gx, 25.20.-x, 25.40.Cm, 25.60.Gc, 25.85.Jg

## SUMMARY

The results considered in this review were obtained for nucleon-nucleon (NN) system, when the tensor forces and the forbidden states (FSs), those forbidden by the Pauli principle, exist in the interaction potentials. At first some variants of pure central NN interactions with FS will be considered. Furthermore, it will be shown that for a description of significant amounts of NN data it is possible to use the exponential potential (EP) taking into account one-pion exchange (OPEP), which contains only two phenomenological parameters and a certain integer constant *N*, characterizing the power of cutoff factor of the potential at low distances. Note that the three adjustable parameters are needed besides *N* for the Gaussian Moscow potential (MP) with FS.

---

* E-mail: dubovichenko@mail.ru.

Furthermore, an interesting dependence of the EP parameters between each other will be shown, which is inherent to the MP too. The technique of exclusion of the unobserved node in the *D*-wave, which exists in the MP and in the first variant of the EP is demonstrated.

The influence of the πNN coupling constant value on the deuteron characteristics, on the NN elastic scattering phase shifts up to 500 MeV, on the tensor and vector polarizations in the elastic $e^2H$ scattering will be demonstrated. The results obtained here generally agree with the known experimental data on tensor polarization $T_{ij}$ up to 3 – 4 $fm^{-1}$, obtained in the Jefferson physical laboratory, located in the USA, in the late nineties. In addition, certain results for Nijmegen potentials, polarizations and other characteristics of the NN system for variants of the Argonne v18 and the CD-Bonn potentials of nineties of XX century in comparison with the results for the EP are given.

The timing from the eighties of XX century to 2002 is conditioned by the fact that exactly in that period all given here results were published for different variants of the EP with FS, whereon, evidently, the history of the development of this type of potentials was finished. The review was written in 2002 and placed as a chapter into the book of Dubovichenko "Characteristics of light atomic nuclei in the potential cluster model", 2004 year, published in Russian in Almaty and uploaded at http://arxiv.org/abs/ 1006.4944. These results do not undergo essential changes since then, and this issue is the first survey review in English.

## 1. INTRODUCTION

It has long been known that a potential description of the cluster-cluster scattering can be based on the deep interactions with the forbidden states [1]. Due to the existence of such states the scattering phases follow the generalized Levinson theorem and become positive at all energies, vanishing in the Born range of big energies. This approach allows one to construct a variety of potentials for cluster systems like $^4He^2H$, $^4He^3H$, $^3He^3H$ [2], which describe correctly not only elastic scattering phase shifts at low energies, but the main characteristics of light nuclei $^6Li$ and $^7Li$ too, including photonuclear reactions in the corresponding cluster models [2].

Potentials with forbidden states were used to describe the characteristics of nucleon-nucleon interactions. It seems likely that for the first time the conception of the forbidden states was applied towards the nucleon-nucleon interactions in the early seventies of the last century, such as in Ref. [3], where the singlet nucleon-nucleon Yukawa potentials, in general, correctly describing *S*, *D*, and *G* scattering phase shifts at energies up to 300 MeV were appropriate. Subsequently, the central singlet and triplet Gaussian interactions were parameterized and it was shown that triplet potentials with forbidden states could be matched simultaneously with the scattering phases up to 250 – 300 MeV, characteristics at low energies, and some properties of deuteron [4].

These results were further developed in Ref. [5] where the Gaussian singlet and triplet potentials with the tensor component and taking into account one-pion exchange OPEP, allow one to obtain a good description of the scattering phases of the nucleon-nucleon interactions and deuteron characteristics. The wave functions of deuteron and the scattering wave functions in such potentials have nodes at distances of about 0.56 fm, practically immovable for energy changes. Central Gaussian interactions with the FSs that correctly describe the general behavior of the scattering phases up to 3 GeV and the detailed justification of the deep nucleon-nucleon potentials in terms of nucleon quark structure have been previously suggested in [6]. The general qualitative comparison of the results with some quark models, where there are nodes in wave functions, is given. As was shown in Refs. [3–6], the potentials with forbidden states allow, essentially, a reduction in the number of the varied adjustable parameters.

In this way, the triplet potential from Ref. [5] has only four parameters, while the core interactions, for example, Reid triplet potential [7] contains about ten adjustable values. In these works generally the Gaussian form of interactions were used. Therefore, it is interesting to consider some other forms of central potentials, for example exponential and Yukawa potentials, for comparison in the calculation of different scattering characteristics and characteristics of the bound deuteron state and to compare them with the results for the known potentials with core.

## 2. Central Nucleon-Nucleon Interactions

If one considers deuteron as a six-quark system then the orbital symmetry {42}, consistent with $L = 0, 2$, corresponds to the ground state, and the bound state with orbital symmetry {6} is forbidden.

The symmetry {51} is also forbidden in the odd waves [5,6]. This means that the potential of the triplet $S$-wave interaction has the lower forbidden state, and the ground state of deuteron corresponds to the second bound state in such a potential. There is a bound but forbidden state in the $P$-wave interaction too. Therefore, the triplet $S$ scattering phase shifts, in accordance with the Levinson theorem, starting with 360º and singlet with 180º rising approximately up to 220º. The triplet and singlet $P$-phases also have to start with 180º. There is one bound allowed state in the $D$-wave which, together with the $S$-wave, determines the ground state of deuteron, and the forbidden state is absent, i.e., there is not node in the $D$ component of the deuteron wave function (WF).

In calculations of the characteristics of the nucleon-nucleon system for central forces, it is possible to use the general radial Schrödinger equation in the form:

$$\chi''_L(r) + [k^2 - V(r) - V_{\mathrm{Coul}}(r) - L(L+1)/r^2]\,\chi_L(r) = 0$$

with the asymptotics of the scattering functions, expressed in terms of Coulomb type functions

$$\chi_L(r) \rightarrow N\,[F_L(kr)\cos(\delta_{LJ}) + G_L(kr)\sin(\delta_{LJ})],$$

where $F_L(kr)$ and $G_L(kr)$ are the spherical Coulomb functions, $\delta_{LJ}$ are the scattering phase shifts, $k$ is the wave number $k^2 = 2\mu E_s/\hbar^2$, and $E_s$ is the scattering energy.

The asymptotics of the wave function for the bound states is presented in the general form [8]

$$\chi_L(r) = \sqrt{2k_0}C_W W_{-\eta L+1/2}(2k_0 r),$$

where $C_W$ is the asymptotic constant (AC), $W_{-\eta L+1/2}$ is the Whittaker constant, η is the Coulomb parameter, $k_0$ is the wave number $k_0^2 = 2\mu E_d/\hbar^2$, and $E_d$ is the module of the deuteron energy.

The numbering wave function $\chi_L$ matches the asymptotics for the bound state (BS) or with scattering processes at distances about 10 – 20 fm, and it allows one to find the asymptotic constant or scattering phases $\delta_{LJ}$.

There are three types of two-parameter central potentials with FSs that were used in the preliminary calculations [9]:

Gaussian

$$V(r) = -V_g \exp(-\alpha_g r^2), \tag{1}$$

exponential

$$V(r) = -V_e \exp(-\beta_e r), \tag{2}$$

Yukawa

$$V(r) = -V_y \exp(-\gamma_y r)/r. \tag{3}$$

The experimental data on the nucleon-nucleon scattering phase shifts were taken from works by Brown and Jackson [10]. The potential parameters were chosen in such a way that one set of parameters ought to describe one partial scattering phase with the given $L$ at energies up to 400–500 MeV. The scattering length $a_0$ [9] is defined as

$$1/a_0 = -\lim_{k\to 0}(k \cot \delta_0) \tag{4}$$

and the triplet effective radius $r_0$ is expressed in terms of the binding energy of deuteron $k_0$ and the scattering length $a_0$ [14]

$$1/a_0 = k_0 - 1/2\, k_0^2\, r_0. \tag{5}$$

The parameters of potentials obtained in this way and the energies of the forbidden states are given in Table 1 [9]. There are two variants of parameters for the Gaussian $S$ interactions. The first of them is obtained on the basis of the phases at high energies, and the second one is obtained from the scattering characteristics at low energies only. The wrong form of the "tail" of the Gaussian potential, which determines the asymptotic behavior of the wave function, does not allow one to match the description of phases at low and high energies on the basis of unified parameters. The repulsive potentials for the triplet $^3D_1$ phase with $J = 1$ are used because the tensor component of the interaction is not taken into account.

**Table 1. Parameters of the np potentials of different types and energies of the forbidden states [9]. Here, $E_{FS}$ is the energy of the FS**

| $L_J$ | $V_g$ (MeV) | $\alpha_g$ (fm$^{-2}$) | $E_{FS}$ (MeV) | $V_e$ (MeV) | $\beta_e$ (fm$^{-1}$) | $E_{FS}$ (MeV) | $V_y$ (MeV) | $\gamma_y$ (fm$^{-1}$) | $E_{FS}$ (MeV) |
|---|---|---|---|---|---|---|---|---|---|
| $^1$S (1) | 1128.6 | 1.55 | 448.3 | 2990 | 3.1 | 514.5 | 3141.5 | 3.0 | 951.6 |
| $^1$S (2) | 1131.5 | 1.55 | 447 | – | – | – | – | – | – |
| $^1$P | 1050 | 1.3 | 126.7 | 3600 | 3.1 | 78.0 | 625 | 1.4 | 37.9 |
| $^1$D | 70 | 0.6 | – | 320 | 2.0 | – | 200 | 1.5 | – |
| $^3$S (1) | 1406.1 | 1.8 | 581.8 | 3094.7 | 3.0 | 581 | 3360.8 | 3.1 | 1197.6 |
| $^3$S (2) | 949.7 | 1.2 | 395.9 | – | – | – | – | – | – |
| $^3P_0$ | 830 | 0.8 | 160.9 | 2500 | 2.3 | 125.5 | 660 | 1.3 | 82.5 |
| $^3P_1$ | 990 | 1.2 | 125.6 | 4400 | 3.5 | 75.7 | 660 | 1.5 | 38.1 |
| $^3P_2$ | 3200 | 3.0 | 46.5 | 12300 | 5.0 | 686.6 | 3750 | 3.8 | 875.9 |
| $^3D_1$ | –120 | 0.4 | – | –530 | 1.6 | – | –360 | 1.1 | – |
| $^3D_2$ | 63 | 0.3 | – | 160 | 1.15 | – | 125 | 0.7 | – |
| $^3D_3$ | 11 | 0.3 | – | 55 | 1.5 | – | 14 | 0.4 | – |

The Schrödinger equation is solved for bound states of the Yukawa potentials starting with 0.003 fm, which, evidently, could be considered as an implementation of the cutoff radius, so all results essentially depend on this value. The step of 0.01 fm is used with the number of steps for finite-difference mesh equaling 2500. The changing of the step from 0.01 fm to 0.005 fm does not essentially influence the results of calculations of the scattering phases and changes the deuteron energy no more than 0.3% – 0.5%. The asymptotic constant $C_0$ (for the exponential asymptotic value) for wave functions of bound states is constant in the range of 6 – 25 fm [9]. The deuteron characteristics and low energy parameters, calculated with the potentials described above, are listed in Table 2 together with the data [10,11]. The values $R$, given in the last column, are the distances for which the wave function of the ground state has the node given in Ref. [9], $R_r$ is the deuteron radius obtained from the cluster expression and $R_f$ is the deuteron radius obtained from the deuteron Coulomb form factor [9].

The comparison of results of the calculation of phases for the Reid potential, given in Ref. [7] and obtained for the same potential in Ref. [9] on the basis of using finite-difference method, is listed in Table 3. This comparison demonstrates the accuracy of the used phase shift calculation method, which was applied in Ref. [9] for finding of WFs of BSs that were used later for calculations of radii and ACs (Table 1.2).

**Table 2. Characteristics of deuteron and parameters of the low energy np scattering [9]**

| Potentials | $E_d$ (MeV) | $R_r$ (fm) | $R_f$ (fm) | $C_0$ (fm$^{1/2}$) | $a_{0t}$ (fm) | $r_{0t}$ (fm) | $a_{0S}$ (fm) | $R$ (fm) |
|---|---|---|---|---|---|---|---|---|
| Gauss. 1 | –2.223 | 1.95 | 1.99 | 1.23 | 5.25 | 1.49 | –18.6 | 0.60 |
| Gauss. 2 | –2.223 | 2.02 | 2.06 | 1.29 | 5.41 | 1.75 | –23.1 | 0.71 |
| Expon. | –2.225 | 2.01 | 2.05 | 1.28 | 5.37 | 1.69 | –24.5 | 0.53 |
| Yukawa | –2.223 | 2.00 | 1.73 | 1.27 | 5.33 | 1.64 | –23.6 | 0.51 |
| Reid [7] | –2.2246 | – | – | – | 5.39 | 1.72 | –17.1 | – |
| Experiment [10,11] | –2.22464 | 1.9660(68) | – | 1.30(2) | 5.414(5) | 1.750(5) | –23.719(13) | – |

**Table 3. The comparison of the np scattering phase shifts for the Reid potential from Ref. [7] and calculated in Ref. [9]**

| $L_J$ | $E_{lab}$ (MeV) | Calculations from Ref. [7], (rad) | Calculations from Ref. [9], (rad) |
|---|---|---|---|
| $^1S_0$ | 48 | 0.696 | 0.687 |
| | 144 | 0.277 | 0.265 |
| | 208 | 0.093 | 0.082 |
| | 352 | –0.205 | –0.215 |
| $^1P_1$ | 48 | –0.071 | –0.072 |
| | 144 | 0.277 | 0.265 |
| | 208 | –0.456 | –0.458 |
| | 352 | –0.708 | –0.710 |
| $^1D_2$ | 48 | 0.027 | 0.028 |
| | 144 | 0.089 | 0.091 |
| | 208 | 0.123 | 0.125 |
| | 352 | 0.164 | 0.166 |
| $^3P_0$ | 48 | 0.198 | 0.206 |
| | 144 | 0.105 | 0.107 |
| | 208 | –0.012 | –0.010 |
| | 352 | –0.264 | –0.264 |
| $^3P_1$ | 48 | –0.133 | –0.139 |
| | 144 | –0.304 | –0.312 |
| | 208 | –0.386 | –0.393 |
| | 352 | –0.518 | –0.525 |

The calculation results of the scattering phase shifts in the triplet state are presented in Figures 1 – 3 (solid lines) for the first variant of Gaussian, exponential and Yukawa interactions, respectively. The dashed line in Figure 1a shows the triplet phase shift for the second variant of the Gaussian *S* potential, which describes low energy parameters better, but the calculated phases at high energies lay slightly lower than experimental data. The triplet *P*-phases and singlet *D*-phases of the Reid potential with soft core [7], calculated in Ref. [9], are shown in Figures 2.

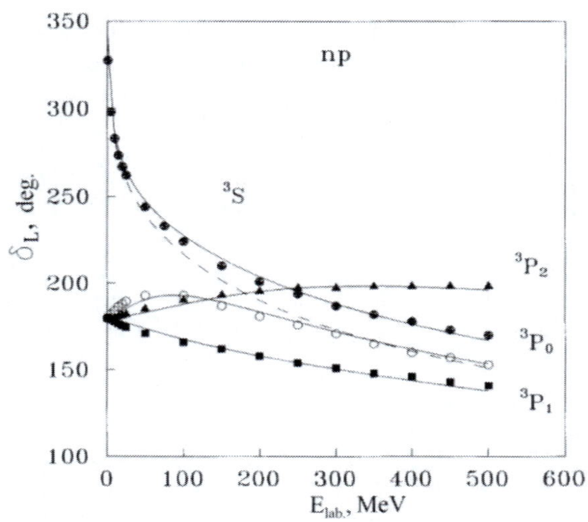


Figure 1a. The phase shifts of the np scattering. Lines are calculations for the Gaussian potentials from Table 1. Circles, triangles, circles, and squares are experimental data from Ref. [10].

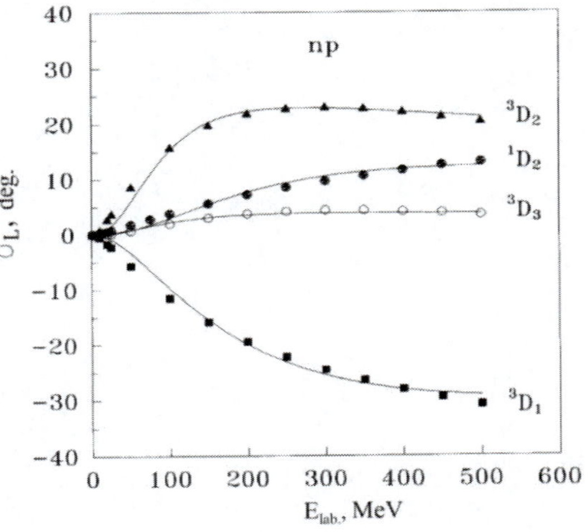


Figure 1b. The phase shifts of the np scattering. Lines are calculations for the Gaussian potentials from Table 1. Circles, triangles, circles, and squares are experimental data from Ref. [10].

The singlet phase shifts of the second Gaussian potential are shown in Figure 4a by the solid line, the results for exponential interaction are given by the dashed line, and the phase shifts for the Yukawa potential coincide with the solid line. The phases for the Reid singlet potentials with soft core [7] are shown by the dotted line.

The upper solid line in Figure 4b shows the calculation results of the phase shifts at low energies for the first singlet Gaussian *S*-wave potential, and the dashed line for the exponential one. The results for the Reid potential are presented by the solid line and for the Yukawa potential they are given by the dotted line. The dot-dashed line in Figure 4b shows the phases of the second singlet Gaussian potential and the dot-dot-dashed line shows the *S*-phase of the Gaussian potential with the OPEP addition and with cutoff at low distances from Refs. [5].

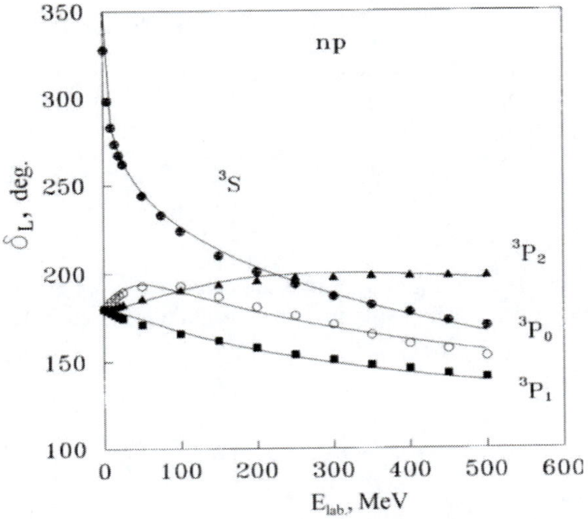


Figure 2a. The phase shifts of the np scattering. Solid (dashed) lines are calculations for the exponential potentials from Table 1 (the Reid potential with soft core [7]).

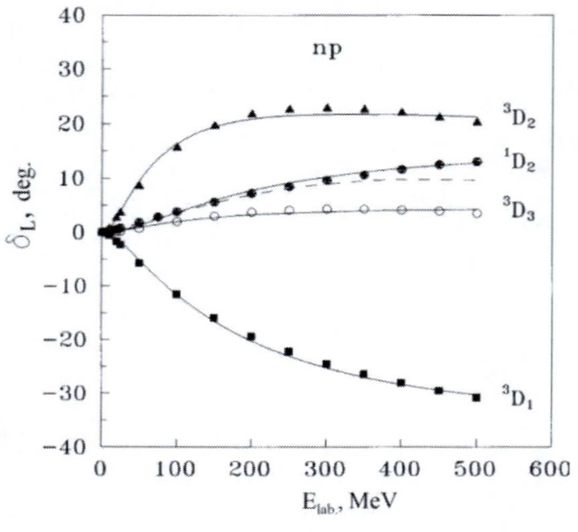


Figure 2b. The phase shifts of the np scattering. Solid (dashed) lines are calculations for the exponential potentials from Table 1 (the Reid potential with soft core [7]).

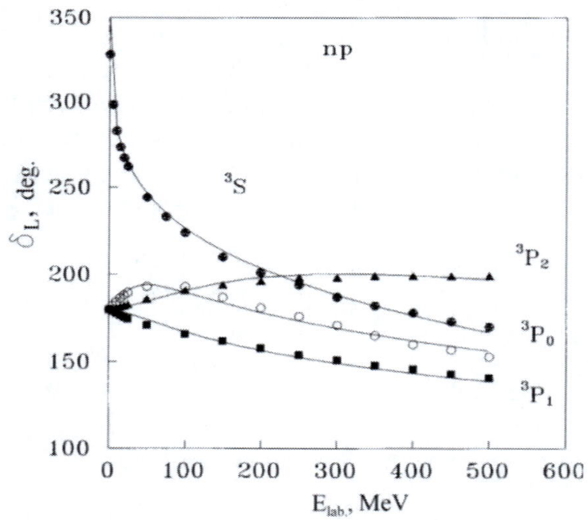


Figure 3a. The phase shifts of the np scattering. Lines are calculations for the Yukawa potentials from Table 1. Circles, triangles, circles, and squares are experimental data from Ref. [10].

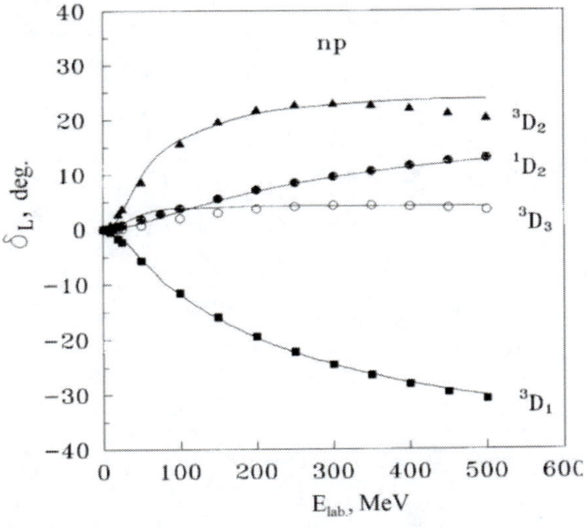


Figure 3b. The phase shifts of the np scattering. Lines are calculations for the Yukawa potentials from Table 1. Circles, triangles, circles, and squares are experimental data from Ref. [10].

Phase shifts of this interaction at energies above 100 MeV coincide with the solid line in Figure 4a. That potential is the best one to describe the phase shifts for energies 1–10 MeV at the expense of true behavior of ``tail" of the interaction, determined by taking into account the OPEP [5]. The Reid potential and the first Gaussian potential slightly underestimate the singlet scattering length, as is well seen in Table 2 and in Figure 4b.

The calculated deuteron form factor [11] for all above considered types of potentials describes the experimental data of Refs. [12, 13] only up to 3 $fm^{-1}$, because the *D*-component of the wave function is not taken into account here. The Gaussian parameterization with $\gamma = 0.0864\ fm^2$ was used as the proton form factor, and the radius of proton took on the value of 0.805(11) fm.

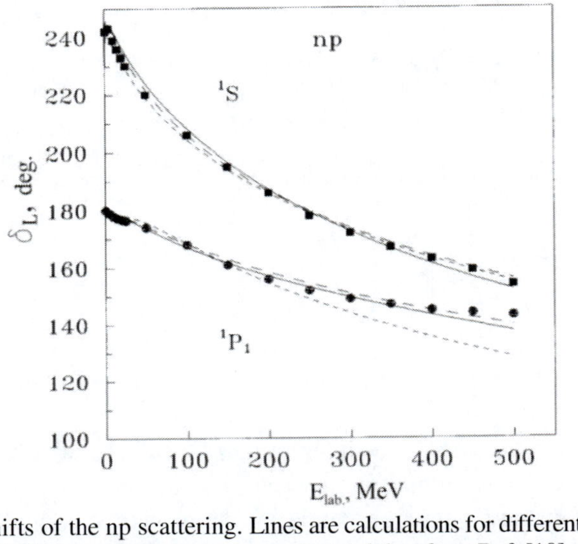


Figure 4a. The phase shifts of the np scattering. Lines are calculations for different potentials from Table 1, and for the Reid potential. Circles and squares are experimental data from Ref. [10].

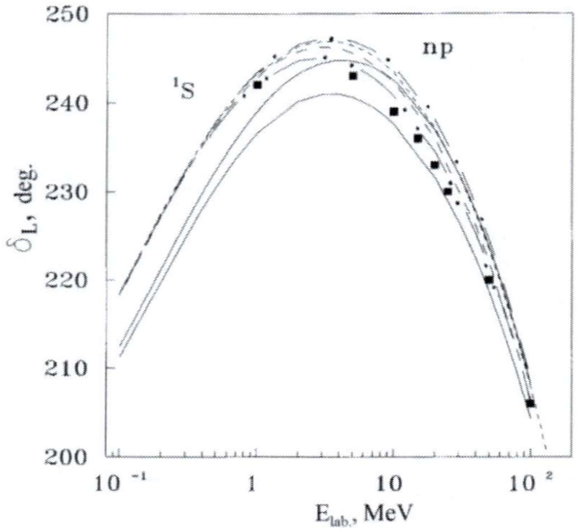


Figure 4b. The phase shifts of the np scattering. Lines are calculations for different potentials from Table 1, for the Reid potential, and for the interaction from Refs. [5]. Circles and squares are experimental data from Ref. [10].

The expression for the total capture cross section in the form [9] was used under consideration of the electromagnetic $E1$ transitions in deuteron

$$\sigma_c(NJ,J_f)=\frac{8\pi Ke^2}{\hbar^2 q^3}\frac{\mu}{(2S_1+1)(2S_2+1)}\frac{J+1}{J[(2J+1)!!]^2}A_J^2(NJ,K)\sum_{L_i,J_i}P_J^2(NJ,J_f,J_i)I_J^2(J_f,J_i), \tag{6}$$

where the matrix elements of the $E1$ transition have the form [9]

$$A_J(EJ,K)=K^J\mu^J\left(\frac{Z_1}{m^J_1}+(-1)^J\frac{Z_2}{m^J_2}\right),\quad I_J(J_f,J_i)=\langle\chi_f|R^J|\chi_i\rangle$$

$$P_J^2(EJ,J_f,J_i)=\delta_{S_iS_f}\left[(2J+1)(2L_i+1)(2J_i+1)(2J_f+1)\right](L_i0J0|L_f0)^2\begin{Bmatrix}L_i & S & J_i\\ J_f & J & L_f\end{Bmatrix}^2,$$

and the approximate formula of dipole transition at photodisintegration is [11]:

$$\sigma_c(E1)=\frac{8\pi}{3}\frac{e^2}{\hbar c}\frac{\hbar^2}{m}\frac{W^{1/2}E^{3/2}}{(E+W)}\left(\frac{1}{1-k_0r_0}\right), \tag{7}$$

where $W$ is the binding energy of deuteron and $E$ is the energy of nucleons in continuous spectrum.

The calculation results of the $E1$ disintegration of deuteron are shown in Figure 5. The dashed line shows the cross section obtained from the approximate formula, the solid one – for the second variant of the Gaussian potential of the ground state and the dotted line – for the first variant, calculated with the help of Eq. (6).

Because the first variant of the Gaussian potential leads to slightly underestimated effective radius, the disintegration cross section lays lower than the experiment given in Refs. [14] and the approximated results. The calculated cross sections reproduce the experimental data quite well up to 70 MeV for the second Gaussian interaction at relatively correct values of low energy parameters.

The calculation results of the pp scattering that are shown in Figure 6 were obtained on the basis of the parameters of the exponential potential taking into account Coulomb interaction, given in Table 1. It can be seen that the potential allows one to obtain acceptable description the experimental data from Ref. [10] and in this case too.

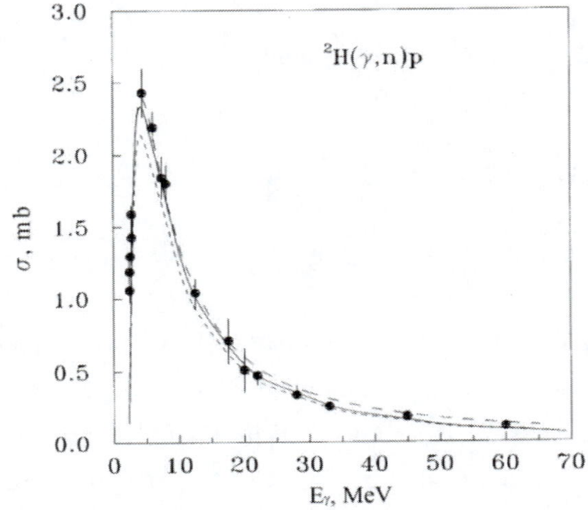


Figure 5. The total cross sections of the deuteron photodisintegration for potentials from Table 1. Experimental data are from Refs. [14].

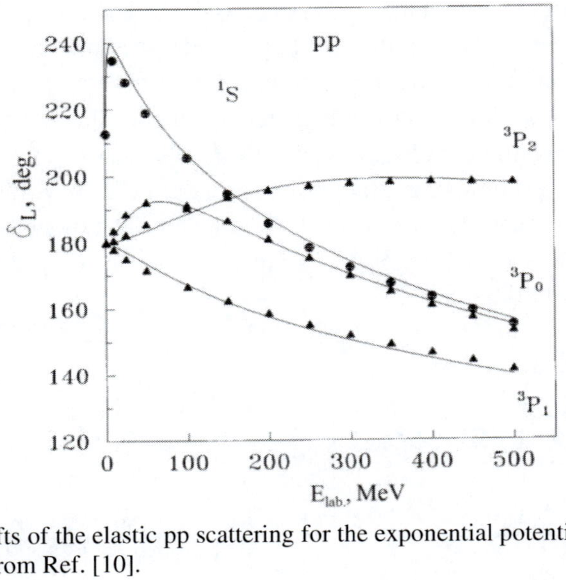


Figure 6. The phase shifts of the elastic pp scattering for the exponential potential from Table 1. Experimental data are from Ref. [10].

Thereby, three studied forms of potentials give practically identical description of the experimental scattering phases for all considered energies. However, the exponential form of interaction leads to the slightly better description of the experimental data at low energies than for Gaussian potential. It is possible to match parameters with phases at high energies with the characteristics of low energy scattering. The Yukawa type potential, coinciding in shape with the OPEP, allows a correct reproduction of the *S*- and *P*-phase shifts, although the description of the *D*-phases is obtained slightly worse.

## 3. Nucleon-Nucleon Forces with Tensor Component

We now examine the variants of deep phenomenological np and pp interaction for the Schrödinger equation with the tensor part

$$u''(r) + [k^2 - V_c(r) - V_{Coul}(r)]\, u(r) = \sqrt{8}\, V_t(r)\, w(r),$$

$$w''(r) + [k^2 - V_c(r) - 6/r^2 - V_{Coul}(r) + 2\, V_t(r)]\, w(r) = \sqrt{8}\, V_t(r)\, u(r), \qquad (8)$$

where $V_{Coul}(\mathrm{r}) = 2\mu Z_1 Z_2 / r\hbar^2$ is the Coulomb potential, μ is the reduced mass of nucleus in the considered cluster channel, the constant $\hbar^2/\mathrm{m}_0$ usually equals 41.4686 or 41.47 MeV fm$^2$, $k^2 = 2\mu E/\hbar^2$ is the wave number of the cluster relative motion, $V_c$ is the central part of the potential, $V_t$ is the tensor part of the interaction potential, *u* and *w* are the scattering wave functions with the orbital moments $L = 0$ and 2.

As a potential we use the exponential form with tensor part in the triplet spin channel and the OPEP that plays the basic role at long distances. The dependence of the parameters α and β was observed in Refs. [5] for the Gaussian potential, the last of which is the cutoff OPEP parameter. By fixing one of them within certain bounds, it is practically always possible to find the values of another one, which reproduce experimental data. Consequently, the value of 3.2433 fm$^{-1}$ was proposed for the cutoff parameter β, which, as it will be seen later, turns out to be close to the obtained here value α, so far as in this case α = β. Thereby by using the exponential form of interaction, it is possible to exclude the third parameter and reduce the potential to the two-parameter form Ref. [15]. As a result the exponential potential has only two adjustable parameters $V_0$ and α, and it is presented in the form that is analogous to the Gaussian interaction from Refs. [5].

$$V(r) = V_C(r) + V_T(r)\, S_{12},$$

$$V_C(r) = -V_0 \exp(-\alpha r) + V_{0C}(r)\, g(r),$$

$$V_T(r) = V_{0T}(r)\, g^3(r). \qquad (9)$$

The potentials of one-pion exchange have the form:

$$V_{0C}(r) = -V_1 \exp(-\mu r)/(\mu r),\ V_{0T}(r) = -V_1[1 + 3/(\mu r) + 3/(\mu r)^2] \exp(-\mu r)/(\mu r). \quad (10)$$

These are cutoff at low distances via factor $g(r) = 1 - \exp(-\alpha r)$ [5]. As opposed to the Gaussian interaction, the cutoff OPEP parameter $\alpha$ coincides with the width parameter of the exponential potential.

The average mass of $\pi$ mesons taken as the value of $m_\pi = 138.03$ MeV [5, 15] leads to value of $\mu = m_\pi/(\hbar c) = 0.6995\ \text{fm}^{-1}$. Constants are equal to $\hbar c = 197.327$ MeV fm and $\hbar^2/m_N = 41.47\ \text{MeV fm}^2$. If we use the experimental value of the $\pi$NN coupling constant $f^2 = 0.0776(9)$ [16], then for the depth of the OPEP we will obtain the value $V_1 = m_\pi f^2 = 10.71(12)$ MeV. Note that in Ref. [5] slightly other values are used, notably $V_1 = 10.69$ MeV and $f^2 = 0.07745$.

**Table 4. The comparison of the characteristics for deuteron and np scattering for the Reid soft core alternative potential (RSCA) and the Moscow potential (MP)**

| Deuteron characteristics | Calculation from the work [7] (Reid) | Calculation from the work [15] (Reid) | Calculation from the work [5] (MP) | Calculation from the work [15] (MP) |
|---|---|---|---|---|
| $E_d$ (MeV) | –2.22464 | –2.22458 | –2.2246 | –2.22456 |
| $Q_d$ (fm$^2$) | 0.2762 | 0.276 | 0.2860 | 0.285 |
| $P_D$ (%) | 6.217 | 6.22 | 6.78 | 6.78 |
| $A_S$ | 0.87758 | 0.875(2) | 0.8814 | 0.879(1) |
| $\eta = A_D/A_S$ | 0.02596 | 0.0260(2) | 0.0269 | 0.0270(2) |
| $a_t$ (fm) | 5.390 | 5.39 | 5.40 | 5.40 |
| $r_t$ (fm) | 1.720 | 1.72 | 1.73 | 1.73 |
| $a_s$ (fm) | –17.1 | –17.1 | –23.73 | –23.70 |
| $r_s$ (fm) | 2.80 | 2.81 | 2.61 | 2.62 |
| $R_d$ (fm) | 1.956 | 1.95 | 1.961 | 1.95 |

The use of methods of calculation of the deuteron characteristics and the np system were verified on the Reid soft core alternative potential (RSCA) [7] and on the Moscow potential [5]. The comparison of the results obtained for these potentials in Refs. [5, 7] with calculations [15] are listed in Table 4. The errors in the calculated asymptotic constants are caused by the averaging of their values over distances ranging from 7 – 15 fm. Consequently, Table 4 demonstrates the accuracy of methods of calculation used in Ref. [15].

Furthermore, parameters of the exponential potentials were matched exclusively on the basis of the description of the low energy characteristics of the np scattering. *The binding energy of deuteron, and then the high energy phase shifts and mixing parameters* [15] *were calculated with the obtained potential*, the experimental data are given in Ref. [10]. The same value of 10.69 MeV, as in Ref. [5], was used at first for the depth of the OPEP $V_1$. Consequently, it was found that the experiment can be described if $\alpha = 3.25\ \text{fm}^{-1}$ and with the depth of the central part of 1033.885 MeV – we will consider these parameters as the first variant of triplet potential. The scattering phase shifts and the mixing parameter of the first triplet potential are shown in Figs. 7 by the dotted line, which for the $S$-wave merges with the solid one.

The calculated scattering length and the effective radius are equal to $a_t = 5.410$ fm and $r_t = 1.741$ fm, and are in good agreement with experiment [16]. The value of –2.2246 MeV was obtained for the deuteron binding energy, the quadruple moment equals 0.285 $fm^2$ and the probability of the *D*-state equals 6.78% [15], which also agrees with the data from works such as [16]. Potentials, such as in Ref. [5], in the triplet and singlet variants have deep underlying unobservable forbidden states, leading to a node in the wave functions of deuteron and scattering processes. The location of such node approximately corresponds to the distance at which potentials with a core usually lead to the "extinction" of the relative cluster motion wave function.

We have to specifically note that the two-parameter form allows one to fix potential parameters unambiguously. The changing of the α parameter, on all sides from the given above value, gives overestimated or underestimated values of the effective radius, even in this case, it is possible correctly reproduce the scattering length. The parameter α for the triplet potential should to be in the range of 3.2 – 3.3 $fm^{-1}$ for the correct and simultaneous description of these two values, so that their values must be within the range of experimental errors. This interval is also equal to the value of 3.2 – 3.3 $fm^{-1}$ for the singlet potential. Because the errors for the singlet effective radius are relatively large, then, generally, it is always possible to find a variant of singlet and triplet potentials with the same width parameter α, which equals about 3.22 – 3.23 $fm^{-1}$.

For example, if one uses the value of 10.71 MeV as the depth of the OPEP $V_1$, then it is possible to find $\alpha = 3.20$ $fm^{-1}$ with the depth of the central part as 1062.81 MeV for parameters of the triplet potential – this is the second variant for the triplet potential. The next parameters were obtained for the singlet potential (the first variant of the singlet potential): $\alpha = 3.30$ $fm^{-1}$ and $V_0 = 3170.72$ MeV, which leads to the effective radius of 2.732 fm and the scattering length of –23.716 fm, and is in good agreement with the experimental data [16]. If one wants to coordinate the singlet potential with the experimental data from Ref. [11], then the parameters $\alpha = 3.23$ $fm^{-1}$ and $V_0 = 3037.4$ MeV (the second variant of the singlet potential) have to be used. These parameters lead to the effective radius of 2.768 fm and the scattering length equals –23.721 fm.

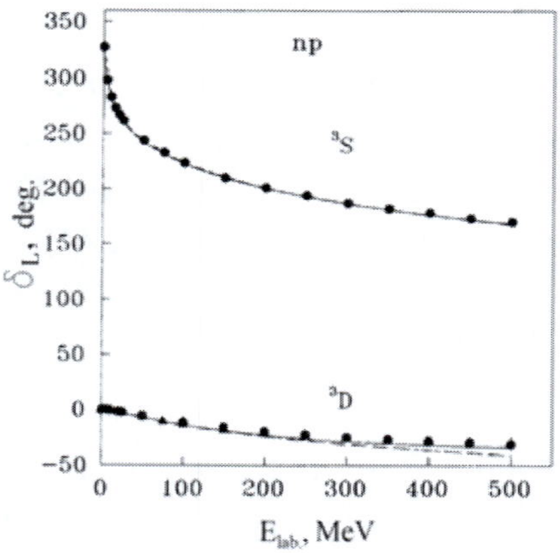


Figure 7a. The triplet phase shifts of the np scattering. Lines are calculations with different potentials. Experimental data are from Ref. [10]. Dashed lines show the phases of the MP [5].

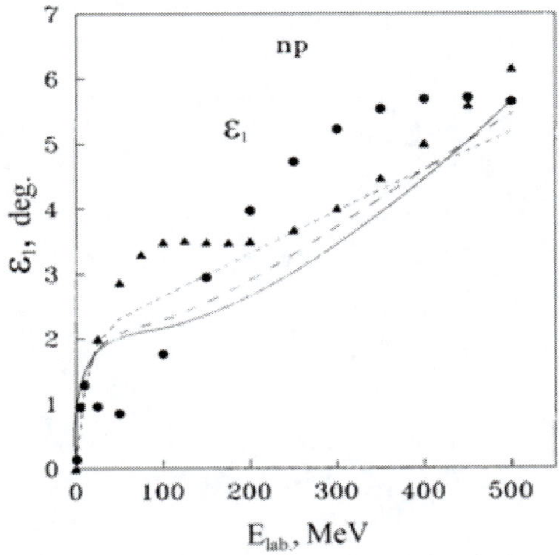


Figure 7b. The mixing parameters for the triplet scattering phase shifts. Lines are calculations with different potentials. Experimental data are from Ref. [10]. Dashed lines show the phases of the MP [5].

The characteristics of deuteron and np scattering, calculated with these potentials, are listed below in Table 5 – the first variant of the singlet exponential potential, the second variant of triplet, and the modified variant of the MP were used (see below).

The singlet scattering phase shifts of the first variant of the potential with $\alpha = 3.30$ fm$^{-1}$ are shown in Figures 8 by the solid line. Dashed lines in Figures 8 show the phases of the MP [5]. The scattering wave functions for the first variant of singlet interaction at different energies are shown in Figure 9. The deuteron form factors, obtained for the above given second variant of the triplet potential with $V_1 = 10.71$ MeV, are shown in Figures 10 by the solid line. The form factor for the Reid potential [7] is shown by the dotted line, and the results for the Moscow potential [5] are shown by the dashed line. The experimental data on form factor are given in Refs. [12, 13].

**Table 5. The characteristics of deuteron and np scattering for the exponential potential and the modified variant of the Moscow potential (MP)**

| Deuteron characteristics | Exponential potential | Calculation for the modified MP | Experiment from work [16] | Experiment from work [11] |
|---|---|---|---|---|
| $E_d$ (MeV) | –2.2246 | –2.22453 | –2.224579(9) | –2.22452(20) |
| $Q_d$ (fm$^2$) | 0.286 | 0.286 | 0.2859(3) | 0.286 |
| $P_D$ (%) | 6.736 | 6.776 | – | – |
| $A_S$ | 0.881(1) | 0.879(1) | 0.8802(20) | – |
| $\eta = A_D/A_S$ | 0.0269(1) | 0.0269(1) | 0.0271(4) | – |
| $a_t$ (fm) | 5.418 | 5.401 | 5.419(7) | 5.414(5) |
| $r_t$ (fm) | 1.752 | 1.729 | 1.754(8) | 1.750(5) |
| $a_s$ (fm) | –23.716 | – | –23.715(15) | –23.719(13) |
| $r_s$ (fm) | 2.732 | – | 2.73(3) | 2.76(5) |
| $R_d$ (fm) | 1.961 | 1.956 | 1.9560 (68) | – |

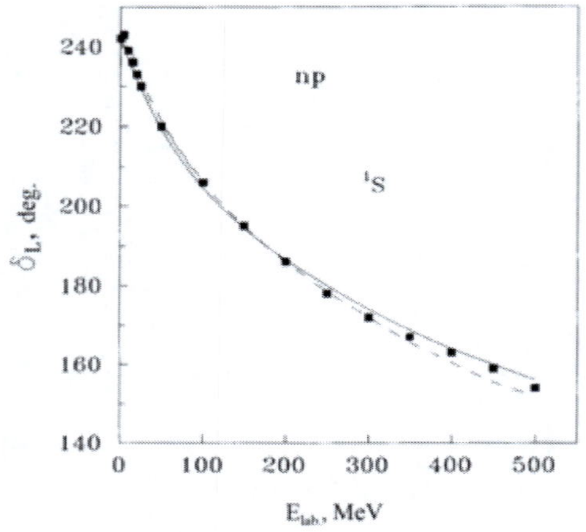


Figure 8a. The singlet phase shifts of the np scattering. Lines are calculations with different potentials. Experimental data are from Ref. [10].

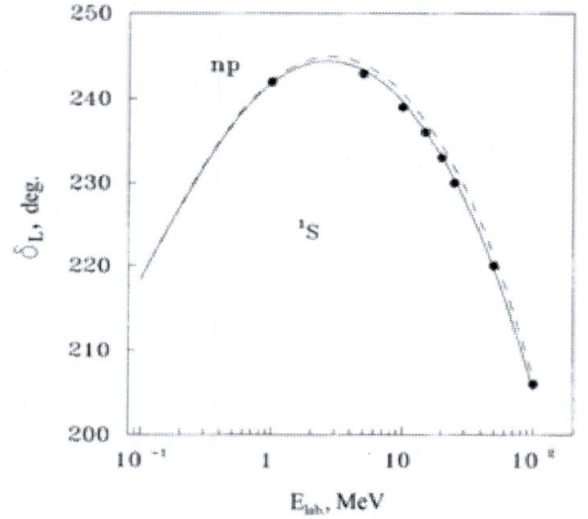


Figure 8b. The singlet phase shifts of the np scattering. Lines are calculations with different potentials. Experimental data are from Ref. [10].

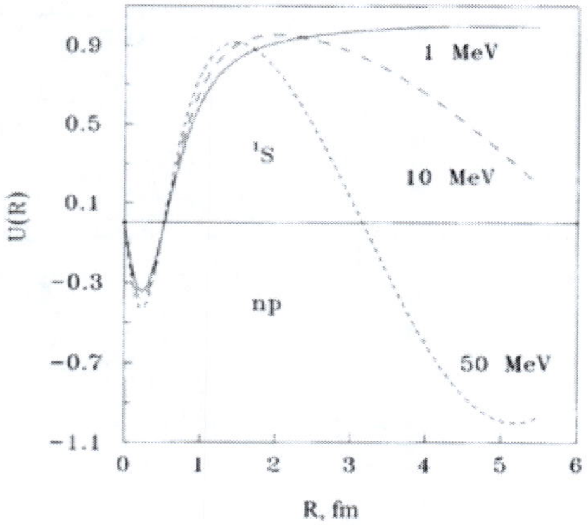


Figure 9. The scattering wave functions for the singlet exponential potential at different energies.

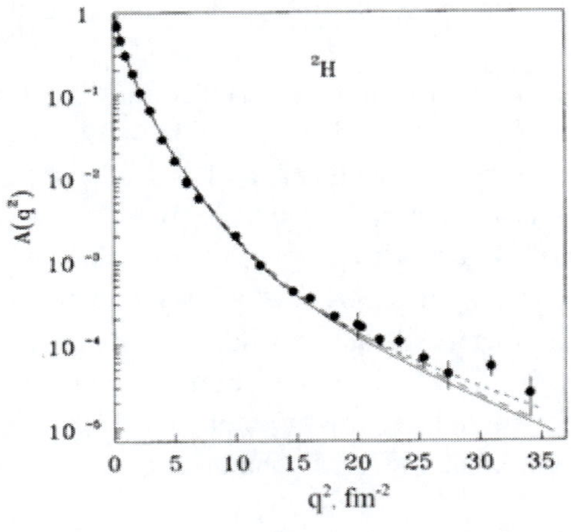


Figure 10a. The deuteron form factor A(q) for different potentials. Experimental data for the form factor are from Refs. [12, 13].

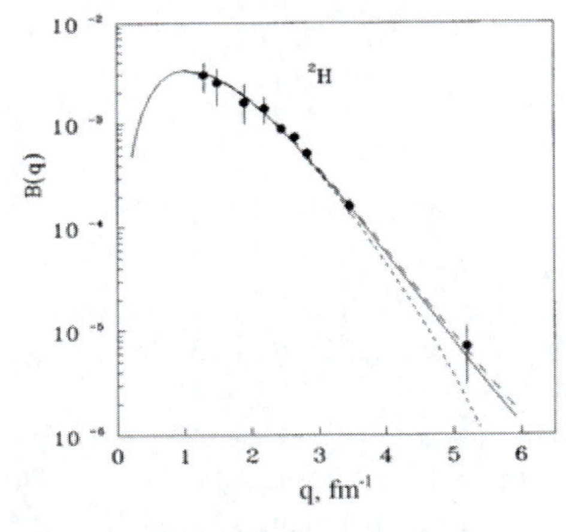


Figure 10b. The deuteron form factor B(q) for different potentials. Experimental data for the form factor are from Refs. [12, 13].

## 4. Interdependence of the Potential Parameters

Because the parameters of the triplet potentials are fixed quite unambiguously, it is possible to find the interconnection between them. Particularly, let us assume $V_0 = AV_1$ and $\alpha = B\mu$, where $A$ and $B$ are unknown dimensionless parameters, and $V_1 = 10.71$ MeV and $\mu = 0.6995$ fm$^{-1}$. Using such a form of parameter notation, one manages to find that the parameter $B$, determining the potential radius, might be presented in the form $B = A^{1/3}$. Then, the value of the depth parameter $A$ has to be equal to 97.765 and equivalently $V_0 = 1047.06$ MeV and $\alpha = 3.2224$ fm$^{-1}$. This potential leads to the scattering length of 5.415 fm and to the effective radius of 1.748 fm, and its phase shift practically does not differ from the results of the previous variant of triplet interaction. The value of $-2.2245$ MeV was obtained for the deuteron binding energy, the value of 0.285 $fm^2$ for the quadruple moment, and 6.76% for the probability of the *D*-state.

Since the πNN coupling constant is known within certain experimental error, so the value of the parameter *A* approximately lies in the range of 96–99. This interval can slightly be amplified if one takes into account the fact that $a_t$ and $r_t$ also have certain experimental errors. Obviously that use of such constant interconnection is no more successful than the suggestion connecting the phenomenological contact between them and has no microscopic justification. However, the use of the form of the potential allows one to obtain quite unambiguous correspondence with the parameters $V_0$ and α, and, consequently between *A* and *B*, at the limit of experimental errors. Besides, as is seen from the obtained results, it is possible to use the same width parameter for singlet and triplet potentials. But in this case, the value located on the upper limit of the experimental errors is obtained, and the value of the triplet effective radius is on the lower limit.

Allowing that a certain functional dependence between potential parameters always exists, independently of their specific form, it is possible to consider modification of the triplet interaction of the Gaussian form of work such as [5]

$$V(r) = V_C(r) + V_T(r)\, S_{12},$$

$$V_C(r) = -V_0 \exp(-\alpha r^2) + V_{0T}(r)\, g(r),$$

$$V_T(r) = V_{0T}(r)\, g^3(r),$$

with one-pion potentials, given in Eq. (10) and with the cutoff of the form $g(r) = 1 - \exp(-\beta r)$. The following parameters of the potential in Ref. [5] were suggested: $V_0 = 349.023$ MeV, $\alpha = 1.6163$ $fm^{-2}$, and $\beta = 3.2433$ $fm^{-1}$.

In trying to find the functional dependence between these parameters similarly to the exponential interaction, it is possible to assume $V_0 = AV_1$, $\alpha = B\mu^2$, and $\beta = C\mu$, where *A*, *B* and *C* are unknown dimensionless parameters, and $V_1 = 10.69$ MeV and $\mu = 0.6995$ $fm^{-1}$. Using such parameters, it is possible to find parameters *B* and *C* and to determine the radius of the potential and the cutoff OPEP factor which can be presented in the form $B = (1/3A)^{1/2}$ and $C = (2/3A)^{1/2}$. So, the value of the depth parameter *A* should be equal to 32.2875, which is equivalent to $V_0 = 345.1534$ MeV, $\alpha = 1.6052$ $fm^{-2}$, and $\beta = 3.2453$ $fm^{-1}$. The potential parameters obtained in such a way differ little from the parameters given above. The value of the parameter *A* is located in the interval of 31 – 33, because of the existence of experimental errors of the πNN coupled constant. The calculated results of deuteron characteristics and np scattering for this modified potential are listed in Table 5. The phase shifts, the mixing parameters, the wave functions and the form factor practically coincide with the results for the initial potential of Ref. [5], shown in Figures 7 by the dashed line. Thereby, one can see that the modified variant of the Moscow potential leads to the same results of the interaction with the initial parameters. However, in this case we have only one adjustable parameter *A*, and others are connected with it by a certain empirical dependence. Meanwhile, the value of the sole parameter varied on experimental data is determined unambiguously.

## 5. The NN Potential Without Node in the *D*-Wave

All above considered NN potentials (exponential [15] and Gaussian [5]) lead to a node in the *S*- and *D*-components of the deuteron wave function and point to the presence of a forbidden level in the *D*-wave. However, this does not agree with the given above classification, because tableaux {6} and {42} allow only one bound *D*-state, which corresponds to the ground state of deuteron [6].

Therefore, let us consider another variant of the deep np exponential interaction with the tensor part and the OPEP with depth $V_1 = 10.71$ MeV, which does not have a node in the *D*-wave and does not lead to an unnecessary forbidden state. The potential has the form (9) similar to one given in Ref. [15], however the cutoff factor in the tensor part raises not to the third but to the 20$^{th}$ power which allows one to avoid the appearance of a node in the *D*-component of the deuteron wave function. The potential parameters, as usual, are chosen on the basis of the description of low energy scattering characteristics and binding energy of deuteron. Furthermore, the other high energy scattering phases, mixing parameters and other deuteron characteristics are calculated with these parameters, the detailed description of them together with the experimental data is possible to find, for example, in reviews [16]. Consequently, it was found that the correct description of the experiment can be obtained, if one accepts for the triplet potential $\alpha = 3.9$ fm$^{-1}$ with the depth of the central part 3652.86 MeV [17] – third variant of the triplet potential. The resulting calculations of different deuteron characteristics and np scattering are listed in Table 6, and scattering phases are shown in Figures 7 by the solid lines.

The wave function in the *S*-wave has the node at 0.5 fm, and the function in the *D*-wave is nodeless, as is shown in Figure 11 by the solid line. The dashed line shows the results for the Gaussian potential with two nodes from Refs. [5]. The deuteron form factors calculated with these wave functions practically do not differ from the results shown in Figs. 10 by solid line. The momentum distributions of nucleons in deuteron are shown in Figure 12 for the Reid potential by dotted line, for the given above exponential interaction by solid line, and for the Moscow potential by dashed line. Thus, one can see that all considered characteristics of the np system can be successfully described on the basis of the two-parameter deep exponential potential with the forbidden bound state in the *S*-wave, which does not have unobserved forbidden *D*-level.

**Table 6. The comparison of the characteristics of deuteron and np scattering calculated for the exponential potential without node in the *D*-wave with the experimental data**

| Characteristics of deuteron | The exponential potential without node in the *D*-wave | Experiment from Ref. [16] |
|---|---|---|
| $E_d$ (MeV) | –2.2246 | –2.224579(9) |
| $Q_d$ (fm$^2$) | 0.283 | 0.2859(3) |
| $P_D$ (%) | 6.22 | – |
| $A_S$ | 0.881(1) | 0.8802(20) |
| $\eta = A_D/A_S$ | 0.0265(2) | 0.0271(4) |
| $a_t$ (fm) | 5.417 | 5.419(7) |
| $r_t$ (fm) | 1.751 | 1.754(8) |
| $R_d$ (fm) | 1.960 | 1.9560 (68) |

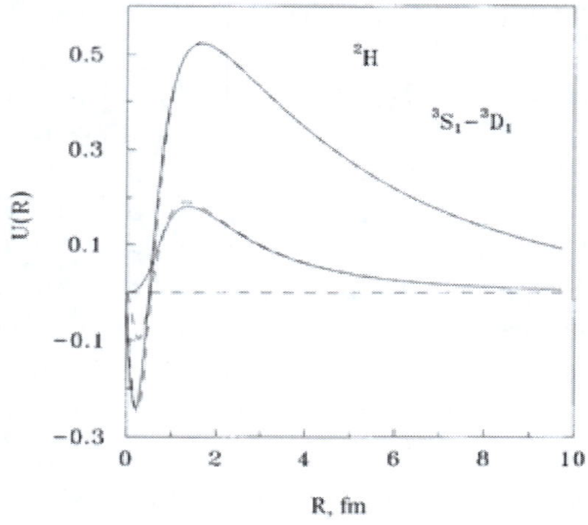


Figure 11. The wave functions for the bound triplet state of deuteron for the Gaussian [5] and exponential np-potentials.

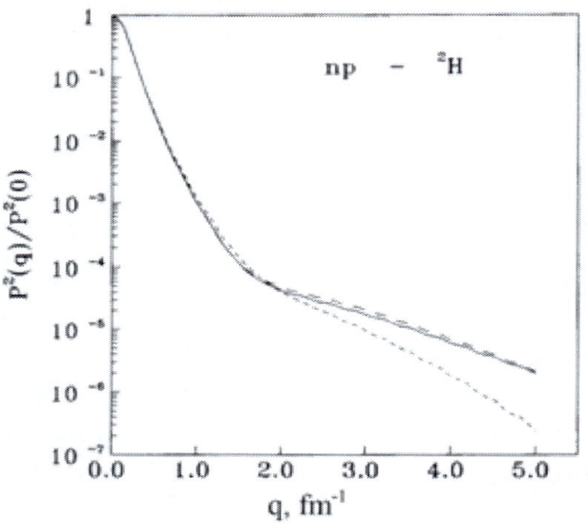


Figure 12. Momentum distributions of nucleons in deuteron for different nucleon-nucleon potentials.

# 6. The Deuteron Characteristics for Nijmegen Potentials

Several new variants of the phenomenological nucleon-nucleon potentials [18] with repulsive cores were recently suggested. They were parameterized on the basis of the phase shift analysis carried out by the Nijmegen group [19]. At that time, the variants of potentials Nijm-1, Nijm-2 and Reid-93 were some of the best from all suggested NN interactions, because they ensure $\chi^2$ about 1.03 in the energy range up to 350 MeV. For example, the classical variant of the Reid-68 [7] or the Paris-80 [20] potentials lead to the $\chi^2$ about 2 – 3 in the energy range up to 300 MeV.

The fourth variant of the Nijmegen potential Nijm-93 gives approximately the same value $\chi^2 = 1.9$, but at energies up to 350 MeV [18, 21]. As was shown in work [21], only Argonne potential, from 1984 [22], gives an approximately small value of $\chi^2 = 3.3$ in the range up to 350 MeV. Other potentials like Johnson-62, Reid-68, Urban-81, Bonn-89 etc. lead to large values of $\chi^2$ in this energy range, because they are parameterized on the basis of a more narrow energy range.

The phase shift analysis [19] was done in the comparatively limited energy range 0–350 MeV, but, generally, is in a good agreement with other known results [23], where the phase shift analysis was done in the energy range up to 1600 MeV and [24] for the range of 0–2500 MeV. Some characteristics of deuteron and low energy np scattering [25] were adjusted by the Nijmegen group on the basis of the phase shift analysis [19].

It should be mentioned that for the parameterization of all variants of Nijmegen potentials the coupling πNN constant $f^2 = 0.074$ was used, which is noticeably smaller than results given in Ref. [16] 0.0776(9) and Ref. [26] 0.0803(14). The used value of the coupling constant more exactly agrees with the data [27] where the value of 0.0760(8) was obtained and with results of Ref. [28] where the value of 0.0760(2) is given. The detailed analysis of many NN characteristics and properties of deuteron [19] was done with the obtained in Ref. [18] interactions; the comparison of the results with the other known potentials [21] was done. The three-body calculations of the binding energy of $^3$H were carried out and they lead, for different variants of interactions, to the value from –7.7 MeV to –7.6 MeV [29], at the experimental binding energy of –8.48 MeV.

However, the deuteron form factors were not considered in the above noted works. Such calculations can be carried out on the basis of wave functions that were obtained in works [18], and the numerical results were given in Ref. [30]. But the wave functions in the range 0–25 fm from Ref. [30] are given with the variable-pitch and this complicates the calculations. Therefore, we have to approximate the numerical wave function [30] for obtaining its values in arbitrary points. For the approximation of the numerical wave functions of the Nijmegen potential, we have used the expansion into a series on gaussoids of the form:

$$R_L(r) = r^L \sum_k C_k \exp(-\alpha_k r^2), \tag{12}$$

where $C_k$ are coefficients and $\alpha_k$ are expansion parameters of the total radial function, which for $L = 0$ and 2 is represented as

$R_0(r) = u(r)/r$, $R_2 = w(r)/r$,

here $u(r)$ and $w(r)$ are the solutions to the radial Schrödinger Eq. (8) for the $S$- and $D$-waves. The summation was carried out up to $N = 13$ which allows one to approximate the numerical function at the range 0–10 fm rather accurately. The excitation coefficients in the process of approximation were found so that the total wave function should be normalized to unity. For example, the normalization for the variant Nijm-1 is equal to 1.00002, i.e., the error of normalization is equal to 0.002% only. The approximation wave function (AWF) at the calculation of the deuteron characteristics was joined at long distances with the asymptotics of the form [18, 25]

$u(r) \to A_s \exp(-r/R)$,

$w(r) \rightarrow A_d\,[1 + 3R/r + 3(R/r)^2]\exp(-r/R).$

Here $R = 4.319$ fm and $A_d = \eta A_s$ are with the asymptotic constants $A_s = 0.8845(8)$ and $\eta = 0.0253(2)$ [25]. The difference between the accurate wave function and its asymptotics in the $S$ wave at the distance 9 – 10 fm, for example, for the variant Nijm-1 is equal to $10^{-4}$, for the $D$-wave about $10^{-3}$. Therefore, it is quite possible to use the above asymptotics of the wave function at such distances. The calculated results of deuteron properties with the AWF and a comparison with accurate results [18] are given in Appendix 1 together with the relative errors, which are shown as the deviation of the approximation function from the numerical results [30]. One can see good agreement of the characteristics obtained with the approximation wave functions and the results of Refs. [18]. A difference is observed only for the variant Nijm-1, which can be explained by the bad behavior of the $D$ asymptotic wave function at long distances. Meanwhile, in this case, the increasing of the dimension of the series (12) up to 15 does not lead to a visible improvement of the results.

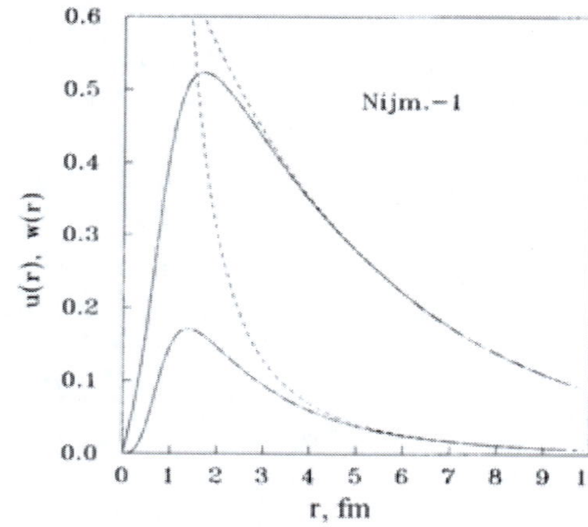


Figure 13. The wave functions for the Nijm-1 potential and their asymptotics.

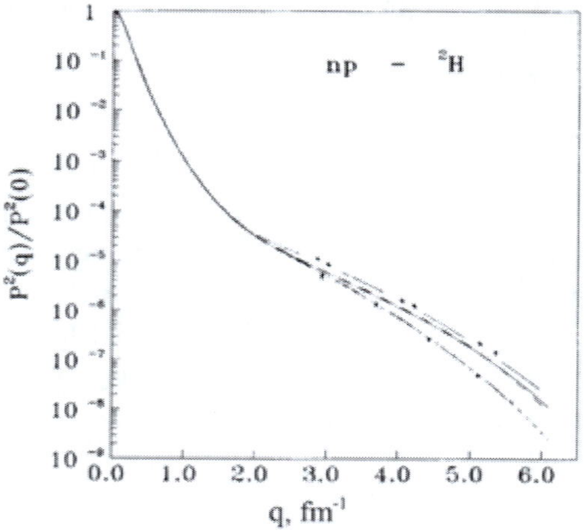


Figure 14. Momentum distributions for different variants of the Nijmegen potentials.

The accurate deuteron wave functions for the variant Nijm-1 are shown in Figure 13 by the solid lines [18,30], and the dashed lines that practically merge with the solid ones which are their approximations. The asymptotics of these wave functions are shown by the dotted lines. One can see that already at 9 – 10 fm the asymptotic lines totally merge with the accurate solution and approximation, as for the *S*- and for the *D*-wave merge too.

Instead of the above considered characteristics, it is possible to compare the results for momentum distributions, given in Ref. [30] for different variants of the potential. The momentum distributions calculated with the AWFs and with accurate WFs [30] are normalized to unity at the zero momentum transfer. The momentum distributions of nucleons in deuteron for the variants of Nijm-1 (dot-dashed line) and Nijm-93 (dashed line) obtained with the AWF are shown in Figure 14 in comparison with the accurate results of Ref. [30] – dotted and solid lines, respectively. One can see that a difference exists only for the potential Nijm-93, but for the first variant of Nijm-1 the lines coincide completely. The results for the alternative Reid potential with the soft core [7] are shown by the dot-dot-dashed line.

As is clear from the given results, it is possible to match the deuteron characteristics calculated with the AWFs and with the accurate numerical functions [30] for all considered variants of the potentials. Therefore, it is practical to use the obtained AWFs for calculation of the deuteron form factors.

$$\frac{d\sigma}{d\Omega} = \left(\frac{d\sigma_M}{d\Omega}\right)\left[A + B\tan^2\left(\frac{\theta}{2}\right)\right],$$

$$A = G_0^2 + G_2^2 + \frac{2}{3}\eta(1+\eta)G_M^2, \quad B = \frac{4}{3}\eta(1+\eta)^2 G_M^2,$$

$$G_0 = 2G_E C_E, \quad G_2 = 2G_E C_Q, \quad G_M = \frac{M_d}{M_P}\left(2G_{M0}C_S + G_E C_L\right),$$

$$2G_E = G_{Ep} + G_{En}, \qquad 2G_{M0} = G_{Mp} + G_{Mn},$$

$$\eta = \frac{(\hbar cq)^2}{4M_d^2} = 0.002767q^2,$$

$$C_E = \int\left(u^2 + w^2\right)j_0(x)dr,$$

$$C_Q = 2\int w\left(u - \frac{w}{\sqrt{8}}\right)j_2(x)dr,$$

$$C_L = \frac{3}{2}\int w^2\left(j_0(x) + j_2(x)\right)dr, \quad x = \frac{qr}{2},$$

$$C_S = \int\left(u^2 - \frac{w^2}{2}\right)j_0(x)dr + \frac{1}{\sqrt{2}}\int w\left(u + \frac{w}{\sqrt{2}}\right)j_2(x)dr.$$

Here $u(r)$ and $w(r)$ are wave functions of the bound state, and $j_i$ are $i$-order Bessel functions. One can use the following values: $M_p = 938.28$ MeV and $M_n = 939.57$ MeV [16] for the nucleon masses, the deuteron mass is equal to 1875.63 MeV. The charged form factor of neutron can be set to zero, and the parameterization [31] is used in the capacity of the charged form factor of proton:

$$G_{Ep} = \frac{1}{\left(1 + 0.054844q^2\right)^2}.$$

Here $q$ is the momentum transfer measured in $fm^{-1}$. The magnetic form factors of nucleons were obtained on the basis of the ``scale law" [31]

$$G_{Mp} = \mu_p G_{Ep},\ G_{Mn} = \mu_n G_{Ep},$$

The following values are used in the capacity of the magnetic moments of nucleons:

$$\mu_p = 2.7928\ \mu_0,\ \mu_n = -1.9131\ \mu_0.$$

The calculated results of the form factors are shown in Figs. 15 together with the experimental data of Ref. [13]. The dotted line in Figure 15a shows the results for the Reid-68 potential [7], and the dashed one – for the variant Nijm-1. The results for the potentials Nijm-2 and Nijm-93 practically coincide and are shown by the solid line, and the Reid-93 is presented by the dot-dashed line that merges with the solid line. The results for the Reid-93 in Figure 15b coincide with the classical Reid-68 and are shown by the dotted line. The dashed line shows the variant Nijm-2, results for the variant Nijm-93 are shown by the solid line, and the dot-dashed line shows the variant Nijm-1, which effectively coincides with the solid line.

Non-relativistic formulae for the form factors, which were used in these calculations, were used earlier in Refs. [5, 13]. In Ref. [32], the comparison of relativistic and non-relativistic calculation methods for the Paris and the Argonne potentials was presented and it was shown that in the range of 3.5 – 4.0 $fm^{-1}$ these results practically coincide. The relativistic effects become appreciable at high momentum transfers of 5.0 – 6.0 $fm^{-1}$, but do not change the shape of form factors essentially and do not play a dominant role. These relativistic corrections were examined in Ref. [33] and it may be considered that they do not give a decisive contribution up to 5.0 – 6.0 $fm^{-1}$, and their value does not exceed the available experimental errors. Thereby, one can see from the given results that the obtained AWFs are quite able to correctly describe the behavior of the numerical functions for all variants of the Nijmegen potentials and allow one to obtain a good agreement of the considered characteristics of deuteron with the calculations for the accurate wave functions. The deuteron form factors obtained on the basis of AWFs are in full agreement with the known experimental data and do not have significant difference between each other in the considered range of momentum transfers [34].

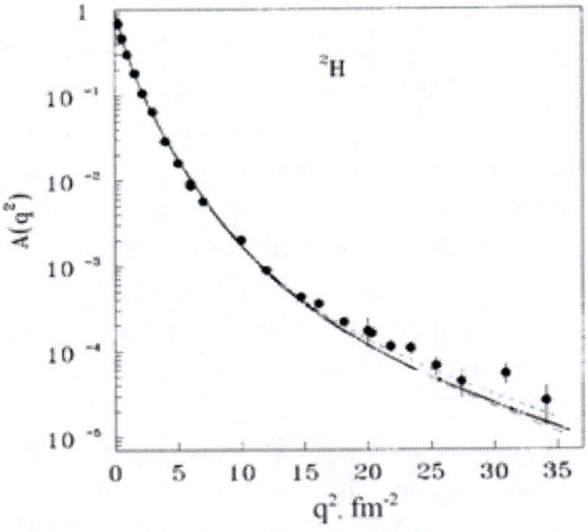


Figure 15a. The deuteron form factor A(q) for the Nijmegen potentials.

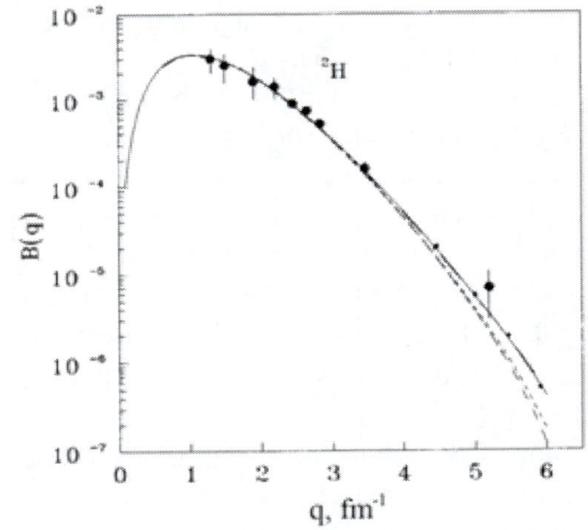


Figure 15b. The deuteron form factor B(q) for the Nijmegen potentials.

## 7. The Influence of the Coupling πNN Constant on the Deuteron Characteristics

The above obtained deep exponential NN potential with a node only in the *S* wave based on the coupling πNN constant $f^2 = 0.0776$, which was found from the analysis of the NN scattering characteristics in [16]. At the same time, as it was said before, there are other results: particularly in works of the Nijmegen group [25] the value of $f^2 = 0.074$ was obtained, and in the researches carried out by the GWU/VPI [27] the value of 0.076 was found.

Therefore, it is interesting to clear up the dependence of parameters of the deep potential with the forbidden states and the quality of the description the results of the NN phase shift analysis with this potential and the properties of deuteron from the value of coupling πNN constant. We will use here the same form of the local NN potential, as in Eq. (9) with the cutoff in the 20th power and with the OPEP of the form (10). The parameters of the potential for each $f^2$ were matched in terms of description of the effective radius $r_0$, the scattering length $a_0$ and the binding energy of deuteron $E_d$, and then along with them were calculated the np scattering phase shifts up to 500 MeV, the mixing parameters and the other characteristics of deuteron. The recent data on the NN phase shifts are given in Refs. [24,25], and the other deuteron characteristics can be found, for example, in Refs. [16]. The calculated results of different deuteron characteristics and np scattering for the obtained variants of potentials and the parameters of these potentials with the value of constant $f^2$ are given in the Appendix 2 [35]. One can see that in all cases it is possible to reproduce all considered characteristics, and the value of the quadruple moment slightly decreases with the reducing of $f^2$, and it is between values of Refs. [16,18].

The power of the cutoff factor of the OPEP, equaling $N = 20$, was chosen in terms of necessity to remove node in the *D*-wave. The small value of $N = 3 - 5$ does not allow one to get rid of the node [5,15], and only increasing *N* up to 15 gradually leads to its disappearance. In general, the value of the power of the cutoff has a quite weak influence on the results and can be approximately in the range of 15–30. Further increasing of this value leads to the degradation of the results of the np scattering phase shifts, although the general form of the deuteron wave function essentially does not change.

The deuteron wave function for any variant of the potentials has the node in the *S* wave approximately in the range 0.50 – 0.52 fm, and the *D*-wave has no node, as is shown in Figure11 by the solid line for the variant of the potential with $f^2 = 0.074$. The dashed line shows the wave function of the Gaussian potential from Ref. [5]. The scattering phases and mixing parameters, obtained with the given in Appendix 2 parameters for the variant $f^2 = 0.074$, are shown in Figures 16 by the solid line. The dot-dashed line shows the results of the phase shift analysis SM97 [24] carried out by the group from the GWU/VPI, and the dotted line shows the phases obtained by the Nijmegen group [25]. The dashed line shows the phases of the potential of Refs. [5].

The quality of description of the triplet phase shifts for $f^2$ value more than 0.076, for example, becomes appreciably worse for 0.0776 – 0.0803 [16,26], in particular the *D* phase slopes down quicker than it follows from the results of the phase shift analysis [24]. The dashed line shows it well in Figure 16b for the Gaussian potential with $f^2 = 0.07745$ from Ref. [5]. In terms of the best description of the *D* phase shift the coupling πNN constant value equals 0.074 is more preferable, as it was suggested in Refs. [25].

The deuteron form factors calculated with the wave function of the ground state of the exponential interaction at $f^2 = 0.074$ are shown in Figures 17 by the solid lines; the dashed line shows the results for the potential from Refs. [5]. The dotted line shows the form factor for the Reid potential [7], the dot-dashed line is for the interaction Nijm-93 [18]. The experimental data are from Refs. [13].

It was shown in Refs. [36] that the potentials with the node in the *D* wave, obtained in Refs. [5], lead to the strongly overestimated cross sections of the disintegration of deuteron by protons, and the $p^2H$ backward scattering at the big momentum transfers. This behavior of the cross sections connects with the relatively high value of the momentum transfers for the potentials that lead to two nodes in the wave function from Refs. [5, 15].

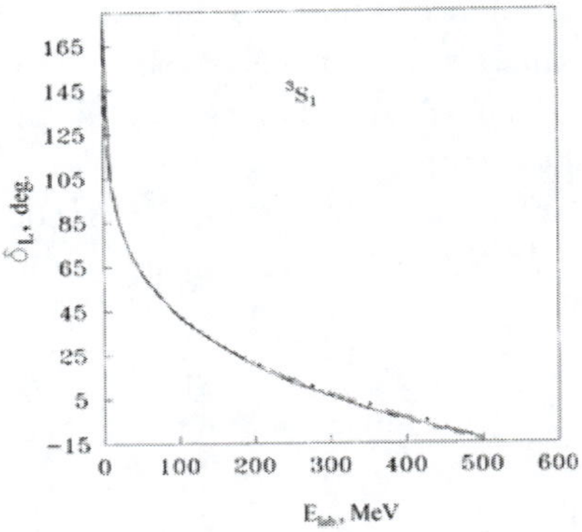


Figure 16a. The triplet *S* phase shift of the NN scattering is solid line. Dot-dashed line is results of Ref. [24], dotted line is the phase shifts obtained in Ref. [25], and dashed line shows the phases of the potential of Refs. [5].

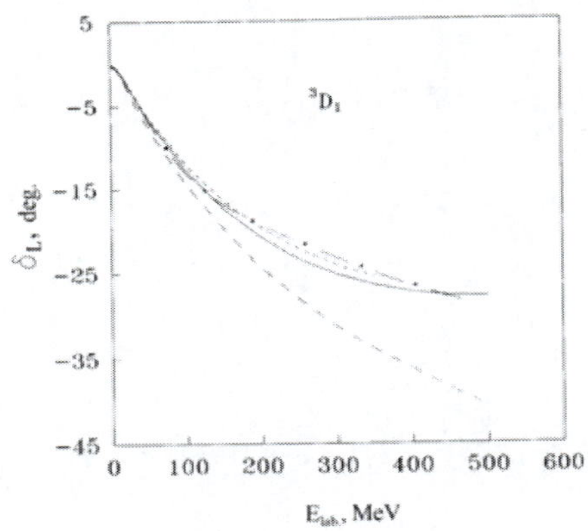


Figure 16b. The triplet *D* phase shift of the NN scattering is solid line. Dot-dashed line is results of Ref. [24], dotted line is the phase shifts obtained in Ref. [25], and dashed line shows the phases of the potential of Refs. [5].

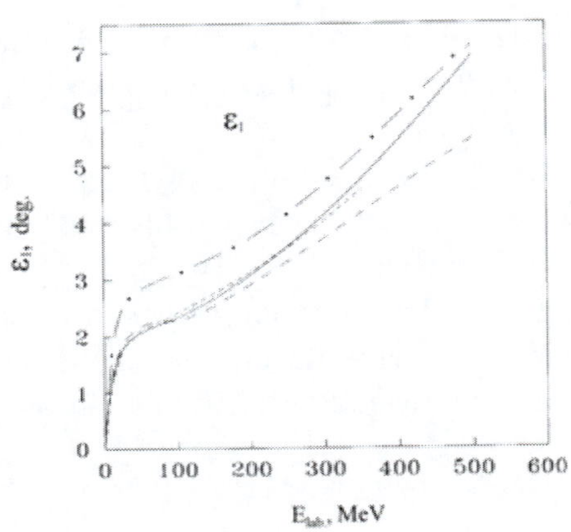


Figure 16c. The mixing parameters for the triplet phase shifts of the NN scattering is solid line. Dot-dashed line is results of Ref. [24], dotted line is the phase shifts obtained in Ref. [25], dashed line shows the phases of the potential of Refs. [5].

The decrease of the momentum distribution's value for the momentum transfers is about 3 $fm^{-1}$ which, obtained for the interaction with the node only in the *S*-wave, can lead to the improvement of the behavior of the calculated cross sections of the elastic $p^2H$ scattering and to the deuteron disintegration by protons.

The tensor and vector polarizations at the elastic $e^2H$ scattering, the definitions of which can be found, for example, in Refs. [32, 37], were considered for solving these problems with the obtained here potentials.

$$d\sigma/d\Omega = S\,(d\sigma/d\Omega)_{Mott},$$

$$S = A(q) + B(q)\tan^2(\theta_e/2),$$

$$A(q) = G_c^2(q) + (8\eta^2/9)\,G_q^2(q) + (2\eta/3)\,G_m^2(q),$$

$$B(q) = (4\eta/3)(1+\eta)\,G_m^2(q),$$

$$t_{20} = -1/(\sqrt{2}\,S)\{(8\eta/3)G_cG_q + (8\eta^2/9)G_q^2 + (\eta/3)[1 + 2(1+\eta)\tan^2(\theta_e/2)]\,G_m^2\},$$

$$t_{21} = 2\eta/(\sqrt{3}\,S)\,[\eta + \eta^2\sin^2(\theta_e/2)]^{1/2}\,G_m G_q \sec(\theta_e/2),$$

$$t_{22} = -\,\eta/(2\sqrt{3}\,S)G_m^2,$$

$$t_{10} = \eta/S\sqrt{\frac{2}{3}}\,[(1+\eta)(1+\eta\sin^2(\theta_e/2))]^{1/2}\tan(\theta_e/2)\sec(\theta_e/2)\,G_m^2,$$

$$t_{11} = 2/(\sqrt{3}\,S)\,[\eta(1+\eta)]^{1/2}\tan(\theta_e/2)\,G_m(G_c + (\eta/3)\,G_q),$$

$$G_c(0) = 1,\ G_q(0) = M_d^2 Q_d = 25.83,\ G_m(0) = (M_d/M_p)\,\mu_d = 1.714, \qquad (13)$$

were A(q) and B(q) are the deuteron form factors, $\eta$ is the Coulomb parameter and the definitions of structure functions $G_c$, $G_q$, and $G_m$ through wave functions are given in Refs. [32, 37].

In the review [32], it was shown by way of the example of the Argonne v14 potential that the relativistic effects (using the relativistic impulse approximation) become visible only at comparatively high momentum transfers of 4 – 6 $fm^{-1}$, slightly changing the value of the polarization $T_{20}$ and form factors of deuteron. But the next accounting of the meson $\rho\pi\gamma$ and $\omega\sigma\gamma$ currents in common with the relativistic impulse approximation acts in the opposite direction, nearly compensating the influence of relativistic effects in this range of momentum transfers.

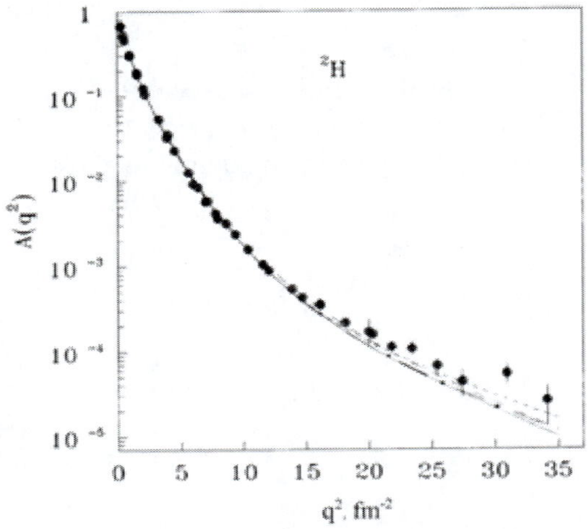


Figure 17a. The deuteron form factor A(q) for different NN potentials.

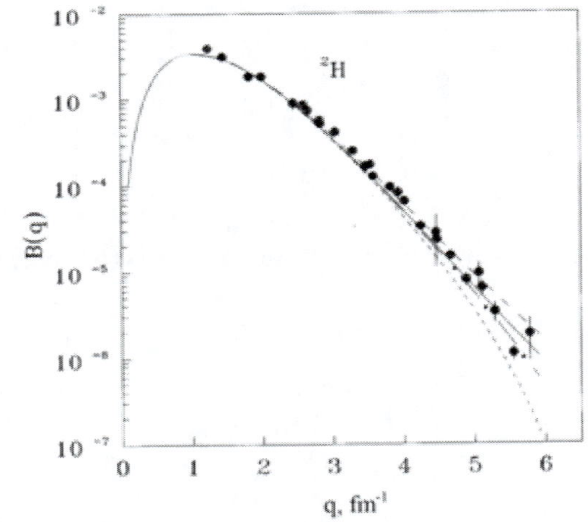


Figure 17b. The deuteron form factor B(q) for different NN potentials.

Consequently, the changes of form factors and polarizations that lead to the simultaneous accounting of such effects do not exceed the values of relatively large experimental errors. Therefore, in our calculations we use the given above nonrelativistic expressions (13).

The calculation results of tensor and vector polarizations for the variant of the potential with $f^2 = 0.074$ are shown in Figures 18 and 20 by the solid lines, and the long dashed lines show the results for potential from [5]. The dot-dashed lines show the polarization for the Nijmegen potential Nijm-1 [18], the dotted line that for the Reid-68 potential [7]. The dot-dot-dashed lines show results for the Nijm-2 [18] and the short dashes that for the Paris interaction [20]. The experimental data on tensor polarizations from Refs. [32, 37] are given by points. It is clear that the results for different potentials practically coincide at small momentum transfers. At $q > 15\ \text{fm}^{-1}$, there is a noticeable difference between results for the potentials with core and forbidden states. However, the absence of data at large momentum transfers and the experimental errors do not allow one to come to final conclusions in favor of any form of the NN interactions.

As it is clear now, the potential from Ref. [5] leads to the understated value of $T_{20}$ at momentum values above 20 $fm^{-2}$. The variant of the potential at $f^2 = 0.074$ suggested here gives slightly better results, although they lie distinctly below the available experimental data.

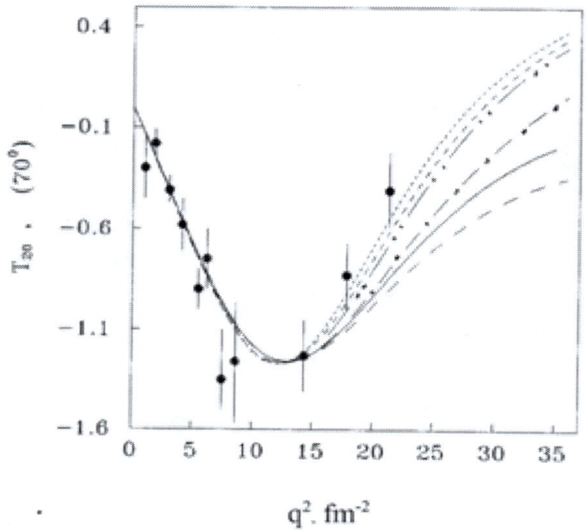


Figure 18a. The tensor polarizations in the elastic $e^2H$ scattering for different NN potentials.

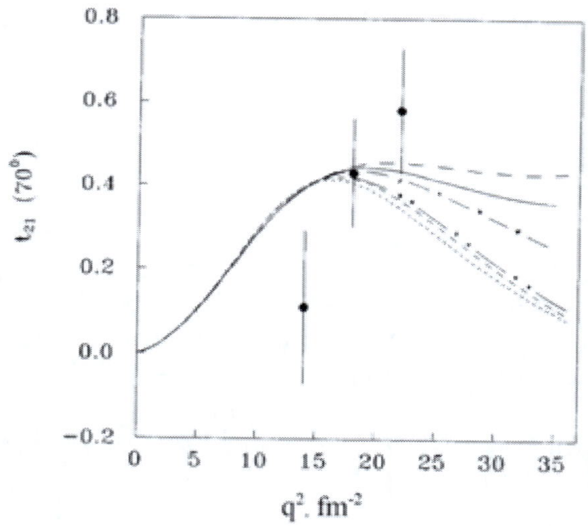


Figure 18b. The tensor polarizations in the elastic $e^2H$ scattering for different NN potentials.

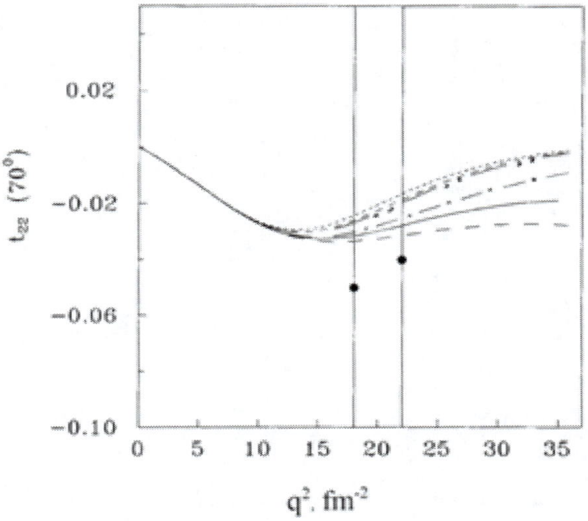


Figure 18c. The tensor polarizations in the elastic $e^2H$ scattering for different NN potentials.

There are quite a few experimental data for other tensor characteristics $t_{21}$ and $t_{22}$, and the errors of such measured data are too large to use them in choosing the type of potential. The experimental data on vector polarizations are absent and it is only possible to compare calculation results for different NN potentials. Thereby, the small value of the coupling πNN constant, which allows one to describe the available experimental data on deuteron and np scattering, turns out to be preferable for the considered type of interaction, the same as for the potentials of the Nijmegen group [18].

## 8. The Last Results for the EP and for New NN Potentials

Recently, in the second half of $90^{th}$, the new experimental data on tensor polarization in the $e^2H$ scattering [38] have appeared. Therefore, it is interesting to recalculate results for the EP from the start. Let us consider tensor polarizations of deuteron in the elastic $e^2H$ scattering for the described EP and compare them with the results for some new NN potentials [39]. The calculation results for the EP at $f^2 = 0.074$ (LP2 from the Attachment 2) are listed in Table 7 together with the known data [16, 18] and with the results for new variant of the Argonne potential v18 [40].

The calculated results of tensor polarization of deuteron for the EP [39] are shown in Figures 19 by solid lines. Dot-dot-dashed line shows the results for the Nijmegen potential Nijm-2 [18]. Dot-dashed lines show the polarizations for a new variant of the Bonn potential CD-Bonn [41], and dotted lines are for a variant of the Argonne v18 [40]. Short dashes lines show the results for the Paris potential [20], and long dashes lines show polarizations for the Gaussian potential with nodes in *S*- and *D*-waves from Refs. [5], which practically coincide with the results for two-nodal exponential potential [15]. The experimental data from Refs. [32, 37, 38] are shown by circles.

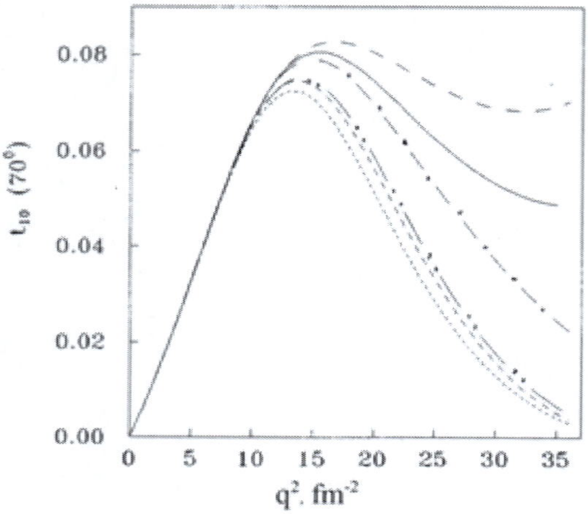


Figure 19a. The vector $t_{10}$ polarization in the elastic $e^2H$ scattering for different NN potentials.

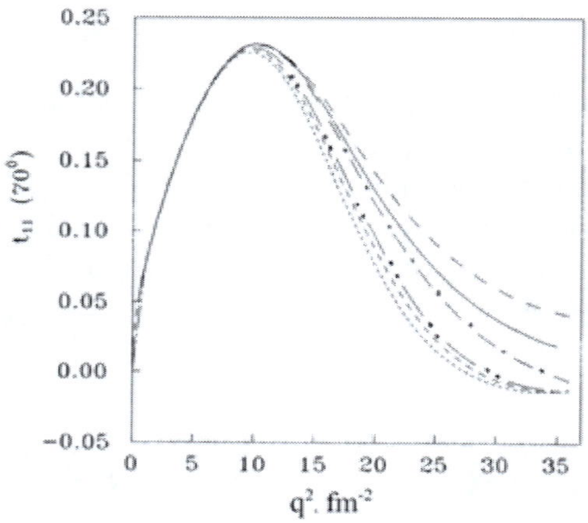


Figure 19b. The vector $t_{11}$ polarization in the elastic $e^2H$ scattering for different NN potentials.

**Table 7. The characteristics of deuteron and np scattering for different variants of the NN potentials**

| Characteristics of deuteron and np scattering | LP2 | Nijm.93 [18] | CERN [16] | Argonne v18 [40] |
|---|---|---|---|---|
| $E_d$ (MeV) | –2.2246 | –2.224575(9) | –2.224579(9) | –2.224575 |
| $Q_d$ (fm$^2$) | 0.274 | 0.271(1) | 0.2859(3) | 0,270 |
| $P_D$ (%) | 5.75 | 5.67 | – | 5,76 |
| $A_S$ | 0.884(1) | 0.8845(8) | 0.8802(20) | 0,8850 |
| $\eta = A_S/A_D$ | 0.0256(1) | 0.0253(2) | 0.0271(4) | 0,0250 |
| $a_t$ (fm) | 5.417 | 5.4194(20) | 5.419(7) | 5,419 |
| $r_t$ (fm) | 1.753 | 1.7536(25) | 1.754(8) | 1,753 |
| $R_d$ (fm) | 1.966 | 1.9676(10) | 1.9560(68) | 1,967 |
| $V_1$ (MeV) | 10.214 | – | 10.71(12) | – |
| $f^2_{\pi NN}$ | 0.0740 | 0.074 | 0.0776 | – |
| $\alpha$ (fm$^{-1}$) | 3.72 | – | – | – |
| $V_0$ (MeV) | 3524.172 | – | – | – |

There is apparent difference between results for two groups of potentials at $q > 15$ fm$^{-2}$. The Argonne, the Paris and the Nijm-2 potentials are the part of the first, and the second one consists of the Bonn and the EP from Refs. [35, 39] and the MP from work [5]. Dividing them into two groups becomes clear, because the WF of deuteron for the Bonn potential has the node in the *D*-wave at the distances lower 0.22 fm, and it immediately leads to the results similar to that obtained for the EP and the MP potentials with FSs.

As one can see from Figure 20a, the best results are given by the potentials with core from the first group, and the curves for potentials, leading to the node in the WF, lie slightly lower than the available data. However, the influence of the relativistic effects and the meson currents on the calculation results of different characteristics, form factors and polarizations of deuteron in the elastic $e^2H$ scattering for the potentials with FSs were still not considered, which can be slightly different than for the interactions with core [32]. As one can see from Figure 20a, the absence of the node in the *D* wave for the EP potential from Refs, [35,39] leads to the results that are slightly better (the solid line) than the interaction with two nodes (long dashes) [5], which can be compared with the results for the CD-Bonn potential. Besides, none of the considered potentials describe the last two or even three points of the experimental polarizations $T_{20}$ above 20 fm$^{-2}$.

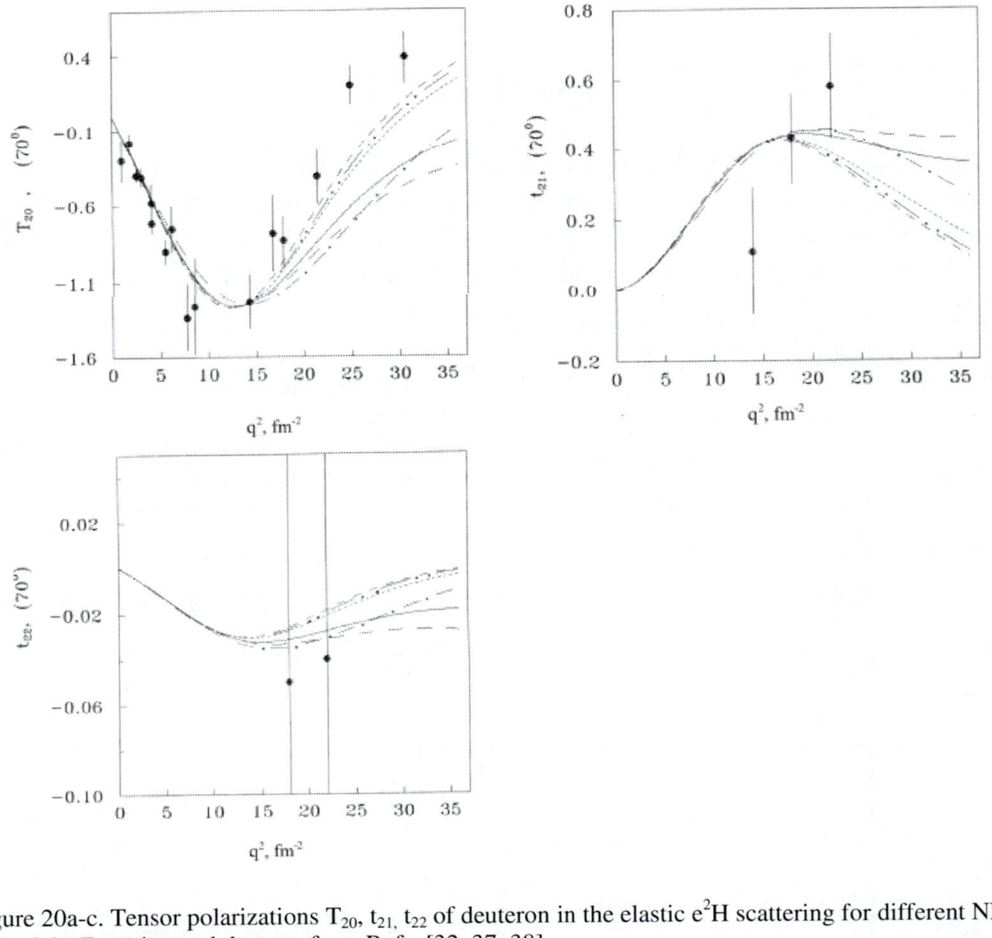


Figure 20a-c. Tensor polarizations $T_{20}$, $t_{21}$, $t_{22}$ of deuteron in the elastic $e^2H$ scattering for different NN potentials. Experimental data are from Refs. [32, 37, 38].

# CONCLUSION

The basic facilities of the Gaussian and the exponential potentials with FSs, their positive and negative features, and, most importantly, the domain of applicability were shown in this review. The simplicity of the EP, having only three parameters ($V$, $\alpha$, $N = 20$), its consistency with the general-theoretical ideas about structure of FSs and the possibility to describe all considered characteristics of the NN system at the energy up to 500 MeV and the momentum transfers up to 15 – 20 $fm^{-2}$, undoubtedly are attractive features.

The incorrect description of the tensor polarizations in the $e^2H$ scattering at high momentum transfers, specifically, higher than 4 $fm^{-1}$, can be caused by the absence of taking into account the relativistic effects and the meson currents, and, possibly, some other effects for the potentials of such type, i.e., NN interactions that contain states forbidden by the Pauli principle.

These studies, as far as is known to authors of the EP, have not been updated and their results can differ from the earlier obtained conclusions for the potentials with core [32].

In conclusion, the author expresses his deep gratitude to I.I. Strakovsky for direct participation in the final part of this work and A.V. Dzhazairov-Kakhramanov for extremely useful discussion of certain parts of this work and assistance in its translation.

# APPENDIX 1

The comparison of the deuteron characteristics, calculated with the accurate and the approximation wave functions for different variants of the Nijmegen potentials. Here ΔES and ΔED are the average relative error of approximation for the WFs for $S$ and $D$ waves. The results for the Nijmegen potentials and the experimental data are taken from [18, 19].

| Characteristics of deuteron | Results for Nijm-1 | Results for Nijm-1 with AWF | Results for Nijm-2 | Results for Nijm-2 with AWF | Results for Nijm93 | Results for Nijm93 with AWF | Results for Reid93 | Results for Reid93 with AWF | Exp. results [18,19] |
|---|---|---|---|---|---|---|---|---|---|
| $P_D$ (%) | 5.664 | 5.648 | 5.635 | 5.637 | 5.754 | 5.740 | 5.699 | 5.698 | 5.67 |
| $Q_d$ ($fm^2$) | 0.2719 | 0.2707 | 0.2707 | 0.2708 | 0.2706 | 0.2700 | 0.2703 | 0.2706 | 0.271(1) |
| $R_d$ (fm) | 1.967 | 1.966 | 1.968 | 1.967 | 1.966 | 1.965 | 1.969 | 1.967 | 1.9676(10) |
| ΔES | | $2\ 10^{-3}$ | | $4\ 10^{-3}$ | | $2.5\ 10^{-3}$ | | $2.6\ 10^{-3}$ | |
| ΔED | | $7\ 10^{-3}$ | | $10^{-2}$ | | $8.0\ 10^{-3}$ | | $1.2\ 10^{-2}$ | |

# Appendix 2

Characteristics of deuteron and np scattering for different variants of NN potentials.

| Characteristics of deuteron and np scattering | LP1 | **LP2** | LP3 | LP4 | LP5 | Nijm.93 [25] | CERN [16] |
|---|---|---|---|---|---|---|---|
| $E_d$ (MeV) | 2.2246 | **2.2246** | 2.22458 | 2.2246 | 2.2246 | 2.224575(9) | 2.224579(9) |
| $Q_d$ ($fm^2$) | 0.271 | **0.274** | 0.279 | 0.285 | 0.290 | 0.271(1) | 0.2859(3) |
| $P_D$ (%) | 5.62 | **5.75** | 6.00 | 6.23 | 6.56 | 5.67 | --- |
| $A_S$ | 0.884(1) | **0.884(1)** | 0.884(1) | 0.884(1) | 0.884(1) | 0.8845(8) | 0.8802(20) |
| $\eta=A_S/A_D$ | 0.0253(1) | **0.0256(1)** | 0.0261(1) | 0.0266(1) | 0.0273(1) | 0.0253(2) | 0.0271(4) |
| $a_t$ (fm) | 5.417 | **5.417** | 5.419 | 5.417 | 5.419 | 5.4194(20) | 5.419(7) |
| $r_t$ (fm) | 1.753 | **1.753** | 1.754 | 1.751 | 1.752 | 1.7536(25) | 1.754(8) |
| $R_d$ (fm) | 1.965 | **1.966** | 1.967 | 1.968 | 1.968 | 1.9676(10) | 1.9560(68) |
| $V_1$ (MeV) | 10.076 | **10.214** | 10.490 | 10.710 | 11.080 | --- | 10.71(12) |
| $f^2_{\pi NN}$ | 0.0730 | **0.0740** | 0.0760 | 0.0776 | 0.0803 | 0.074 | 0.0776 |
| $\alpha$ ($fm^{-1}$) | 3.68 | **3.72** | 3.80 | 3.90 | 4.05 | --- | --- |
| $V_0$ (MeV) | 3495.641 | **3524.172** | 3575.435 | 3652.860 | 3744.980 | --- | --- |

# References


[1] Neudatchin, V. G. et al. *Lett. Nuovo Cim.* 1972, *5*, 834; *Phys. Lett. B* 1971, *34*, 581; Kurdyumov, I. V. et al. *Phys. Lett. B* 1972, *40*, 607.

[2] Neudatchin, V. G. et al. *Phys. Rev. C* 1992, *45*, 1512; Neudatchin, V. G.; Sakharuk, A. A.; Smirnov, Yu. F. *Fiz. Elem. Chast. Atom. Yadra* 1993, *23*, 480; Dubovichenko, S. B.; Dzhazairov-Kakhramanov, A. V. *Yad. Fiz.* 1993, *56*, 87; *Phys. Atom. Nucl.* 1994, *57*, 733; *Phys. Atom. Nucl.* 1995, *58*, 579; *Phys. Atom. Nucl.* 1995, *58*, 788; Neudatchin, V. G.; Sakharuk, A. A.; Dubovichenko, S. B. *Few Body Sys.* 1995, *18*, 159; Dubovichenko, S. B. *Phys. Atom. Nucl.* 1995, *58*, 1174; *Phys. Atom. Nucl.* 1995, *58*, 1295; *Phys. Atom. Nucl.* 1995, *58*, 1866.

[3] Neudatchin, V. G.; Obukhovsky, I. T.; Smirnov, Yu. F. *Phys. Lett. B* 1973, *43*, 13; Neudatchin, V. G. et al. *Phys. Rev. C* 1975, *11*, 128.

[4] Dubovichenko, S. B.; Zhusupov M. A. *Bull. Sci. Acad. KazSSR phys.-math.* 1982, *6*, 34; In: *Proc. Int. Conf. on Nucl. Phys.*, Florence, 1983, *1*, 46.

[5] Kukulin, V. I.; Pomerantsev, V. N.; Krasnopol'sky, V. M.; Sazonov, P. B. *Phys. Lett. B* 1984, *135*, 20; Krasnopol'sky, V. M.; Kukulin, V. I.; Pomerantsev, V. N.; Sazonov, P. B. et al. *Phys. Lett. B* 1985, *165*, 7; Kukulin, V. I.; Pomerantsev, V. N.; Faessler, A.; Buchmann, A. J.; Tursunov, E. M. Phys. Rev. C, 1998, *57*, 535; Kukulin, V. I.; Pomerantsev, V. N. *Nucl. Phys. A* 1998 *631* 456.

[6] Dorodnyukh, Yu. L.; Neudatchin, V. G.; Yudin, N. P. *Yad. Fiz.* 1988, *48*, 1796; *Preprint Inst. Nuc. Res. Acad. Sci USSR* 1988, *P-0595*, pp 11.

[7] Reid, R. V. *Ann. Phys.* 1968, *50*, 411.

[8] Plattner, G. R.; Viollier, R. D. *Nucl. Phys. A* 1981, *365*, 8.

[9] Dubovichenko, S. B. *Phys. Atom. Nucl.* 1997, *60*, 425.

[10] Arndt, R. A. et al. *Phys. Rev. D* 1983, *28*, 97; Mac Gregor, M. et al. *Phys. Rev.* 1969, *182*, 1714; Arndt, R. A.; Strakovsky, I. I.; Workman, R. L. *Phys. Rev. C* 1996, *53*, 430.

[11] Brown, G. E.; Jackson, A. D. *The nucleon-nucleon interaction*; Publisher: North-Holland Pablishing Company, Amsterdam, 1976.

[12] Lomon, E. et al. *Phys. Rev. C* 1974, *9*, 1329; Platner, D. *Europ. Few Body Probl. Nucl. Part. Phys.* Sesimbra, Portugal, 1980, 31.

[13] Buchanan, C. D.; Yearian, M. R. *Phys. Rev. Lett.* 1965, *15*, 303; Ellias, J. I. et al. *Phys. Rev.* 1969, *177*, 2075; Arnold, R. G. et al. *Phys. Rev. Lett.* 1975, 35, 776; Cramer, R. et al. *Z. Phys. C* 1985, *29*, 513; Platchkov, S. et al. *Nucl. Phys. A* 1990, *508*, 343; Auffret, S. et al. *Phys. Rev. Lett.* 1985, *54*, 649; Bosted, P. et al. *Phys. Rev. C* 1990, 42, 38; Drickey, D. J.; Hand, L. N. *Phys. Rev. Lett.* 1962, *9*, 521.

[14] Hulten, L.; Sugavara, M. In: *Structure of atomic nuclei*; Ed. Flugge S., Publisher: Springer-Verlag, Berlin-Gottingen-Heidelberg, 1957, pp 1-143.

[15] Dubovichenko, S. B. *Phys. Atom. Nucl.* 1997, *60*, 621.

[16] Ericson, T. E. *Nucl. Phys. A* 1984, *416*, 281; Ericson, T. E.; Rosa-Costa, M. *Nucl. Phys. A* 1983, *405*, 497; *Ann. Rev. Nucl. Part. Sci.* 1985, *35*, 271.

[17] Dubovichenko, S. B. In: *XVI-th European Conference on Few - Body Problems in Physics.*, 1-6 June, 1998, Autrans, France, 49.

[18] Stoks, V. G. J. et al. *Phys. Rev. C* 1994, *49*. 2950; de Swart, J. J. et al. *Few Body Sys. Sypl.* 1995, *8*, 437.

[19] Klomp, A. M. M.; Stoks, V. G. J.; de Swart, J. J. *Phys. Rev. C* 1991, *44*, R1258; Stoks, V. G. J. et al. *Phys. Rev. C* 1993, *48*, 792.

[20] Lacombe, M. et al. *Phys. Rev. C* 1980, *21*, 861.

[21] Stoks, V.; de Swart, J. J. *Phys. Rev. C* 1995, *52*, 1698.

[22] Wiringa, R. B.; Smith, R. A.; Ainsworth, T. L. *Phys. Rev. C* 1984, *29*, 1207.

[23] Arndt, R. A.; Strakovsky, I. I.; Workman, R. L. *Phys. Rev. C* 1994, *50*, 2731.

[24] Arndt, R. A. et al. *Phys. Rev. C* 1997, *56*, 3005.

[25] de Swart, J. J.; Terheggen, C. P. F.; Stoks, V. G. J. In: *Proc. 3-rd. Int. Symp. "Deuteron 95"*, Dubna, Russia, 1995; Stoks, V. G. J.; Timmermans, R.; de Swart, J. J. *Phys. Rev. C* 1993, *47*, 512; *Phys. Rev. C* 1993, *48*, 792.

[26] Rahm, J. et al. *Phys. Rev. C* 1998, *57*, 1077; Höhler, G. *Pion-Nucleon Scattering, Tables of Data*; Landoldt-Bornstein, Vol. I/9b2, Ed. H. Schopper, Publisher: Springer-Verlag, Berlin, 1983.

[27] Arndt, R. A. et al. *Phys. Rev. C* 1995, *52*, 2120.

[28] Arndt, R. A.; Strakovsky, I. I.; Workman, R. L. *Phys. Rev. C* 1995, *52*, 2246.

[29] Friar, J. L. et al. *Phys. Lett. B* 1993, *311*, 4.

[30] Stoks, V. G. J. et al. http://nn-online.sci.kun.nl/NN/inde.html

[31] Baldin, A. M. *Electromagnetic interactions and particle structure*; Publisher: Mir, Moscow, 1969.

[32] Garson, M. et al. *Phys. Rev. C* 1994, *49*, 2516.

[33] Phillips, D. R.; Wallace, S. J.; Devine, N. K. *Phys. Rev. C* 1998, *58*, 2261.

[34] Dubovichenko, S. B. *Phys. Atom. Nucl.* 2000, *63*, 734.

[35] Dubovichenko, S. B.; Strakovsky, I. I. *Phys. Atom. Nucl.* 2000, *63*, 582.

[36] Perdrisat, C. F.; Punjabi, V. *Phys. Rev. C* 1990, *42*, 1899; Imambekov, O.; Uzikov, Yu. N.; Schevchenko, L. V. *Z. Phys. A* 1989, *332*, 349.

[37] Voitsekhovsky, B. B. et al. *Pis'ma v ZhETF* 1985, *43*, 567; Gilman, R. et al. *Phys. Rev. Lett*. 1990, *65*, 1733; Schulze, M. E. et al. *Phys. Rev. Lett.* 1984, *52*, 597; The, I. et al. *Phys. Rev. Lett.* 1991, *67*, 173.

[38] Eerro-Luzzi, M. et al. *Phys. Rev. Lett.* 1996, *77*, 2630; Dmitriev, V. F. et al. *Phys. Lett.* B 1985, *157*, 143; Furget, C. et al. *MESON98*, Cracow. Poland. May 30 - June 2. 1998; *Acta. Phys. Polon. B* 1998, *29*, 3301.

[39] Dubovichenko, S. B.; Strakovsky, I. I. – *Bull. Rus. Acad. Sci. Ser. Fiz.* 2001, *65*, 746.

[40] Wiringa, R. B.; Stoks, V. G. J.; Schiavilla, R. *Phys. Rev. C* 1995, *51*, 38; Wiringa, R. B.; Smith, R. A.; Ainsworth, T. L. *Phys. Rev. C* 1984, *29*, 1207.

[41] Machleidt, R.; Sammaruca, F.; Song, Y. *Phys. Rev. C* 1995, *53*, R1483.

In: NN and 3N Interactions
Editors: Leonid Blokhintsev and Igor Strakovsky
ISBN: 978-1-63321-053-0


*Chapter 4*

# TWO- AND THREE-NUCLEON FORCES IN THE COVARIANT SPECTATOR THEORY

*Franz Gross*[1*] *and Alfred Stadler*[2†]

[1]Thomas Jefferson National Accelerator Facility, Newport News, VA, US
and College of William and Mary, Williamsburg, VA, US
[2]Departamento de Física, Universidade de Évora, Évora, Portugal
and CFTP, Instituto Superior Técnico,
Universidade de Lisboa, Lisboa, Portugal

**Abstract**

This chapter provides a short overview of the Covariant Spectator Theory (CST) and its applications to the study of two- and three-nucleon forces. The basic ideas are illustrated using a simple model where two scalar "nucleons" exchange a scalar meson. Then, the equations needed for the realistic description of $NN$ and $3N$ scattering and bound states are introduced, and a kernel (or relativistic potential) for use in these equations is chosen. Results recently obtained by adjusting the free parameters in this kernel in order to give high-precision fits to the two-nucleon scattering data are presented. These fits automatically give the correct three-body binding energy without the introduction of any additional forces or parameters. The chapter concludes with a discussion of the implications of these results, and an assessment of our current understanding of two- and three-nucleon forces.

## 1. Introduction and Theoretical Overview

### 1.1. Background

Nonrelativistic (NR) quantum mechanics has been used to describe nuclear forces since the earliest days of the study of nuclear physics (for a very early discussion, see the review by Bethe and Bacher [1], and the nice introductory books by Moravcsik [2] and Brown and

*E-mail address: gross@jlab.org
†E-mail address: stadler@uevora.pt

Jackson [3]). Early nucleon-nucleon (*NN*) potentials were constructed from 10 different spin and angular momentum dependent terms, and soon it was realized that all of these degrees of freedom were required in order to quantitatively explain the scattering data. At first the data were sparse and the structure of the potential was ambiguous, but as more measurements were made it became possible to constrain the complex structure of the nuclear potential. The 1993 phase shift analysis by the Nijmegen group [4] set the standard for precision fits to the data (a "precision" fit has $\chi^2$/datum $\simeq 1$). In addition to the data maintained by the Nijmegen group [5], R. A. Arndt and the GWU group maintains the *NN* data and publishes new fits periodically [6]. Precision fits by the Argonne group using the V18 potential [7] and R. Machleidt using the CD-Bonn potential [8] soon followed. All of these fits are limited to data below a *NN* lab energy of 350 MeV; above this energy the scattering is no longer elastic (actually, pion production begins at about 290 MeV, but inelastic effects are quite small below 350 MeV) and therefore cannot be described using real phase shifts and mixing parameters, a very useful constraint. These models include some relativistic corrections, but are largely NR. They show that many different *NN* potentials can be found which give precision fits to the data.

Recently the study of the *NN* force has evolved into two quite different directions. These use either (i) chiral perturbation theory (ChPT), well described elsewhere in this volume, or (ii) manifestly covariant equations, represented by the covariant spectator theory (CST), which is the subject of this chapter. Before turning to the CST, we offer a few remarks about ChPT.

## 1.2. Chiral Perturbation Theory

ChPT exploits the fact that the interactions of quantum chromodynamics (QCD – believed to be the fundamental theory of the strong interactions from which the nuclear force must be derived) can be written as an effective theory of nucleons interacting with pions. From this point of view the small pion mass ($m_\pi \simeq 139$ MeV, much smaller that the masses of other mesons and baryons) is a signature of the dynamical breaking of chiral symmetry, and allows the dynamics of QCD, in the *NN* and 3*N* sectors, to be written as a perturbation series in which the pion mass and momenta are expressed in terms of the small quantities $m_\pi/\Lambda_\chi$ and $p/\Lambda_\chi$, where $p$ is a typical momentum and $\Lambda_\chi \simeq 1$ GeV is the hadronic scale at which the chiral perturbation series breaks down.

In implementing ChPT, the *NN* potential is expanded in a power series in the small parameter $Q/\Lambda_\chi$ (where $Q$ is the generic name for momenta $p$ or the pion mass $m_\pi$). At each order $\nu$ in the $(Q/\Lambda_\chi)^\nu$ expansion, there are only a finite number of diagrams that contribute to the potential, so that it is possible to compute all of the terms that contribute to a given order $\nu$. At each order constants are introduced that effectively represent short range (high momentum) contributions that cannot be computed. These constants can, in principle, be computed from QCD, but in the absence of such a calculation the constants are determined by fits to the data. At present, calculations have been done up to $N^3LO$, where $\nu = 4$ and there are a total of 35 undetermined parameters that must be fit to the *NN* data. Using this method, a precision fit to the database below 290 MeV was made by Entem and Machleidt [9].

The use of ChPT to describe nuclear forces enjoys many advantages which make it very popular:

- It provides a systematic power series expansion of the *NN* potential.
- A theoretical error can be estimated by looking at the size of the next order term in the series.
- The rules for computing the $\pi N$ interactions can be applied to a wide variety of physical process from $\gamma N \to X$, $\pi N \to X$ and 3*N* interactions, and some of the constants that contribute to one class of interactions can be used in another, reducing the number of unknown constants.

The method also has some disadvantages, which will be discussed in the final Reflections section of this chapter.

## 1.3. Manifestly Covariant Methods

The laws of nature are constrained by the principle of relativity, i.e.,, they should be *covariant*. Some would argue that the size of "relativistic effects" are small: typical average velocities of nucleons in nuclei lead to an estimate that relativistic corrections to their *kinetic energy* should not exceed a few percent. Why, then, is it not sufficient to treat these effects as relativistic "corrections?"

Here are a few reasons why it may be useful to construct a manifestly covariant theory, even if it turns out to be substantially more complicated than a NR theory:

- Few-nucleon physics has entered an era of high-precision experiments. Even if relativity should turn out to lead to only small differences compared to nonrelativistic theory, these differences might be experimentally detectable and therefore significant. For example, there are some discrepancies between measured observables and NR theoretical calculations with the most accurate nucleon-nucleon (*NN*) potentials that remain still unexplained. Examples are the well-known $A_y$-puzzle in *Nd* scattering, and the underprediction of the three-nucleon (3N) binding energy (usually fixed by the introduction of three-body forces) and the deuteron quadrupole moment. It is important to know if relativity can explain these discrepancies, at least in part.
- Experience with the Dirac equation has taught us that spin-dependent observables can be especially well described by a relativistic treatment, which can sometimes unify phenomena that might appear to be otherwise unrelated when studied from a NR point of view. A good example is the case of hydrogen-like atoms, where four relativistic corrections to the NR Hamiltonian, including the Darwin term and the spin-orbit interaction with its famous gyromagnetic ratio $g = 2$ for the electron, were originally viewed as unrelated and therefore computed independently. The Dirac equation revealed that they all have a common origin in the equation's relativistic structure, and that they are in fact completely determined by it. In this chapter we emphasize another striking advantage: a *manifestly covariant* calculation (one based on the use of Feynman diagrams) automatically includes far more physics than other relativistic methods that use *time-ordered* diagrams.

- Eventually, relativity becomes unavoidable: for example, when the structure of light nuclei is investigated by means of scattering with hadronic or electromagnetic probes, the transferred momentum can become so large that NR quantum mechanics is no longer sufficient to describe the high-velocity final nuclear state. The validity of adding relativistic corrections perturbatively to NR calculations can only be tested when it can be compared to truly covariant calculations.

- Finally, the study of relativistic equations is interesting from an intellectual point of view. There are several ways to extend NR calculations to include the requirements of special relativity. They can be divided into two groups: formalisms in which the number of particles is fixed, referred to collectively as Relativistic Hamiltonian Dynamics [10], and others based on Relativistic Quantum Field Theory. There are many interesting conceptional questions about how the various forms of relativistic theories within each group are related to each other, and perhaps even more about the relations between the two groups.

Only through the comparison of a manifestly covariant treatment of the *NN* interaction with modern alternative approaches can one assess the importance of these issues. The purpose of this chapter is to report on recent progress with one of these methods, the CST.

## 1.4. Foundations of the CST Equations

### 1.4.1. Feynman vs time-ordered diagrams

We begin the discussion with a brief overview of the kinds of interactions that are used in the description of the nuclear force. In order to be specific, consider a field theory for three distinct neutral (real) scalar fields called $\Psi$, $\psi$, and $\phi$, with masses $m_1$, $m_2$, and $m_b$, respectively, described by the Lagrangian

$$\mathcal{L}(x) = \mathcal{L}^0_{m_1}(\Psi) + \mathcal{L}^0_{m_2}(\psi) + \mathcal{L}^0_{m_b}(\phi) - g\left[\Psi^2(x)\,\phi(x) + \psi^2(x)\,\phi(x)\right] \tag{1}$$

where the free Lagrangians are

$$\mathcal{L}^0_m(\phi) = \frac{1}{2}\left[\frac{\partial\phi(x)}{\partial x^\mu}\frac{\partial\phi(x)}{\partial x_\mu} - m^2\phi^2(x)\right] \tag{2}$$

and the term $g\left[\Psi^2(x)\,\phi(x) + \psi^2(x)\,\phi(x)\right]$ describes the interactions of the fields at any space-time point $x^\mu$. Because there are three fields interacting at a point, this is referred to as a $\phi^3$-type theory.

One way to calculate the amplitude for scattering of $\Psi$ from $\psi$ particles implied by the Lagrangian (1) is to express it as an infinite sum of Feynman diagrams which describe physically and mathematically how the interactions take place. Since the $\Psi$ and $\psi$ fields do not interact with each other at the same point, the simplest (lowest order in the coupling constant, $g$) interaction between them is the exchange of the boson $\phi$, represented by the Feynman diagram shown in Fig. 1.

The rules for calculating Feynman diagrams usually include one stating that external lines represent particles on their mass shells. However, when summing Feynman diagrams by means of integral equations one has to consider diagrams that will be combined with

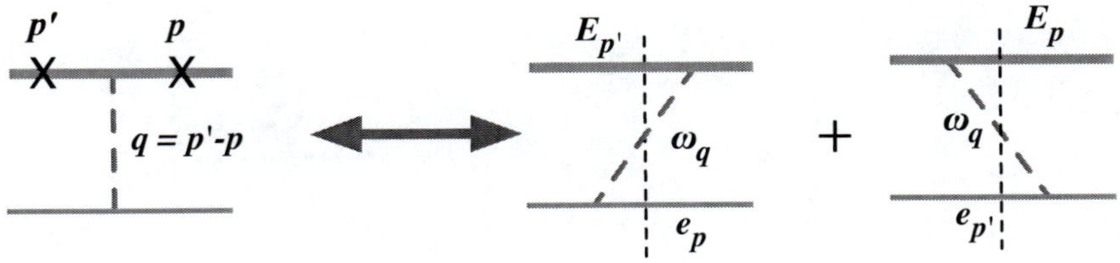


Figure 1. A single Feynman diagram (left figure) is equivalent to a sum of time-ordered diagrams (the two on the right). We adopt the convention that time flows from the right to the left. The vertical dashed line identifies the particles that contribute to the total energy of the intermediate state. A cross on a line indicates that the particle is on mass shell.

others to form larger diagrams. In these cases, some external lines eventually become internal lines of other diagrams and can therefore be off mass shell. For this reason, in the following we do not automatically restrict external particle lines to be on mass shell, but we will explicitly indicate when this is the case.

If the particle with mass $m_1$ (which we will think of as the heavy particle) is on-shell, the diagram in Fig. 1 has the mathematical form

$$\begin{aligned} M_2 &= \frac{g^2}{m_b^2 - q^2} = \frac{g^2}{\omega_q^2 - (E_{p'} - E_p)^2} = \frac{g^2}{2\omega_q}\left[\frac{1}{\omega_q + E_{p'} - E_p} + \frac{1}{\omega_q - E_{p'} + E_p}\right] \\ &= \frac{g^2}{2\omega_q}\left[\frac{1}{\omega_q + E_{p'} + e_p - W_p} + \frac{1}{\omega_q + e_{p'} + E_p - W_{p'}}\right] \end{aligned} \tag{3}$$

where $q = p' - p$, the relativistic on-shell energies of the particles are $E_p = \sqrt{m_1^2 + \mathbf{p}^2}$, $e_p = \sqrt{m_2^2 + \mathbf{p}^2}$, $\omega_q = \sqrt{m_b^2 + \mathbf{q}^2}$, and $W_p = E_p + e_p$. This Feynman diagram is illustrated in the left panel of Fig. 1. (We will return to a discussion of the second line in Eq. (3) in a moment.)

Another way to solve for the scattering amplitude is to construct the Hamiltonian from the Lagrangian and use the generalized Born series. Recall that the second order Born approximation to the scattering amplitude can be written

$$\begin{aligned} \mathcal{M}_2(W) &\simeq \sum_n \frac{\langle f | H_{\rm int} | n \rangle \langle n | H_{\rm int} | i \rangle}{W_n - W} \\ &= \frac{g^2}{2\omega_q}\left[\frac{1}{\omega_q + E_{p'} + e_p - W} + \frac{1}{\omega_q + e_{p'} + E_p - W}\right] \end{aligned} \tag{4}$$

where $W$ is the total energy and the sum is over all intermediate states $n$ with energy $W_n$. The matrix elements of the interaction Hamiltonian $H_{\rm int}$ which describes the production of a boson with energy $\omega$ are proportional to $g$, and there are two intermediate states describing different *time-orderings* for the production of the intermediate state. In one the boson is emitted by the heavy particle and absorbed by the lighter; in the other the roles of the two particles are reversed. The sum of these two terms is displayed in the second line of Eq. (4), and illustrated by the two right-hand panels of Fig. 1. Note here that the sum of the individual particle energies in the intermediate states is not equal to the energy $W$ of the

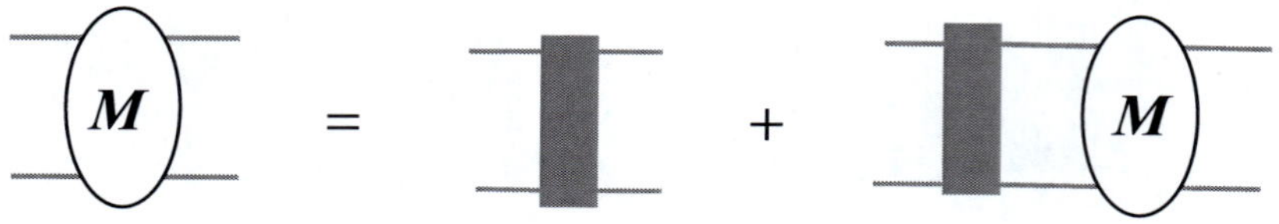


Figure 2. Diagrammatic representation of the BS equation. In its most general form, the intermediate nucleon propagators are dressed by the self-energy as given in Eq. (8).

initial and final states; one says that the particles are "on mass shell" but the states are "off energy shell."

Now compare the second line of Eq. (3) with the time ordered result Eq. (4). We see that they are very similar, and would be identical if we insisted that the initial and final states have energy $W_p = W_{p'} = W$. In this case $|\mathbf{p}| = |\mathbf{p}'|$, and both expressions collapse to the *instantaneous* result

$$M_2 = \mathcal{M}_2(W) = \frac{g^2}{\omega_q^2}. \tag{5}$$

The Feynman diagram is not identical to the second order Born approximation when external particles are off mass shell (but in our example the heavy particle was on mass shell). As mentioned above, this feature allows the Feynman diagram to be used as part of a higher order contribution to the scattering series. Furthermore, since all time orderings are automatically included in each Feynman diagram, they are distinguished only by their topologies; two diagrams which can be deformed into each other by sliding the vertices to the left or right (without crossing another vertex) are identical.

In summary, it can be shown that the sum of all time-ordered diagrams is equivalent to the Feynman series. Figure 1 illustrates this correspondence to second order where a single one boson exchange (OBE) Feynman diagram is equivalent to two *time-ordered* diagrams. In general, an $n$th order Feynman diagram is equivalent to the $n!$ time ordered diagrams derived by time ordering the $n$ vertices in all possible ways. Since the Feynman diagram allows the external particles to be off mass shell, and the time-ordered series does not, these two expressions are identical only when all external particles in the Feynman diagram are placed on mass shell. In second order this gives the instantaneous interaction usually used in NR potential theory.

We conclude this brief introduction with a warning: because Feynman diagrams include all possible time orderings, they contain much more information than might appear to an unwary reader. We will return to this point in subsequent discussion.

### 1.4.2. Bethe-Salpeter equation

We are now ready to review the well-known Bethe-Salpeter (BS) equation introduced by Salpeter and Bethe in 1951 [11]. This equation provides a manifestly covariant way of summing an infinite class of Feynman diagrams. It is illustrated in Fig. 2 and, in its most general form, is written

$$M = V^{\mathrm{BS}} - V^{\mathrm{BS}} G_{BS} M, \tag{6}$$

where $V^{BS}$ is the irreducible kernel (i.e., the sum of all Feynman diagrams that cannot be separated into two parts by cutting *only* two nucleon lines) and $G_{BS}$ is the intermediate state propagator. Its specific form is

$$M_{12}(p,p';P) = V_{12}^{BS}(p,p';P) + i\int \frac{d^4k}{(2\pi)^4} V_{12}^{BS}(p,k;P)G_1(k_1)G_2(k_2)M_{12}(k,p';P)\,, \quad (7)$$

where a compact notation $1 \to \{\alpha\alpha'\}$ and $2 \to \{\beta\beta'\}$ is used to denote the Dirac indices of particles 1 and 2 (they are summed over when repeated, so that $O_1 = A_1B_1$ is a shorthand for $O_{\alpha\alpha'} = A_{\alpha\alpha''}B_{\alpha''\alpha'}$). The total four-momentum $P$ is conserved, and $p, p'$, and $k$ are relative four-momenta related to the momenta of particles 1 and 2 by $p_1 = \frac{1}{2}P + p$, $p_2 = \frac{1}{2}P - p$, and similarly for $p'$ and $k$. Both particles are off-shell: no constraint is placed on the squares of the individual particle momenta, as implied by the four-dimensional integral over the internal relative four momentum $k$. For spin-$\frac{1}{2}$ particles, the dressed propagators are

$$G_i(k_i) = \frac{1}{m_{0i} - \not{k}_i + \Sigma_i(\not{k}_i) - i\varepsilon} \quad (8)$$

where $m_{0i}$ is the bare mass and $\Sigma_i$ the self-energy of particle $i$ (see Ref. [12]). The product of the two one-body propagators, $-iG_1(k_1)G_2(k_2)$, is referred to as the two-body BS propagator.

If the BS kernel and the dressed nucleon propagators were known *exactly*, the solution to the BS equation would give the exact $NN$ scattering amplitude. Unfortunately, there is no known way to find the exact kernel or the exact propagators. The kernel and propagator must be approximated before the equation can be solved. It is often assumed that a suitable approximation is to replace the kernel by the sum of one boson exchanges (OBE) between the two nucleons, and to replace the dressed propagators $G_i(k_i)$ by bare propagators (obtained by setting the self-energy to zero), but with the physical masses $m_i$ in place of the bare masses $m_{0i}$. Substituting this kernel into the equation shown in Fig. 2 gives the famous "ladder sum."

As an example of the rich content of the ladder sum, Fig. 3 shows how the single 4th order ladder diagram (the "box" diagram) is a sum of 24 time-ordered diagrams. The first four [($a_1$) to ($a_4$)] preserve the group orderings $\{t_2,t_4\} > \{t_1,t_3\}$ and are the expected iteration of the two time-ordered OBE diagrams shown in Fig. 1. The box not only includes these, but also an additional 20 diagrams. Largest among these remaining diagrams are the "stretched" boxes, ($a_5$) and ($a_6$), with the time orderings $t_4 > t_3 > t_2 > t_1$ ($a_5$) and $t_2 > t_1 > t_4 > t_3$ ($a_6$). Each of the 24 diagrams has three "cuts" which identify the intermediate states that appear in the 4th order Born expansion, but only the "middle cut" with—except for diagrams (a1) to (a4)—the largest energy intermediate state is identified in the figure. The larger the energy of this state, the smaller its expected contribution, and since the stretched boxes involve an intermediate state with two bosons, they are expected to be smaller than the other (a)-type diagrams. The other diagrams fall into three classes of 6 diagrams each: the ($b_i$) class with an intermediate state with a threshold starting at $m_1 + 3m_2 + 2m_b$ (corresponding to the production of a pair of particles with mass $m_2$, the "Z-graphs" for particle 2), the ($c_i$) class with a threshold starting at $3m_1 + m_2 + 2m_b$ (corresponding to the production of a pair of particles with mass $m_1$, "Z-graphs" for particle 1), and the two-pair class ($d_i$) with a

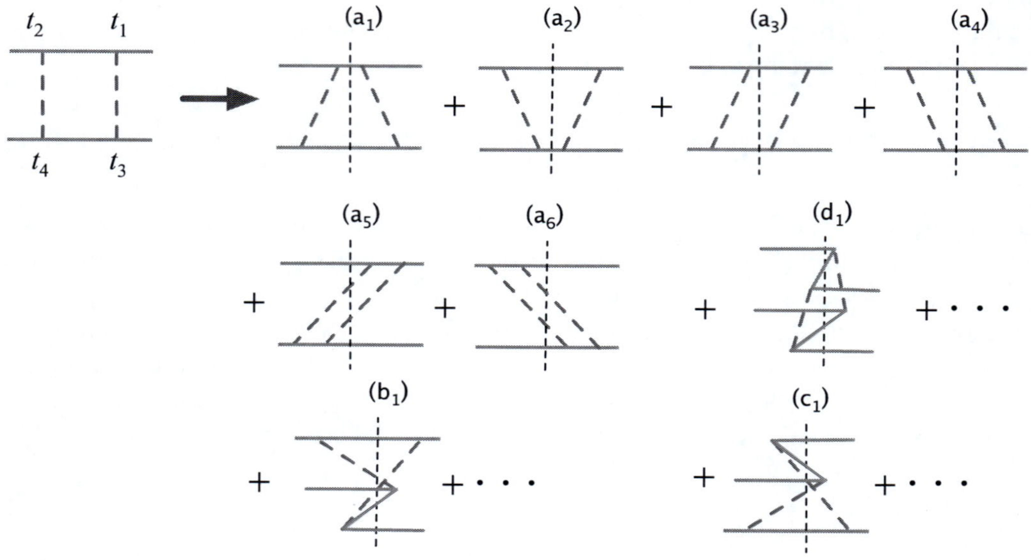


Figure 3. Decomposition of the box diagram into its 24 time-ordered parts, as discussed in the text. The four vertices are labeled by the times $t_i$. The diagram ($b_1$) has the ordering $t_2 > t_3 > t_4 > t_1$, ($c_1$) the ordering $t_4 > t_1 > t_2 > t_3$, and ($d_1$) the ordering $t_3 > t_1 > t_2 > t_4$.

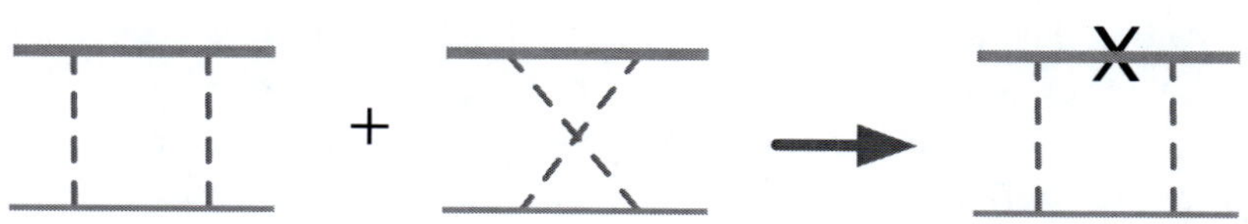

Figure 4. In $\phi^3$-type theories, as the mass of the heavy particle goes to infinity, the sum of box and crossed-box diagrams becomes equal to the box diagram with the heavy particle on mass shell (indicated by a cross on the line).

threshold starting at $3m_1 + 3m_2$. This example shows explicitly how the OBE sum of Feynman diagrams contains far more physics than the iteration of the time-ordered OBE kernel of Eq. (4).

In spite of this rich structure, it can be shown in some cases that the BS ladder sum is *not* the best approximation, leading naturally to a discussion of the CST equation.

### 1.4.3. CST equation

The basic idea of the CST [13] is to reorganize the manifestly covariant BS equation with its complete kernel into an equivalent form, in which a different two-body propagator and an accordingly modified kernel is used. It turns out that this reorganization can be done in such way that the OBE approximation to the new kernel is a better approximation to the exact result, and, at the same time, the new two-body propagator is much simpler to use.

The idea is illustrated in Fig. 4, which shows the 4th order box and crossed box contributions to the scattering amplitude (there are other 4th order diagrams associated with the particle self energies and vertex corrections that are not being considered here). The diagrams can be labeled so that, in both diagrams, the light particle has four-momentum $P-k$, and the momenta of the two bosons are $q_f = k-p$ and $q_i = p'-k$, with $p$ ($p'$) the four-momentum of the heavy outgoing (incoming) particle 1. To illustrate the idea and keep the notation as simple as possible, we limit this discussion to scalar particles interacting at a point (the $\phi^3$-type theory discussed above). Then the sum of the box and crossed box diagrams is

$$M_4 \simeq \int \frac{d^4k}{(2\pi)^4} I(k) \left[ \frac{1}{m^2-k^2-i\varepsilon} + \frac{1}{m^2-(p+p'-k)^2-i\varepsilon} \right], \tag{9}$$

where, for simplicity, $m_1 = m$ is the dressed mass of the heavier particle (and includes the average of the contributions from its self-energy). The factor $I(k)$ includes the propagators of the exchanged mesons and the particle with mass $m_2$. To estimate the behavior of $M_4$, evaluate it at the point where the external incoming and outgoing particles are on shell, and choose the frame where they are at rest, so that $p = p' = \{m, \mathbf{0}\}$. Rewriting the denominators of (9), and defining $E_k = \sqrt{m^2+\mathbf{k}^2}$ gives

$$\begin{aligned} M_4 \quad &\to \quad \int \frac{d^4k}{(2\pi)^4} I(k) \left[ \frac{1}{E_k^2-k_0^2-i\varepsilon} + \frac{1}{E_k^2-(2m-k_0)^2-i\varepsilon} \right] \\ &= \quad \int \frac{d^4k}{(2\pi)^4} I(k) \left[ \frac{1}{E_k-k_0-i\varepsilon} + \frac{1}{E_k-2m+k_0-i\varepsilon} + R_x \right], \end{aligned} \tag{10}$$

where the first term comes from the box and the second term from the crossed box, and the remainder term $R_x$ can be neglected (because the $k_0$ integral of the product $I(k)R_x$ is small everywhere).

Now evaluate the $k_0$ integral by closing the contour in the lower half plane using the residue theorem. The result is a sum of residues at each of the poles of $I(k)$ that lie in the lower half plane, plus the pole at $k_0 = E_k$ coming from the box diagram. It is not necessary to actually evaluate the residues of the poles of $I(k)$ if we observe that the two terms in the square brackets tend to cancel. If the integral over $d^3k$ converges, so that the average value of $\langle \mathbf{k}^2 \rangle \ll m^2$ when $m$ is large, then if one of the poles in $I(k)$ is at $k_0 = m+\delta$,

$$\lim_{m\to\infty} \left[ \frac{1}{E_k-k_0-i\varepsilon} + \frac{1}{E_k-2m+k_0-i\varepsilon} \right] = \frac{1}{\frac{\langle \mathbf{k}^2 \rangle}{2m}-\delta} + \frac{1}{\frac{\langle \mathbf{k}^2 \rangle}{2m}+\delta} \to 0. \tag{11}$$

Hence the contributions from all of the $k_0$ poles in $I(k)$ cancel as $m = m_1 \to \infty$. Because of this cancellation, we may ignore the $k_0$ poles in $I(k)$, and closing the contour in the lower half plane leaves only the positive energy pole of the heavy particle at $k_0 = E_k$, giving the simple result

$$\lim_{m\to\infty} M_4 = \int \frac{d^3k}{(2\pi)^3} \frac{m}{E_k} \left[ \frac{I(\widehat{k})}{m} \right], \tag{12}$$

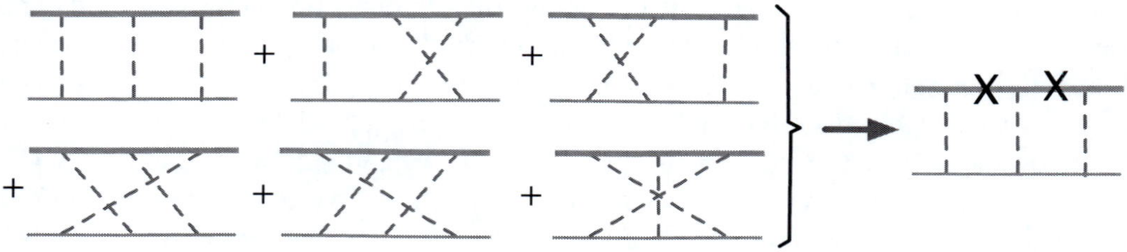

Figure 5. The "generalized ladder sum" to 6th order (two loops). There is one ladder diagram, two iterations of the crossed box, and three irreducible contributions. As $m_1 \longrightarrow \infty$ all of these reduce to a single 6th order ladder with the heavy particle on-shell in each of the loops.

where the notation $\widehat{k}$ is used to denote the on-shell four-vector with $k_0 = E_k$ and $\widehat{k}^2 = m^2$. We refer to this result as the *cancellation theorem*.

This means that, contrary to a commonly held opinion, contributions from the crossed box *cannot be ignored*. As Fig. 4 illustrates, the leading order contribution from the sum of the box and crossed box diagrams in the large $m_1$ limit reduces to the box diagram with the *heavy particle on shell*. Moreover, adding the crossed box to the BS kernel will not preserve the cancellation to higher orders. It can be shown that this cancellation works to *all orders provided all generalized ladder diagrams are included* at each order [12]. Fig. 5 shows how the the cancellation works to 6th order: six diagrams participate, including three new irreducible diagrams which would have to be added to the BS kernel. To preserve the cancellation using the BS equation, an infinite number of irreducible crossed ladder contributions must be added to the original OBE kernel, a formidable task. Yet all of this is automatically accomplished using the CST equation with an OBE kernel.

In summary: as $m \longrightarrow \infty$, the *exact generalized ladder sum reduces to the sum of ladder (OBE) graphs with the heavy particle on shell*. The most general form of the integral equation which sums these ladder graphs is shown in Fig. 6. If the final, outgoing heavy particle is put on-shell (giving the CST equation) the equation is "closed" and can be solved for the one particle on-shell scattering amplitude $M$. Once the on-shell amplitude is known, the generalized equation shown in Fig. 6 (referred to as the CST-BS equation) can be used to extrapolate the amplitude off-shell. (It is important to realize that the BS amplitude obtained in this way is not identical to the amplitude obtained from the BS equation in the OBE approximation.)

There are many ways to rearrange the BS equation leading to a kernel depending on three-momentum of the initial and final state. The resulting equations are called "quasi-potential equations." While the CST belongs in this category, it has properties not all of the other quasi-potential theories possess. For instance, it satisfies the property of cluster-separability [14], without which a calculated two-body scattering amplitude could not be used consistently in the kernel of a three-body equation (see further discussion of this point below). It also has the one-body limit: when one particle is a massive scalar, the two-body equation reduces to a relativistic one-body equation for the light particle moving in an effective potential created by the massive particle.

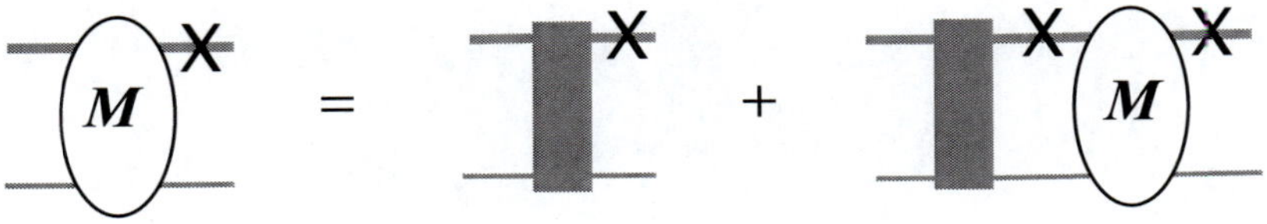


Figure 6. Diagrammatic representation of the CST-BS equation, with both particles off-shell in the final state.

## 2. CST equations for the Two- and Three-Nucleon Systems

### 2.1. Two-Nucleon Equation

The specific form of the CST equation for the $NN$ scattering amplitude $M$, with particle 1 on-shell in the initial state and *both* particles off-shell in the final state, as illustrated in Fig. 6, is

$$M_{12}(p^*,p';P) = \overline{V}_{12}(p^*,p';P) - \int \frac{d^3k}{(2\pi)^3}\frac{m}{E_k}\overline{V}_{12}(p^*,k;P)G_2(k,P)M_{12}(k,p';P)\,, \qquad (13)$$

where, for the remainder of this chapter we use a different notation that assumes particle 1 is on shell (so that in the rest frame where $P=\{W,\mathbf{0}\}$, the relative momentum is $k=\{E_k-\frac{1}{2}W,\mathbf{k}\}$), *unless* the momentum is denoted by a $^*$, in which case $p^*=\{p_0,\mathbf{p}\}$. Since $p^*$ is off-shell in (13), this is the CST-BS equation mentioned above; it gives an expression for the fully off-shell scattering amplitude once the CST amplitude is known.

To find the CST amplitude, one merely sets $p^*\to p$, giving a closed equation for $M$ [15]

$$M_{12}(p,p';P) = \overline{V}_{12}(p,p';P) - \int \frac{d^3k}{(2\pi)^3}\frac{m}{E_k}\overline{V}_{12}(p,k;P)G_2(k,P)M_{12}(k,p';P)\,. \qquad (14)$$

This equation is illustrated in the top panel of Fig. 7. Since particle 1 is on shell throughout, $M_{12}$ may be defined to be the matrix element of the Feynman scattering amplitude $\mathcal{M}$ between positive energy Dirac spinors of particle 1

$$M_{12}(p,p';P) \equiv M_{\lambda\lambda',\beta\beta'}(p,p';P) = \bar{u}_\alpha(\mathbf{p},\lambda)\mathcal{M}_{\alpha\alpha';\beta\beta'}(p,p';P)u_{\alpha'}(\mathbf{p}',\lambda')\,. \qquad (15)$$

The index 2 still refers collectively to the Dirac index of particle 2, $\{\beta\beta'\}$, but index 1 may refer either to the helicities of particle 1 $\{\lambda\lambda'\}$ or the Dirac indices $\{\alpha\alpha'\}$ (the replacement of a Dirac index ($\alpha_i$, $\beta_i$, ...) by a helicity index ($\lambda_i$) always indicates a corresponding contraction with a positive-energy helicity spinor). For applications, it is sufficient to place particle 2 on mass shell in the initial state.

The propagator for the off-shell particle 2 will be dressed by the off-shell nucleon form factor $h$

$$G_2(k,P) \equiv G_{\beta\beta'}(k_2) = \frac{(m+\not{k}_2)_{\beta\beta'}}{m^2-k_2^2-i\varepsilon}h^2(k_2) \qquad (16)$$

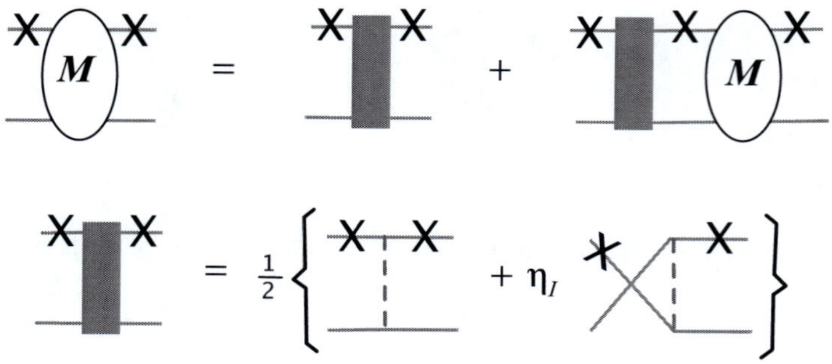


Figure 7. Top panel: diagrammatic representation of the CST equation (14) with particle 1 on-shell. Bottom panel: diagrammatic representation of the definition of the antisymmetrized kernel (18), with $\eta_I = (-)^I$.

with $k_2 = P - k_1$, $k_1^2 = m^2$, and with $h(k_2)$ normalized to unity when $k_2^2 = m^2$. In Ref. [16] we used

$$h(p) = \left[\frac{(\Lambda_N^2 - m^2)^2}{(\Lambda_N^2 - m^2)^2 + (m^2 - p^2)^2}\right]^2, \tag{17}$$

where $\Lambda_N$ is an adjustable cutoff parameter. Since the full nucleon propagator should include the self-energy contribution, $\Sigma(k_2)$, including the form factor $h$ could be regarded as one way to allow for this physics, but its real role is to regularize the high-momentum behavior of the kernel.

Since the propagator of the CST is not symmetric under particle interchange, it is necessary to explicitly antisymmetrize the kernel, as illustrated in the bottom panel of Fig. 7. The Dirac form of the covariant kernel $\overline{V}$ is therefore

$$\overline{V}_{\alpha\alpha';\beta\beta'}(p,k;P) = \frac{1}{2}\left[V_{\alpha\alpha';\beta\beta'}(p,k;P) + (-)^I V_{\beta\alpha';\alpha\beta'}(-p,k;P)\right], \tag{18}$$

where the isospin indices have been suppressed, so that the factor of $(-)^I$ (with $I$=0 or 1 the isospin of the $NN$ state) ensures that the remaining amplitude has the symmetry $(-)^I$ under particle interchange $\{p_1,\alpha\} \leftrightarrow \{p_2,\beta\}$, as required by the generalized Pauli principle. This symmetry guarantees that identical results emerge if a different particle is chosen to be on-shell in either the initial or final state.

The propagator of an off-shell particle with four-momentum $k$ and fixed mass $m$ can be decomposed into positive and negative energy contributions

$$\frac{(m + \not{k})_{\beta\beta'}}{m^2 - k^2 - i\varepsilon} = \frac{m}{E_k}\sum_\lambda \left\{\frac{u_\beta(\mathbf{k},\lambda)\bar{u}_{\beta'}(\mathbf{k},\lambda)}{E_k - k^0} - \frac{v_\beta(-\mathbf{k},\lambda)\bar{v}_{\beta'}(-\mathbf{k},\lambda)}{E_k + k^0}\right\}. \tag{19}$$

Accordingly, the CST equations can be separated into positive- and negative-energy (also called "ρ-spin"- ) channels, which is useful for their numerical solution. However, the decomposition (19) is frame-dependent and thus breaks manifest covariance. This does not

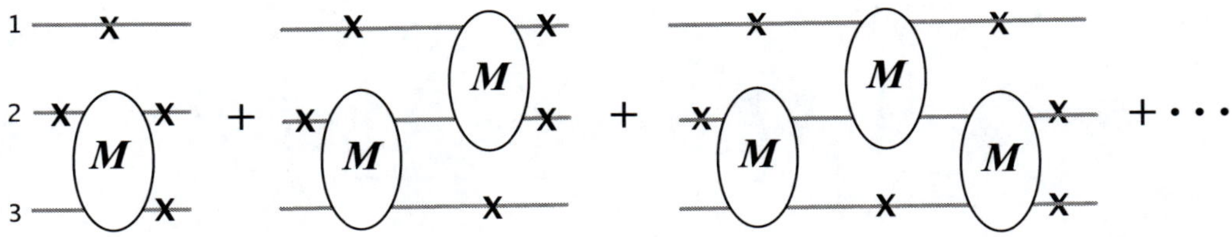


Figure 8. Examples of diagrams contributing to the $3N$ scattering amplitude. In this subset all particles in the initial state are on-shell, particle 1 in the final state was a spectator to the "last" interaction, and particle 2 was chosen to be on-shell in the last interaction. The $NN$ amplitudes marked $M$ are solutions of the CST $NN$ equation (23). The spectator nucleon is always placed on mass shell.

mean that the ρ-spin separated equations are no longer covariant, only that their covariance is no longer explicit as it is in the manifestly covariant form.

Negative-energy states are related to the "Z-graphs" of time-ordered perturbation theory. In this sense, the solutions of Eq. (14) automatically include Z-graphs to all orders, as in the BS theory discussed above. This is useful to keep in mind when CST is compared to other theories in which relativistic corrections are added perturbatively to nonrelativistic calculations.

### 2.2. Three-Nucleon Equation

The CST for $3N$ scattering was formulated for the first time in Ref. [17]. Assuming only two-body interactions, the main idea is to always place the spectators on mass shell. In fact, the name "spectator theory" derives from this principle and is best illustrated in the three-body problem. If this is done consistently, in any intermediate state there are always two nucleons on mass shell, and one off mass shell, as illustrated in Fig. 8. This has the important consequence that the energy-component in all loop integrations is constrained by total energy conservation and two mass-shell conditions, such that all loop integrations are three-dimensional.

Furthermore, in the case of short-range interactions, cluster separability requires that when any one particle is far removed from the remaining pair, the system has to separate into a free particle and an independent two-particle system with internal interactions only. As long as the particles are close together, each of them is sometimes a spectator and sometimes participates in a pair interaction. However, when one particle's distance to the others increases, its interaction with them eventually vanishes completely (only the first diagram in Fig. 8 remains). It becomes a permanent spectator, and therefore in CST it is automatically always on mass shell. This is of course exactly what is required for a free particle.

In Ref. [18] the CST equations for the $3N$ bound state were formulated in a way suitable for a practical solution. Starting from Faddeev-type integral equations for the $3N$ scattering amplitude $\mathcal{M}_{3N}$, one looks for a pole at the sub-threshold bound-state energy $M_t$, i.e., where

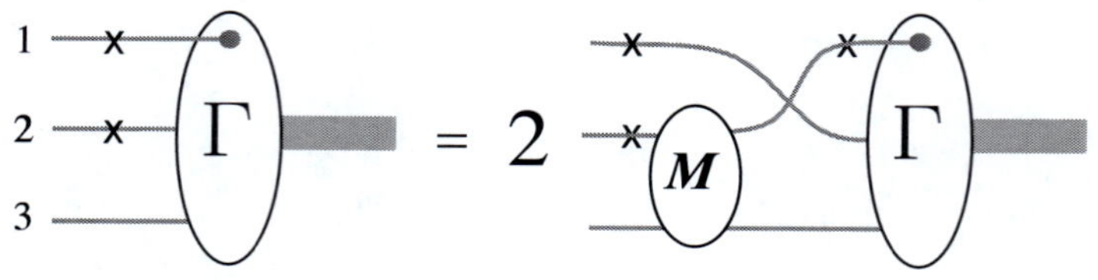


Figure 9. Diagramatic representation of the covariant spectator equation for the three-body bound state vertex function Γ with particles 1 and 2 on-shell (labeled with a ×). Here particle 1 is the spectator to the last two-body interaction between particles 2 and 3, described by the scattering amplitude $M$ with particle 3 off-shell.

it can be written as

$$\mathcal{M}_{3N} = -\frac{\Gamma\bar{\Gamma}}{M_t^2 - P^2} + \mathcal{R}\,. \tag{20}$$

Here, $\mathcal{R}$ is the remaining non-pole contribution, and $P = \widehat{k}_1 + \widehat{k}_2 + k_3$ is the total four-momentum of the 3$N$ system. One obtains a homogeneous equation for the vertex function Γ of the bound state, and suppressing the hat notation (the four-momenta of particles 1 and 2 are always on-shell in this discussion)

$$\Gamma_{\lambda_1\lambda_2\alpha}(k_1,k_2;P) = 2\int \frac{d^3k_2'}{(2\pi)^3}\frac{m}{E_{k_2'}}\sum_{\lambda_2'} M_{\lambda_2\alpha,\lambda_2'\alpha'}(k_2,k_2';P_{23}) \times \mathcal{P}_{12}\, G_{\alpha'\beta}(P_{23}-k_2')\Gamma_{\lambda_1\lambda_2'\beta}(k_1,k_2';P)\,, \tag{21}$$

where the permutation operator $\mathcal{P}_{12}$ interchanges particles 1 and 2 (see also Fig. 9). The particular vertex function in Eq. (21) is for the case that particle 1 was the spectator during the "last" interaction, and particle 2 is the second on-shell particle. For identical particles, all other vertex functions are related to it by exchange symmetry.

As in Eq. (15), we use a shorthand for the contraction of Dirac indices with nucleon helicity spinors with helicity $\lambda_i$,

$$\Gamma_{\lambda_1\lambda_2\alpha'}(k_1,k_2;P_t) \equiv \bar{u}_{\alpha_1}(k_1,\lambda_1)\bar{u}_{\alpha_2}(k_2,\lambda_2)\Gamma_{\alpha_1\alpha_2\alpha'}(k_1,k_2;P_t)\,, \tag{22}$$

and $M_{\lambda_2\alpha,\lambda_2'\alpha'}(k_2,k_2';P_{23})$ is the scattering amplitude of nucleons 2 and 3 with total pair momentum $P_{23}$. It satisfies the CST two-body equation

$$M_{\lambda_2\alpha,\lambda_2'\alpha'}(k_2,k_2';P_{23}) = \overline{V}_{\lambda_2\alpha,\lambda_2'\alpha'}(k_2,k_2';P_{23}) - \int \frac{d^3k_2''}{(2\pi)^3}\frac{m}{E_{k_2''}}\sum_{\lambda_2''} \overline{V}_{\lambda_2\alpha,\lambda_2''\beta}(k_2,k_2'';P_{23})G_{\beta\beta'}(P_{23}-k_2'')M_{\lambda_2''\beta',\lambda_2'\alpha'}(k_2'',k_2';P_{23})\,. \tag{23}$$

Note that the total pair momentum $P_{23} = k_2 + k_3 = P - k_1$ depends now on the spectator momentum $k_1$, which implies that the $NN$ amplitude $M_{\lambda_2\alpha,\lambda_2'\alpha'}(k_2,k_2';P_{23})$ has to be known in arbitrary reference frames. However, because of the CST equation's manifest covariance it is sufficient to calculate the two-body amplitude in its rest frame, and determine it in moving frames through the corresponding boost operations.

If $\Lambda$ is a space-time Lorentz transformation, and $S(\Lambda)$ its representation in a one-particle Dirac space, satisfying

$$S^{-1}(\Lambda)\gamma^{\mu}S(\Lambda) = \Lambda^{\mu}{}_{\nu}\gamma^{\nu}, \tag{24}$$

then $M$ transforms like

$$\begin{aligned} M_{\alpha\alpha',\beta\beta'}&(\Lambda k_2, \Lambda k_2'; \Lambda P_{23}) \\ &= S_{\alpha\alpha_1}(\Lambda)S_{\beta\beta_1}(\Lambda)M_{\alpha_1\alpha_2,\beta_1\beta_2}(k_2,k_2';P_{23})S^{-1}_{\alpha_2\alpha'}(\Lambda)S^{-1}_{\beta_2\beta'}(\Lambda). \end{aligned} \tag{25}$$

All relativistic effects can be calculated *exactly* in CST, and the full Dirac structure of the nucleons is also taken into account.

The $3N$ equation (21) was solved numerically for the first time in Ref. [19], for a family of older OBE potentials. Since then, much progress has been made in the development of more accurate CST $NN$ interaction models, which will be described in the following section.

# 3. Applications

## 3.1. High-Precision *np* Kernels

The first covariant $NN$ OBE kernels in CST, based on the exchange of either four or six mesons, were published in Ref. [15]. After a long process of improvements of the applied numerical techniques, the structure of the kernels, and the enlargement of the $np$ data base, two new models were developed [16] whose precision in representing the most recent world $np$ data is on par with all commonly used "realistic" potentials.

The kernels include a simple one photon exchange contribution (depending on the neutron anomalous magnetic moment) plus a sum of OBE contributions,

$$V_{\alpha\alpha';\beta\beta'}(p,k;P) = \sum_b V_{12}^b(p,k;P). \tag{26}$$

In the notation of Fig. 10, the individual boson contributions are of the form

$$V_{12}^b(p,k;P) = \varepsilon_b \delta \frac{\Lambda_1^b(p_1,k_1)\otimes\Lambda_2^b(p_2,k_2)}{m_b^2+|q^2|} f(\Lambda_b,q). \tag{27}$$

**Table 1. Mathematical forms of the $bNN$ vertex functions, with $\Theta(p) \equiv (m-\not{p})/2m$. The vector propagator is $\Delta_{\mu\nu} = g_{\mu\nu} - q_\mu q_\nu/m_v^2$ with the boson momentum $q = p_1 - k_1 = k_2 - p_2$**

| $J^P(b)$ | $\varepsilon_b$ | $\Lambda_1 \otimes \Lambda_2$ | $\Lambda(p,k)$ or $\Lambda^{\mu}(p,k)$ |
|---|---|---|---|
| $0^+(s)$ | $-$ | $\Lambda_1\Lambda_2$ | $g_s - \nu_s[\Theta(p)+\Theta(k)]$ |
| $0^-(p)$ | $+$ | $\Lambda_1\Lambda_2$ | $g_p\gamma^5 - g_p(1-\lambda_p)\left[\Theta(p)\gamma^5+\gamma^5\Theta(k)\right]$ |
| $1^-(v)$ | $+$ | $\Lambda_1^{\mu}\Lambda_2^{\nu}\Delta_{\mu\nu}$ | $g_v\left[\gamma^{\mu}+\frac{\kappa_v}{2M}i\sigma^{\mu\nu}(p-k)_{\nu}\right]+g_v\nu_v\left[\Theta(p)\gamma^{\mu}+\gamma^{\mu}\Theta(k)\right]$ |
| $1^+(a)$ | $+$ | $\Lambda_1^{\mu}\Lambda_2^{\nu}g_{\mu\nu}$ | $g_a\gamma^5\gamma^{\nu}$ |

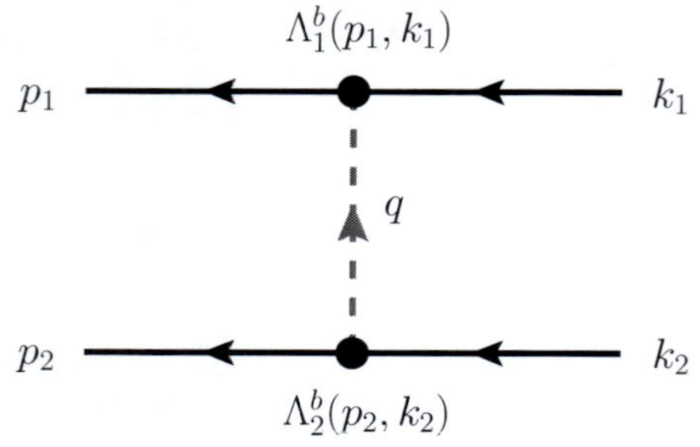


Figure 10. The structure of a OBE kernel.

Here, $b = \{s, p, v, a\}$ denotes the boson type (scalar, pseudoscalar, vector, axial vector), $q = p_1 - k_1 = k_2 - p_2 = p - k$ the momentum transfer, $m_b$ the boson mass, $\varepsilon_b$ a phase factor, and $\delta = 1$ for isoscalar bosons and $\delta = \tau_1 \cdot \tau_2 = -1 - 2(-)^I$ for isovector bosons. All boson form factors, $f$, have the simple form

$$f(\Lambda_b, q) = \left[\frac{\Lambda_b^2}{\Lambda_b^2 + |q^2|}\right]^4, \tag{28}$$

where $\Lambda_b$ is the boson form factor mass. The use of the absolute value $|q^2|$ amounts to a covariant redefinition of the propagators and form factors in the region $q^2 > 0$. This choice removes all singularities, and can be justified by a detailed study of the structure of the exchange diagrams [16]. It is a significant improvement over the choice previously used in Ref. [15].

The axial vector bosons are treated as contact interactions, with a structure as in Eq. (27), but with the propagator replaced by a constant, $m_a^2 + |q^2| \to m^2$, where the nucleon mass sets a convenient scale not related to a boson mass (the effective boson mass in a contact interaction is infinite). The explicit forms of the numerator functions $\Lambda_1^b \otimes \Lambda_2^b$ can be inferred from Table 1. Note that $\lambda_p = 0$ corresponds to pure pseudovector coupling, and that the definitions of the off-shell coupling parameters $\lambda$ or $\nu$ differ for each boson.

The parameters of two models, called WJC-1 and WJC-2, are displayed in Table 2. Model WJC-1 was constructed with the goal to obtain the best possible fit, while the objective of WJC-2 was to use the smallest number of parameters without destroying the quality of the fit significantly.

Table 3 shows that both models provide excellent fits to the most complete *np* scattering data base, and with a considerably smaller number of adjustable parameters than other realistic potential models. In fact, in view of the $\chi^2/N_{\rm data} = 1.06$ of model WJC-1, the corresponding phase shifts can be considered a new phase shift analysis which includes many more data than the "standard" Nijmegen 93 analysis to which all realistic potential models were fitted.

We emphasize that both models are true OBE models, *proving that OBE models not only work, but can provide an excellent description of the data*. This is compatible with the large $N_c$ limit of QCD. It seems likely that the efficiency of these models (precision fits with fewer parameters) may be attributed to the *efficiency of the Dirac equation*, as mentioned in the Introduction.

**Table 2. Values of the 27 parameters for WJC-1 with 7 bosons and 2 axial vector contact interactions, and the 15 parameters for WJC-2 with no axial vectors. All masses and energies are in MeV; other couplings are dimensionless; $G_b = g_b^2/4\pi$. Parameters in bold were varied during the fit; those labeled with an * were constrained to equal the one above. The triton binding energy is $E_t$ (with its experimental value in parentheses)**

| | | Model WJC-1 | | | | | Model WJC-2 | | | | |
|---|---|---|---|---|---|---|---|---|---|---|---|
| $b$ | I | $G_b$ | $m_b$ | $\lambda_b/\nu_b$ | $\kappa_v$ | $\Lambda_b$ | $G_b$ | $m_b$ | $\lambda_b/\nu_b$ | $\kappa_v$ | $\Lambda_b$ |
| $\pi^0$ | 1 | **14.608** | 134.9766 | **0.153** | — | **4400** | **14.038** | 134.9766 | 0.0 | — | **3661** |
| $\pi^\pm$ | 1 | **13.703** | 139.5702 | **−0.312** | — | 4400* | 14.038* | 139.5702 | 0.0 | — | 3661* |
| $\eta$ | 0 | **10.684** | **604** | **0.622** | — | 4400* | **4.386** | 547.51 | 0.0 | — | 3661* |
| $\sigma_0$ | 0 | **2.307** | **429** | **−15.169** | — | **1435** | **4.486** | **478** | **−2.594** | — | 3661* |
| $\sigma_1$ | 1 | **0.539** | **515** | **4.763** | — | 1435* | **0.477** | **454** | **9.875** | — | 3661* |
| $\omega$ | 0 | **3.456** | **657** | **0.843** | **0.048** | **1376** | **8.711** | 782.65 | 0.0 | 0.0 | **1591** |
| $\rho$ | 1 | **0.327** | **787** | **−1.263** | **6.536** | 1376* | **0.626** | 775.50 | **−2.787** | **5.099** | 1591* |
| $h_1$ | 0 | **0.0026** | — | — | — | 1376* | — | | | | |
| $a_1$ | 1 | **−0.436** | — | — | — | 1376* | — | | | | |
| | | $\Lambda_N =$ **1656**; $E_t = -8.48\,(-8.48)$ | | | | | $\Lambda_N =$ **1739**; $E_t = -8.50\,(-8.48)$ | | | | |

The phase shifts for models WJC-1 and WJC-2 are shown in Figs. 11–13, together with those of the Nijmegen phase shift analysis. While overall the three cases lie close together, it is perhaps surprising that in some regions the agreement between the WJC models and the Nijmegen phase shifts is not better. It can be concluded that the new WJC-1 phase shifts provide a superior representation of the data and should be used in future comparisons of model predictions with experiment. Note that the WJC phase shifts can be used outside the framework of CST, just like any other phase shift analysis.

### 3.2. The Deuteron

The bound-state equation is derived from the representation of the CST scattering amplitude near the deuteron pole,

$$M_{\lambda_1\lambda_1',\beta\beta'}(p,p';P) = R_{\lambda_1\lambda_1',\beta\beta'}(p,p';P) - \sum_{\lambda_d} \frac{\mathcal{G}_{\lambda_1\beta}(p,P,\lambda_d)\,\bar{\mathcal{G}}_{\lambda_1'\beta'}(p',P,\lambda_d)}{M_d^2 - P^2 - i\varepsilon}, \tag{29}$$

where $\lambda_1, \lambda_1'$ are the helicities of the outgoing and incoming on-shell particle 1, $\beta, \beta'$ are the Dirac indices of the off-shell particle 2, $\lambda_d$ is the helicity of the deuteron, $\mathcal{G}$ is the *dnp* vertex function [contracted with $\bar{u}(\mathbf{p}_1,\lambda_1)$] describing the coupling of the deuteron to the neutron and proton, and $R$ is a remainder function finite at the pole. The equation for $\mathcal{G}$ is found by substituting (29) into the scattering equation (14) and demanding that it hold at the pole, giving

$$\mathcal{G}_{\lambda_1\beta}(p,P,\lambda_d) = -\int \frac{d^3k_1}{(2\pi)^3}\frac{m}{E_{k_1}}\overline{V}_{\lambda_1\lambda_1',\beta\alpha}(p,k;P)G_{\alpha\beta'}(k_2)\,\mathcal{G}_{\lambda_1'\beta'}(k,P,\lambda_d)\,. \tag{30}$$

The deuteron binding energy was constrained during the fitting of the CST *NN* kernels, and therefore they reproduce the experimental binding energy of $E_d = 2.2246$ MeV automatically.

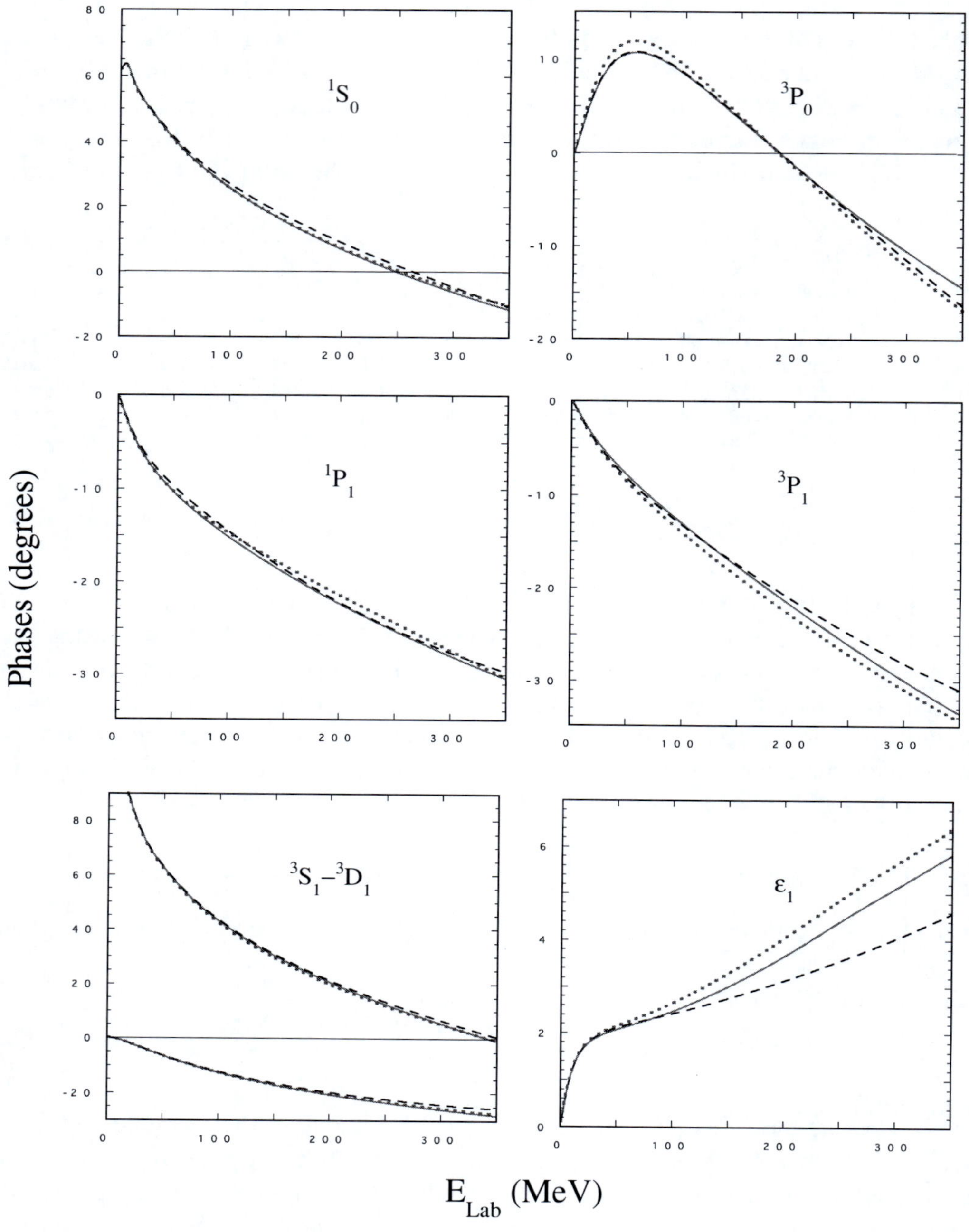


Figure 11. Phase shifts and mixing parameters for all states with $J \leq 1$. Models WJC-1 (solid line) and WJC-2 (dotted line) are compared to the Nijmegen phases (dashed line).

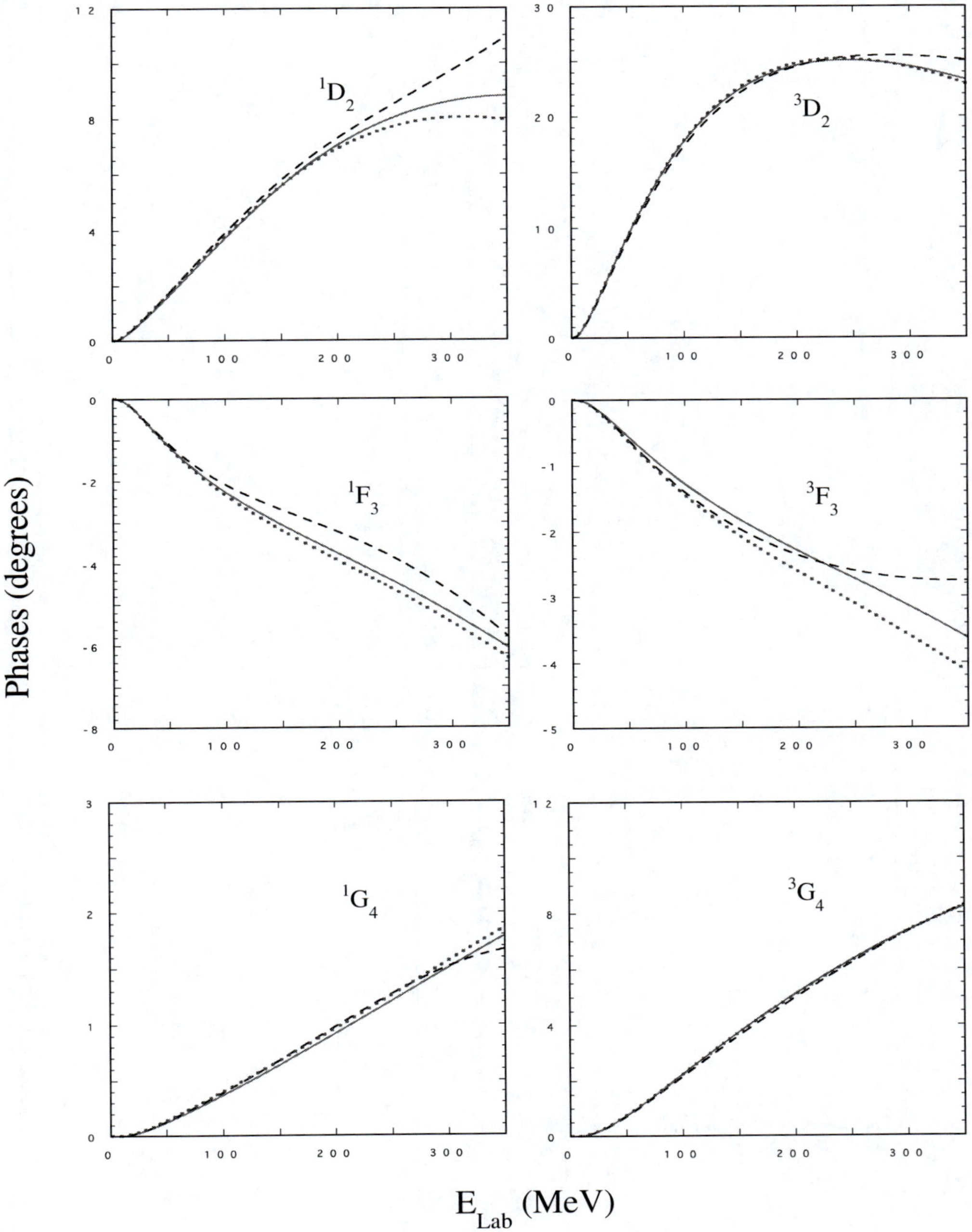


Figure 12. Singlet and uncoupled triplet phase shifts for all states with $2 \leq J \leq 4$. Curves are drawn as in Fig. 11.

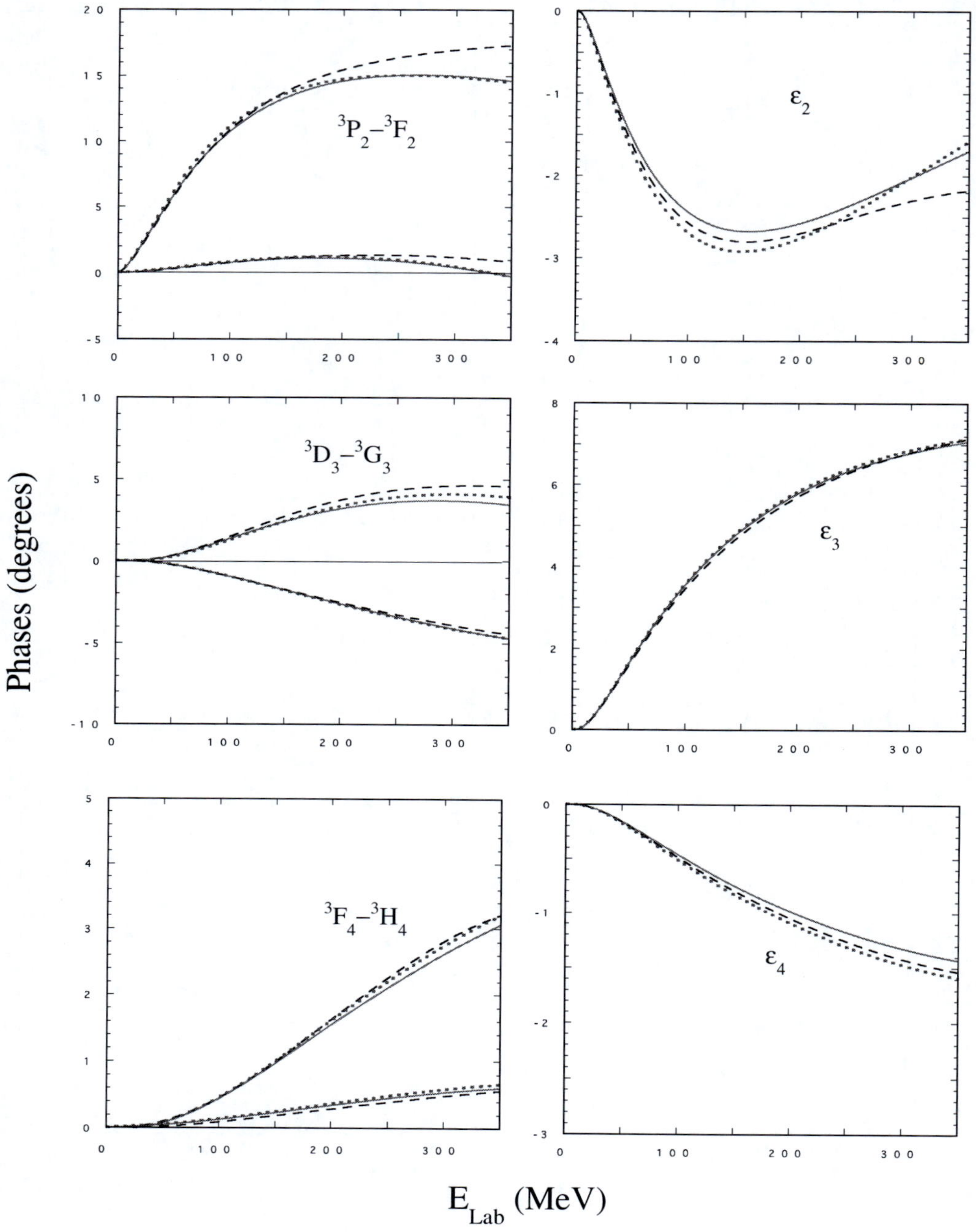


Figure 13. Triplet coupled phase shifts and mixing parameters for states with $2 \le J \le 4$. Curves are drawn as in Fig. 11.

**Table 3. Comparison of precision $np$ models and the 1993 Nijmegen phase shift analysis. The first column specifies the model, the second the number of adjustable parameters (in the case of the first five models for both $np$ and $pp$ data), and the third the year of the data base (data prior to this year are included). Columns four to six are the obtained $\chi^2/N_{\rm data}$ for various data bases (identified by their year), where the number of included data is given in parentheses. Our calculations are in bold face; the others are the author's published results**

| Model | | | $\chi^2/N_{\rm data}(N_{\rm data})$ | | |
|---|---|---|---|---|---|
| Reference | $N_{\rm pars}$ | Year | 1993 | 2000 | 2007 |
| PWA93 [4] | 39 | 1993 | 0.99(2514) | — | — |
| | | | **1.09**(3011) | **1.12**(3336) | **1.13**(3788) |
| Nijm I [20] | 41 | 1993 | 1.03(2514) | — | — |
| AV18 [7] | 40 | 1995 | 1.06(2526) | — | — |
| CD-Bonn [8] | 43 | 2000 | — | 1.02(3058) | — |
| ChPT-Idaho [9] | 35 | 1999 | — | 1.10(2402) | — |
| WJC-1 | 27 | 2007 | **1.03**(3011) | **1.05**(3336) | **1.06**(3788) |
| WJC-2 | 15 | 2007 | **1.09**(3011) | **1.11**(3336) | **1.12**(3788) |

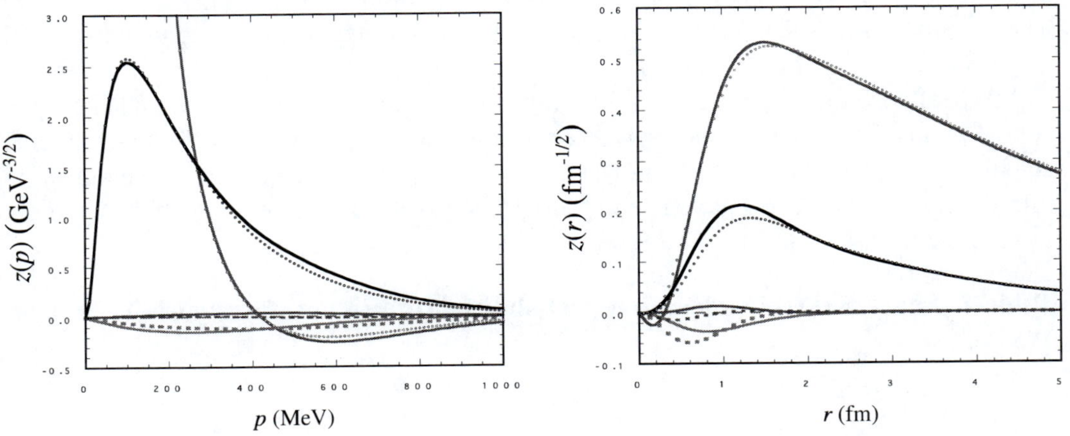


Figure 14. Deuteron wave functions (generically denoted as $z$) in momentum space (left panel) and in coordinate space (right panel). The largest two wave functions are $u$ and $w$ (with $u$ reaching the highest values), with solid lines for WJC-1 and dotted lines for WJC-2. The small P-waves $v_t$ are solid for both models, WJC-1 being negative and WJC-2 positive, whereas both dotted lines are for $v_s$, with WJC-1 negative and WJC-2 closer to zero.

The vertex function $\mathcal{G}$ can be be expanded in terms of four component wave functions: the well-known nonrelativistic S- and D-state deuteron wave functions, $u(p)$ and $w(p)$, and two P-wave functions, the spin singlet $v_s(p)$ and spin triplet $v_t(p)$, which are of purely relativistic origin [21,22]. They satisfy the normalization condition

$$\int_0^\infty p^2 dp \left[u^2(p)+w^2(p)+v_s^2(p)+v_t^2(p)\right]+\langle V'\rangle = 1\,, \tag{31}$$

where $\langle V'\rangle$ is a wave function average of the derivative of the kernel with respect to the total energy, evaluated at the pole energy $M_d$. Table 4 shows the individual contributions to the normalization of Eq. (31) for the two models WJC-1 and WJC-2. In both cases, the derivative of the kernel yields a negative result, so that the sum of the wave function probabilities by themselves exceeds 100%.

The $\langle V'\rangle$ term in the normalization condition (31) is a consequence of electromagnetic interaction currents generated by the momentum dependence of the kernel, and it is inconsistent to compute electromagnetic observables without including these currents. Nevertheless, if interaction currents are ignored, and approximate results are deemed sufficient, the best way to do so appears to be to neglect the derivative term and to rescale the wave functions to a total probability of 100%, as indicated in the Tab. 4. Tables with the numerical values, and convenient parameterizations using analytic functions, both in momentum and coordinate space, are given in Ref. [23].

## 3.3. Three-Nucleon Binding Energy and Three-Nucleon Forces

One of the persistent problems of low-energy few-nucleon physics is the apparent inability of realistic $NN$ potentials to explain the experimental triton binding energy $E_t = 8.48$ MeV. The potentials with the best fit to the $NN$ data yield binding energies roughly between 7.6 and 8 MeV.

To deal with this discrepancy, it is commonly assumed that $3N$ forces must be an essential part of $3N$ dynamics and must therefore be included. One of the features of ChPT calculations is that they predict the process that will generate these forces, requiring the introduction of only 2 new parameters. Models based on a phenomenological $NN$ potential

**Table 4. Exact and scaled deuteron probabilities (in percent) for models WJC-1 and WJC-2**

| | WJC-1 | | WJC-2 | |
|---|---|---|---|---|
| Probability | exact | scaled | exact | scaled |
| $P_s$ | 97.3876 | 92.3330 | 95.7607 | 93.5985 |
| $P_d$ | 7.7452 | 7.3432 | 6.5301 | 6.3827 |
| $P_{v_t}$ | 0.1180 | 0.1119 | 0.0103 | 0.0101 |
| $P_{v_s}$ | 0.2234 | 0.2118 | 0.0090 | 0.0088 |
| $\sum P$ | 105.4743 | 100.0000 | 102.3101 | 100.0000 |
| $\langle V'\rangle$ | −5.4743 | — | −2.3101 | — |
| total | 100.0000 | — | 100.0000 | — |

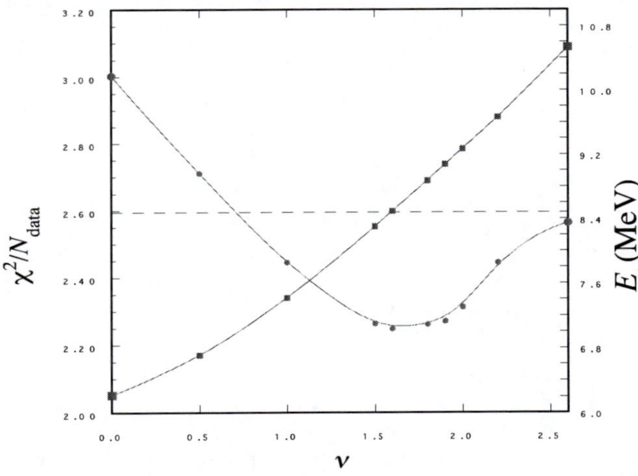


Figure 15. Study of an early family of *NN* models (WNxx) published in 1997 [19]. Here the dependence of $\chi^2/N_{\rm data}$ for a 1994 *np* data base (the curve with the minimum), and the triton binding energy $E_t$ (the curve approximately linear in $\nu$), are shown as a function of the off-shell parameter $\nu$, defined so that $\nu_{\sigma_0} = -0.75\, g_{\sigma_0}\nu$ and $\nu_{\sigma_1} = 2.6\, g_{\sigma_1}\nu$, where $\nu_{\sigma_0}$ and $\nu_{\sigma_1}$ are off-shell couplings of the scalar mesons $\sigma_0$ and $\sigma_1$. Both curves are smooth fits through the points.

must also provide a model for these 3*N* forces with additional parameters. These parameters are usually adjusted to reproduce the triton binding energy, removing the possibility of actually predicting it. Moreover, it raises serious problems of consistency between 2*N* and 3*N* force models. For instance, when the meson-nucleon cut-off parameters employed in realistic potentials are used for the same vertices in 3*N* forces, they lead to very strong overbinding.

Relativity was considered another possible candidate to resolve—at least in part—the triton binding-energy problem. However, the first results with CST OBE potentials were not encouraging, with model IIA of Ref. [15] yielding a binding energy of only $E_t \sim 6$ MeV.

However, the older *NN* kernels of Ref. [15] did not yet exploit the full complexity of boson-nucleon vertices possible in a theory with off-mass-shell nucleons. First steps to explore these additional types of coupling were taken by including terms proportional to $\nu_s$ in the vertices for the coupling of scalar mesons to nucleons [19], in the form

$$g_s\Lambda(p,k) = g_s \quad -\frac{\nu_s}{2m}[(m-\not{p})+(m-\not{k})], \tag{32}$$

as reported in Table 1. (The vector and pseudoscalar meson exchanges also include such terms, fully exploited in WJC-1 fits.) These off-shell couplings contribute only when the incoming or outgoing nucleon is off mass shell, since $(m-\not{p})u(\mathbf{p},\lambda) = 0$ for an on-shell nucleon with momentum $p$.

It turns out that the scalar off-shell coupling terms in Eq. (32) have a remarkable effect, illustrated in Fig. 15 for an early family of models labeled WNxx [19]. Here the parameters $\nu_{\sigma_0}$ and $\nu_{\sigma_1}$ were both fixed (with values depending on the parameter $\nu$ as explained in the figure caption), the other parameters were varied to give the best fit to the data, and the

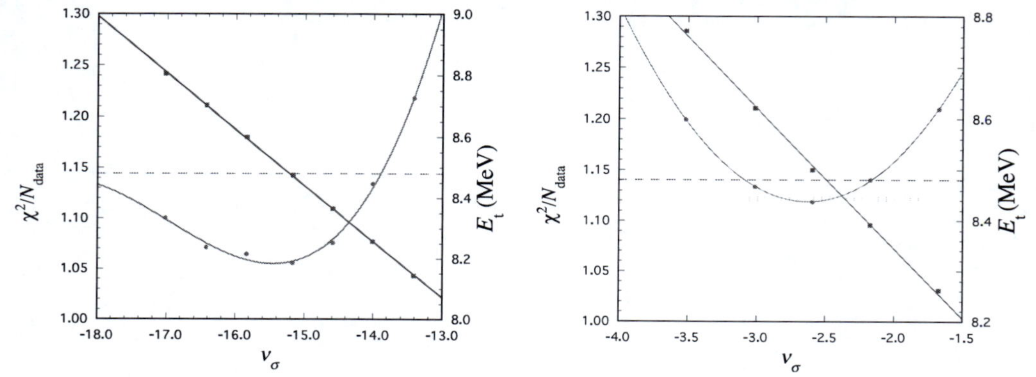


Figure 16. Results for calculations of $\chi^2/N_{\rm data}$ (curves with a minimum and left scale) to a 2007 *np* data base, and triton binding energy $E_t$ (linear curves and right scale) for WJC-1 family (left panel) and for WJC-2 family of models (right panel). The points with the lowest $\chi^2/N_{\rm data}$ are models WJC-1 and WJC-2, respectively. The other models of the two families were obtained by fixing $\nu_\sigma$ at different values and refitting all other parameters. The curves are fits through the actually calculated points. These fits are linear for $E_t$ in both cases, while the fit to $\chi^2/N_{\rm data}$ is cubic for WJC-1 and quadratic for WJC-2. The error in the predicted $E_t$ for model WJC-2 is within the uncertainty in the location of the $\chi^2/N_{\rm data}$, as shown by the grey shaded line.

resulting $\chi^2/N_{\rm data}$ and triton binding energy $E_t$ were determined. The figure shows that both the fit to the *NN* data and the triton binding energy are strongly dependent on the off-shell couplings, and that the *best fit* to the data also gives an *excellent* triton binding energy (correct to 0.01 MeV).

However, the family WNxx does not give a high precision fit; its best $\chi^2/N_{\rm data} \simeq 2.2$, and the fact that the best fit to the data also predicts the correct three-body binding energy might be an accident. The recent models WJC-1 and WJC-2 can also be associated with a family, and the results for the $\chi^2/N_{\rm data}$ and $E_t$ for these families is shown in Fig. 16. Here only the the $\sigma_0$ mixing parameter, $\nu_\sigma = \nu_{\sigma_0}$, was fixed. For each member of the family, characterized by a given $\nu_\sigma$, all other parameters were varied to give the best fit. The same remarkable features occur: within errors, *the best fit to the data also gives the correct triton binding energy*. They again predict the three-body binding energy, even though their detailed structure and parameters differ quite significantly from each other and from the old WNxx models. The result is robust, and it becomes difficult to believe it could possibly be a mere coincidence!

One may ask whether this is really a satisfying result. Don't relativistic 3*N* forces still need to be included, eliminating the nice agreement with the experimental value for $E_t$? The surprising answer is no. In the manifestly covariant OBE theory for the interaction between nucleons no additional irreducible 3*N* forces can be derived. The full three-body interaction is the ladder sum of exchanges between the three nucleons, as illustrated in Fig. 17. As the figure shows, each of these diagrams is a succession of OBE exchanges between only two particles. There are no irreducible three-body forces.

To understand the nuances of this conclusion, consider the time-ordered decomposition of a single $3N$ 4th order Feynman diagram into time-ordered contributions, as sketched in Fig. 18. Only four of the 24 possible diagrams are shown. Diagram18(a) is one of four diagrams resulting from the iteration of the time-ordered OBE diagrams discussed in Fig. 1. The others cannot be generated by iterating the time-ordered OBE contributions, and *in the framework of a time-ordered OBE theory* these are irreducible three-body forces. These include the "stretched" three-body force diagrams (b) similar to the "stretched" two-body forces shown in Fig. 3 ($a_5$ and $a_6$), and contributions from "Z" diagrams associated with particle 2: one with a threshold beginning at $5m$ (diagram c) and another with a threshold at $5m+2m_b$ (diagram d). We repeat: while these are true three-body force diagrams for a time-ordered theory, they are *not three-body forces in the CST*. The conclusion is that the *definition of a three-body force* depends critically on the formalism, and a feature of the CST is that the relativistic OBE generates no three-body forces!

To emphasize the extent to which $3N$ forces (and even the OBE concept itself) are dependent on the theoretical organization of the calculation, turn to Fig. 19, which shows how successive off mass-shell boson couplings (32) can be reinterpreted. As the two left panels show, the effect of the off-shell coupling in one of two successive OBE exchange is to cancel the propagator connecting the two exchanges, generating either a triangle diagram or a three-body contact interaction. As illustrated in the right-hand panels, many more complicated interactions can be generated if more than two off-shell couplings occur in sequence (and the reader may easily write down many more processes that occur in higher order). We are left with an equivalence theorem: an OBE model with *off-shell couplings* (picture #1) is equivalent to another model with *no off-shell couplings supplemented by an infinite number of two- and three-body irreducible diagrams* (picture #2). In picture #2 these extra diagrams describe effective non-OBE contributions to the $NN$ sector and three-body forces in the $3N$ sector.

Remember that these "new" forces are automatically included in an OBE model with off-shell couplings (picture #1). Their appearance in picture #2 shows, however, that the off-shell couplings present in picture #1 give rise to physics that in other models might be interpreted as three-body forces.

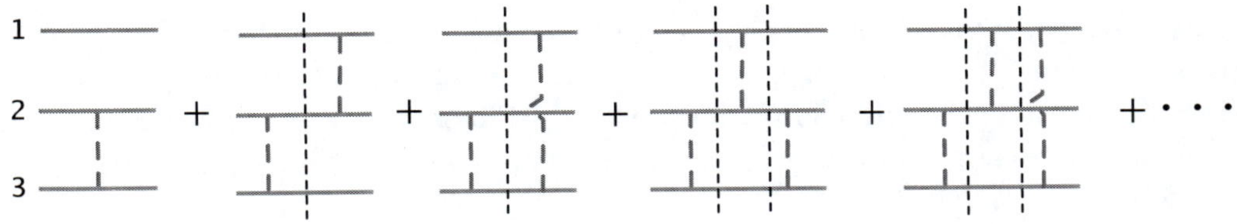


Figure 17. The three-body ladder sum up to 6th order, including only thoses diagrams where particles 2 and 3 are the "last" to interact. The thin vertical dashed line shows how these diagrams can be reduced to successive OBE exchanges between each pair. To allow for multiple exchanges between the same pair, each OBE exchange can be replaced by the two-body scattering amplitude. Similar diagrams exist for the case when the 12 and 13 pairs are the "last" to interact.

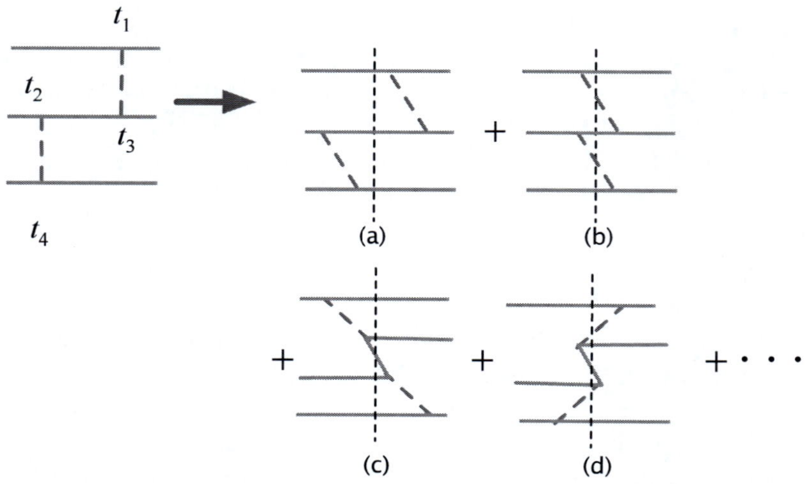


Figure 18. The time-ordered decomposition of a single Feynman diagram. In (a) the time ordering is $t_2 > t_4 > t_1 > t_3$, in (b) $t_2 \sim t_1 > t_3 \sim t_4$, in (c) $t_1 > t_3 > t_2 > t_4$, and in (d) $t_4 > t_3 > t_2 > t_1$.

### 3.4. Corrections to OBE

The cancellation theorem gives strong theoretical justification for considering OBE to be an excellent approximation to the generalized ladder sum. The first corrections to this approximation are the diagrams (b) and (c) shown in Fig. 20. The cancellation theorem (represented diagrammatically in Fig. 4), ensures that this correction approaches zero as the nucleon mass goes to infinity; in the popular language of chiral perturbation theory it could be regarded as the next-to-leading-order (NLO) correction. Results show that this correction is not needed for any of the precision results we have obtained.

Were these corrections to be added to the two-body force, a consistent inclusion of them in the 3$N$ sector would require that all four diagrams shown in Fig. 21 should be treated on the same footing. While diagrams 21(a) and (b) are the reducible counterparts of diagrams 20(b) and (c), the companion diagrams 21(a') and (b') are true three-body forces in the CST. However, since the NLO corrections to the OBE model have been excluded from all considerations, consistency requires that the three-body forces they will generate must also be excluded.

There is a possibility that other degrees of freedom may play a role, such as intermediate-state excitations of the nucleon. Figure 22(a) shows the $2\pi$-exchange 3$N$ force generated by the virtual excitation of a nucleon to a $\Delta$ isobar, the well-known Fujita-Miyazawa force [24]. Such an explicit inclusion of the $\Delta$ or other nucleon resonances goes beyond the current applications of the CST. However, there is some evidence that these kinds of 3$N$ forces would not alter the picture very much.

In their pioneering work on the effects of explicit $\Delta$ isobars in the $NN$- and 3$N$-systems, Hajduk and Sauer [25,26] found that a large part of the Fujita-Miyazawa force of Fig. 22(a) is canceled by the so-called "dispersive" contribution of the $\Delta$ to the $NN$ interaction in the

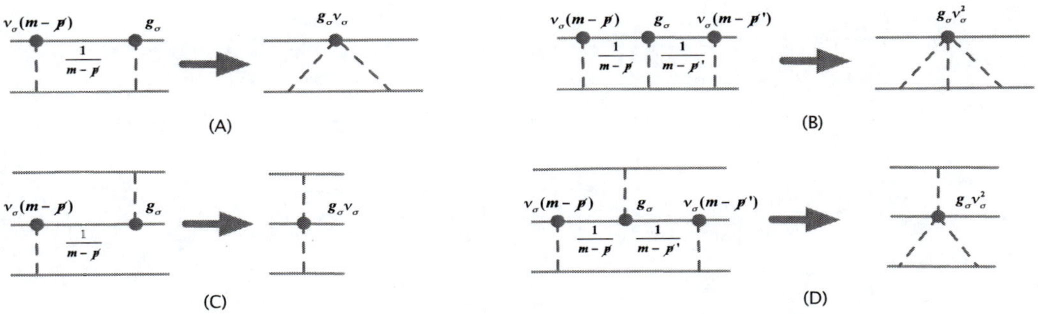


Figure 19. Boson-nucleon vertices with off-shell couplings can generate effective *NN* force diagrams beyond OBE, and effective 3*N* forces. In these examples the off-shell scalar σ meson coupling generates an effective triangle contribution to the *NN* interaction [panel (A)] and a double triangle contribution [panel (B)]. In the 3*N* sector, off-shell exchanges between two different nucleons can also generate effective 3*N* forces [panels (C) and (D)] .

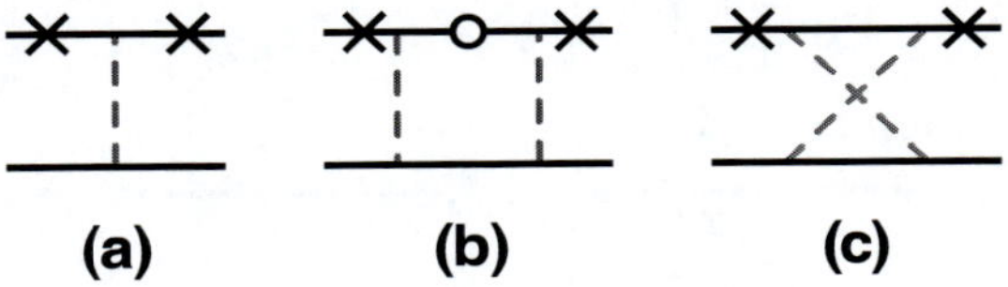


Figure 20. Diagrams (unsymmetrized) for the CST *NN* kernel up to 4nd order. Diagram (a) is the leading OBE approximation, diagrams (b) and (c) are 4th order corrections, not included in an OBE theory. The small circle on the internal nucleon line denotes the subtracted box – the difference between the full box and the iteration of the OBE (the box with the internal particle on-shell).

presence of a spectator nucleon, as shown in Fig. 22(b), leaving only a small remainder. Subsequent work by Picklesimer et al. [27] found an almost vanishing overall effect once multi-Δ configurations are also taken into account.

Peña, Gross, and Surya [28] investigated the 2π-exchange kernel of Fig. 23(a), where a subtracted covariant π*N* scattering amplitude is inserted into a box diagram, containing Δ, Roper, and $D_{13}$ isobars, contact terms, and off-shell (non-pole) dressed nucleon terms, with the spectator nucleon on mass-shell. Embedded in a space of three nucleons, 3*N* forces are generated by this π*N* scattering amplitude which include—but are more general than—the Fujita-Miyazawa force. They found that this 2π-exchange *NN* potential can be very well approximated by the exchange of effective single $\sigma_0$ and $\sigma_1$ mesons. In other words, a simple OBE kernel is able to represent most of the 2π-exchange dynamics, and one might expect the same to be true for the corresponding 3*N* force.

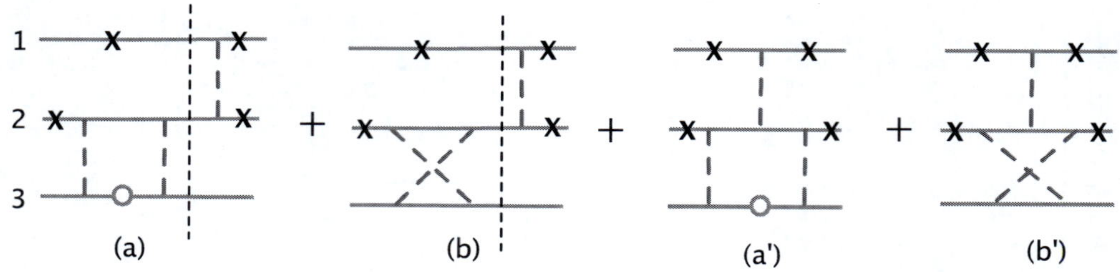


Figure 21. Examples diagrams in the CST that might give rise to $3N$ forces. (As before, the open circle on a nucleon line is the remainder of the diagram once the on-mass-shell contribution has been subtracted.) Diagrams (a) and (b) are reducible and generated automatically through the iteration of the correction kernels 20(b) and (c). Diagrams (a') and (b') are irreducible relativistic $3N$ forces.

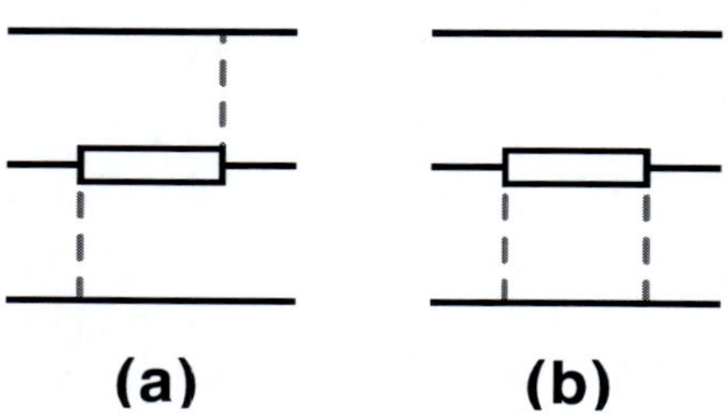


Figure 22. Fujita-Miyazawa $3N$ force (a), and dispersive contribution (b) of the $\Delta$ to the $NN$ interaction in the presence of a spectator nucleon.

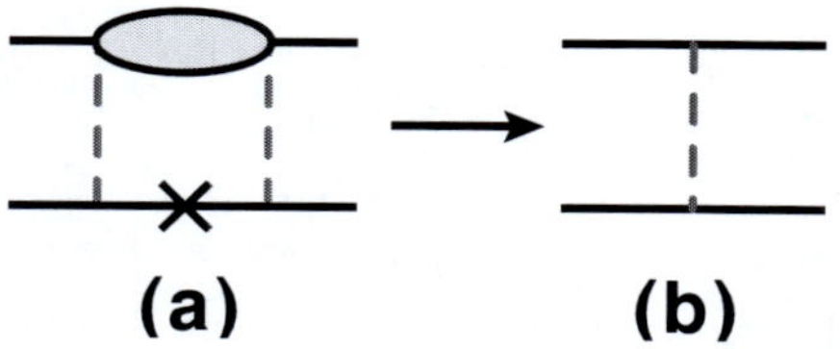


Figure 23. The $2\pi$-exchange potential based on a $\pi N$ scattering amplitude (shaded blob) inserted into a box diagram (a) can be very well approximated by an effective OBE kernel with $\sigma$ and $\delta$ exchange (b).

# 4. Reflections

## 4.1. Overview and Status of CST Applications to *NN* and 3*N* Problems

This chapter reviews the major new insights that emerge from the CST study of nuclear forces. Briefly, these are:

- The covariant OBE model of two- and three-nucleon interactions provides a precision fit to the *np* scattering below 350 MeV lab energy and a precise explanation for the triton binding energy, with *no* additional parameters. To our knowledge, no other method predicts the triton binding without additional parameters that are adjusted to fit it.

- Insight into why the OBE truncation can be so successful follows from the study of a heavy, scalar particle interacting with another particle through a $\phi^3$-like meson exchange. In this case, the infinite sum of all ladder and crossed ladder (generalized ladder) diagrams can be computed *exactly* using the CST equation with the heavy particle on shell, illustrated diagrammatically in Fig. 7. In other cases it has been shown that a scalar exchange effectively approximates many higher order effects [28, 29].

- The remarkable and unexpected discovery that off-shell couplings in the scalar meson exchanges, defined in Eq. (32), not only significantly improve the fits to the *NN* data, but also, at the same time, give the correct triton binding energy, as shown in Figs. 15 and 16. This is a robust result that provides strong evidence that the OBE force is sufficient to describe both two and three-body physics.

- Model WJC-2, with a simple photon exchange and only seven meson exchanges ($\pi^0, \pi^\pm, \eta, \sigma_0, \sigma_1, \omega$, and $\rho$) and only 15 adjustable parameters, shows that the CST can provide a remarkably efficient and simple description of the nuclear force.

Many applications and some development of the CST remain to be done. Among the most important of these is the calculation of the deuteron form factors and static moments, which depend only on the *np* kernels already determined. Early work on the deuteron form factors using dispersion theory [30,31] showed the important role played by the *dnp* vertex function with one nucleon on-shell, providing the initial idea for what became the CST, and an early quantitative calculation using the CST appeared in 1980 [32]. Using deuteron wave functions obtained from the early fits to the *NN* data [15], a successful prediction for the deuteron form factors was made in 1995 [33]. Now this work must be redone because the new precision fits with off-shell scalar (and vector) meson couplings have a momentum dependence that will generate a new class of isoscalar interaction currents never before encountered, and their effect on the deuteron form factors must be studied. This work is expected to be completed in the next year.

The first CST calculations of the three-body form factors were encouraging [34, 35], and a more complete calculation can now be done. In addition, fits to *pp* scattering have yet to be made. Once this is done, it will be possible to predict neutron-neutron scattering, and to calculate the $^3$He binding energy and form factors. While the three-body bound state equation had been solved, the scattering equations have not, and it would be particularly interesting to solve for $Nd$ scattering and look at some of the observables (such as $A_y$) that have defied understanding. Solutions for nuclear matter using a generalized Brueckner-Bethe-Goldstone approach can be readily undertaken, but applications to finite $A \geq 4$ nuclei are much more difficult.

We close this chapter with some reflections on a few issues that provide a contrast between the popular ChPT approach and the CST. This discussion also provides an assessment of the current status of the field.

## 4.2. Renormalization

A successful effective theory of the $NN$ interaction must include a well defined renormalization program. This program will contain a prescription for regularizing the (always present) divergent integrals, i.e., separating them into (i) finite parts describing the low-momentum physics the theory is designed to explain, and (ii) constants parameterizing the unknown short-range (high-momentum) behavior inaccessible to an effective theory. After regularization, all integrals are finite, and any unknown short-range physics has been absorbed into coupling constants and parameters that are determined from experiment. The values of these parameters depend on the renormalization scale, and there should exist renormalization group equations showing how the coupling constants and parameters evolve with this scale.

The gold standard for a successfully renormalized $NN$ theory are the early KSW ChPT results [36], which are unable, however, to completely explain the large and important tensor interaction in the deuteron ${}^3S_1 - {}^3D_1$ channel. The more recent works by Entem and Machleidt [9] and Epelbaum and collaborators [37] have found results insensitive to the regularization parameter as long as it is in the vicinity of 500 to 700 MeV. The CST regularization is contained in the cutoff supplied by the nucleon form factors, $h(p)$, with the masses $\Lambda_N$ shown in Table 2. The fits are very sensitive to the value of this parameter, and, as yet, no attempts have been made to derive equations showing how the other parameters evolve with changes in $\Lambda_N$.

## 4.3. Physics of the $NN$ Interaction

At any energy scale QCD can be described using either confined quark-gluon degrees of freedom or physically observable hadronic (i.e., colorless) degrees of freedom. This choice depends on the scale of the physics being studied. In this respect ChPT and CST are similar: CST uses nucleons and mesons with masses below 1 GeV, while ChPT uses nucleons and pions (only). The great advantage of ChPT is the organizational principle it provides: the existence of a perturbation series for the potential. Unfortunately the scale at which this series diverges is probably less than 1 GeV. Yet we have known for over 20 years that the exchange of mesons with masses in the range of 500 to 700 MeV play a very important role. It is hard to see how these mechanisms can be adequately represented through the perturbation series when the applied cutoff is at the same mass scale as the physics it wants to describe. Is it possible that the need to go to fourth order in ChPT ($N^3LO$) is a reflection of this difficulty?

The organizational principle behind the CST is similar to that in vogue for many years: the exchange of the lightest mesons should account for the longest range force, as suggested by the famous Yukawa potential proportional to

$$V_b(r) \simeq C_b \frac{e^{-m_b r}}{r}. \tag{33}$$

As this suggests, the longest range forces are described by the pion, and the intermediate and shorter range forces described by the exchange of heavier mesons. This organizational principal is less precise and more poorly defined than the chiral perturbation series. Still, the relativistic generalization of the OBE mechanism, using Feynman diagrams to describe the exchanges, works very well. This may be due in part to the fact that scalar, pseudoscalar, and vector meson exchanges have a combined Dirac structure that can reproduce the most general structure of on-shell $NN$ scattering. The covariant OBE model also tells us that there are no many-nucleon forces, confirmed by its ability to predict the three-body binding energy without additional parameters or additional forces. Interaction currents are also comparatively easy to compute using the OBE principle.

Finally, we leave the reader with the observation that the unexpected success of the simple WJC-2 model, with only 15 parameters, shows that, as part of the CST equation, the OBE mechanism works better than any previous study has ever found. It is an attractive alternative to the ChPT.

**Acknowledgements:** F. G. was supported by Jefferson Science Associates, LLC under U.S. DOE Contract No. DE-AC05-06OR23177. A. S. received support from Fundação para a Ciência e a Tecnologia (FCT) under grant No. PTDC/FIS/113940/2009.

## References

[1] H.A. Bethe, R.F. Bacher, Rev. Mod. Phys. **8**, 82 (1936)

[2] M.J. Moravcsik, *The Two-Nucleon Interaction* (Clarendon Press - Oxford, 1963)

[3] G. Brown, A. Jackson, *The Nucleon-Nucleon Interaction* (North Holland - Amsterdam, 1976)

[4] V.G.J. Stoks, R.A.M. Klomp, M.C.M. Rentmeester, J.J. de Swart, Phys. Rev. C **48**, 792 (1993)

[5] The Nijmegen group's NN-OnLine database is found at http://nn-online.org/.

[6] The NN database maintained by the Center for Nuclear Studies (CNS) at the George Washington University is found at http://gwdac.phys.gwu.edu/.

[7] R.B. Wiringa, V.G.J. Stoks, R. Schiavilla, Phys. Rev. C **51**, 38 (1995)

[8] R. Machleidt, Phys. Rev. C **63**, 024001 (2001)

[9] D. Entem, R. Machleidt, Phys. Rev. C **68**, 041001 (2003)

[10] B. Keister, W. Polyzou, Adv.Nucl.Phys. **20**, 225 (1991)

[11] E.E. Salpeter, H.A. Bethe, Phys. Rev. **84**, 1232 (1951)

[12] F. Gross, *Relativistic Quantum Mechanics and Field Theory* (John Wiley & Sons, Inc. – New York, 1993)

[13] F. Gross, Phys. Rev. **186**, 1448 (1969)

[14] S. Weinberg, *The Quantum Theory of Fields, Vol. 1* (Cambridge University Press, 1995)

[15] F. Gross, J.W. Van Orden, K. Holinde, Phys. Rev. C **45**, 2094 (1992)

[16] F. Gross, A. Stadler, Phys. Rev. C **78**, 014005 (2008)

[17] F. Gross, Phys. Rev. C **26**, 2226 (1982)

[18] A. Stadler, F. Gross, M. Frank, Phys. Rev. C **56**, 2396 (1997)

[19] A. Stadler, F. Gross, Phys. Rev. Lett. **78**, 26 (1997)

[20] V.G.J. Stoks, R.A.M. Klomp, C.P.F. Terheggen, J.J. de Swart, Phys. Rev. C **49**, 2950 (1994)

[21] R. Blankenbecler, L.F.C. Jr., Phys. Rev. **119**, 1745 (1960)

[22] W.W. Buck, F. Gross, Phys. Rev. D **20**, 2361 (1979)

[23] F. Gross, A. Stadler, Phys. Rev. C **82**, 034004 (2010). arXiv:1007.0778 [nucl-th]

[24] J. Fujita, H. Miyazawa, Prog. Theor. Phys. **17**, 360 (1957)

[25] C. Hajduk, P.U. Sauer, Nucl. Phys. A **322**, 329 (1979)

[26] C. Hajduk, P.U. Sauer, W. Strueve, Nucl. Phys. A **405**, 581 (1983)

[27] A. Picklesimer, R.A. Rice, R. Brandenburg, Phys. Rev. C **45**, 547 (1992)

[28] M.T. Peña, F. Gross, Y. Surya, Phys. Rev. C **54**, 2235 (1996)

[29] F. Gross, Phys. Rev. C **26**, 2203 (1982)

[30] F. Gross, Phys. Rev. **134**, B405 (1964)

[31] F. Gross, Phys. Rev. **136**, B140 (1964)

[32] R.G. Arnold, C.E. Carlson, F. Gross, Phys. Rev. C **21**, 1426 (1980)

[33] J.W. Van Orden, N. Devine, F. Gross, Phys. Rev. Lett. **75**, 4369 (1995)

[34] S.A. Pinto, A. Stadler, F. Gross, Phys. Rev. C **79**, 054006 (2009)

[35] S.A. Pinto, A. Stadler, F. Gross, Phys. Rev. C **81**, 014007 (2010)

[36] D.B. Kaplan, M.J. Savage, M.B. Wise, Nucl. Phys. **B534**, 329 (1998)

[37] E. Epelbaum, W. Glockle, U.G. Meissner, Nucl. Phys. A **747**, 362 (2005)

In: NN and 3N Interactions
Editors: Leonid Blokhintsev and Igor Strakovsky
ISBN: 978-1-63321-053-0


*Chapter 5*

# Realistic Nucleon-Nucleon Interactions and Two-Nucleon Dynamics in Nuclear Medium

***Claudio Ciofi degli Atti*** *
Istituto Nazionale di Fisica Nucleare, Sezione di Perugia
Department of Physics, University of Perugia, Perugia, Italy

**Abstract**

It is well known that the short-distance behavior of the nucleon-nucleon (NN) interaction cannot be determined uniquely from the analysis of experimental NN elastic phase shifts. At the same time, it is not a priori known whether the two-nucleon interaction in the nuclear medium is the same as the free one, in particular when two nucleons approach a distance of the order of the nucleon radius. The investigation of the short-range two-nucleon dynamics in nuclear medium, commonly referred to as the investigation of short-range correlations (SRCs), is therefore a key topic in nuclear physics. By analyzing recent microscopic many-body calculations in few-nucleon systems and complex nuclei, performed by different groups in terms of realistic NN interactions, it is shown that SRCs exhibit a universal character, in that the correlation hole in nuclei, apart from trivial normalization factors, is almost A-independent and similar to the correlation hole in the deuteron. It seems therefore that in a complex nucleus the short- range two-nucleon dynamics is little affected by the surrounding nucleons. Recent progress in the experimental investigation of SRCs appears to agree with such a picture and with models of the NN interaction exhibiting strong short-range repulsion and intermediate-range tensor attraction. The necessity of more experimental data in this field of research is advocated.

**Keywords**: Many-body approaches; NN interactions; Short Range Correlations; Momentum Distributions

**PACS number:** 21.30.Fe, 21.60.-n, 24.10.Cn, 25.30.-c

---

*ciofi@pg.infn.it

# 1. Introduction

The investigation of the short-range structure of nuclei represents one of the main challenges of modern nuclear physics[1, 2, 3, 4, 5]. The reason is that the detailed theoretical and experimental knowledge of nucleon dynamics at short inter-nucleon distances ($r_{NN} \lesssim 1 \div 1.5\, fm$) could provide decisive answers to longstanding fundamental questions, such as the formation and structure of cold dense matter, possible variations of nucleon properties in the medium, the role of quark -gluon degrees of freedom (d.o.f.) and others

It is well known that many low-energy properties of nuclei can be successfully explained in terms of the independent motion (IM) of nucleons in a mean field (MF) created by their mutual interaction (see, e.g. [6]). These properties however are mainly sensitive to the average inter-nucleon distance ($\bar{r} \simeq 1.5 - 2\, fm^{-1}$); recently it became possible to investigate nuclear structure at high values of energy and momentum transfers, probing inter-nucleon distances of the order of the nucleon radius ($\simeq 1 fm$), a region where serious doubts can be raised as to the validity of a MF description. Here a picture including SRCs is not only more realistic but, as already stressed, it can also provide relevant insight concerning in medium short-range dynamics. The short-range structure of nuclei cannot be inferred uniquely from the knowledge of the short-range behavior of the free NN interaction [7, 8, 9, 10, 11]. As a matter of fact the strong repulsive core in the NN potential, resulting from the analysis of NN elastic scattering data, is introduced by means of various form factors that leave a certain degree of arbitrariness, leading to different short-range behaviors of various NN interaction models. Moreover, elastic on-shell NN scattering cannot in principle determine the details of the NN interaction in medium, because two nucleons that experience interaction with surrounding partners, are off-the-energy shell. As a result, a family of different phase-equivalent potentials can be derived (see, e.g. [12, 13, 14]) that may produce different behaviors of the nuclear wave function at short distances (see e.g. [15]). It should also be stressed that recent *ab initio* many-body approaches (e.g. the Unitary Correlation Operator [16] or the No-Core Shell Model [17] ones) that successfully describe many low-energy properties of nuclei, are based upon various renormalization group (RG) methods (see e.g. Ref. [18, 19, 20]) producing phase equivalent soft NN interactions allowing one to readily diagonalize the many-nucleon Hamiltonian that would be extremely difficult to diagonalize by using the original bare interaction. Due to the soft character of the re-normalized NN interaction, the ground-state wave function produced in these approaches does not contain SRCs. This is however no proof of their absence in nuclei. As a matter of fact, within any RG approach, all necessary operators have to be properly re-normalized, which means that within a discussion of SRCs one faces the problem of evolving high-momentum operators within a low-momentum theory, which is no easy task, though important progress is being done recently [21, 22, 23]. It is not the aim of this review to discuss these approaches, as well those many-body theories (for a recent review see Ref. [24]) based upon effective interactions derived from chiral perturbation theory (see e.g. [12, 13]), where short-range dynamics is described in terms of contact interactions. In the present contribution we focus on the analysis of SRCs produced by realistic many-body calculations performed in terms of realistic NN interactions.

The importance of studying SRCs was stressed more than fifty years ago (see e.g. [25, 26]) but it was only recently that, thanks to the enormous progress made by many-body theories and experimental techniques, the theoretical and experimental studies of SRCs were placed on robust grounds.

In this chapter, a review of recent progress in theoretical and experimental investigations of SRCs will be presented stressing in particular that theoretical calculations demonstrate a universal character of SRCs, in that: (i) in coordinate space they produce a two-nucleon density exhibiting at small relative distances a correlation hole (a region not accessible to nucleons), exhibiting, apart from normalization factors, very mild dependence upon the atomic weight $A$ and essentially resembling the correlation hole in the deuteron; (ii) the correlation hole, in turn, generates in the momentum distributions high-momentum components, missing in MF momentum distributions, and also exhibiting, to a large extent, independence upon A and several interesting scaling properties. Recent experimental evidences supporting such a picture will also be briefly illustrated. The article is organized as follows: in Section 2. a review is presented of modern many-body approaches to the calculation of nuclear properties in terms of realistic NN interactions and their prediction about the short-range structure of nuclei; in Section 3. it is shown how the action of SRCs affects the number of NN pairs in a given spin (S) and isospin (T) state $(ST)$; an exhaustive illustration of the properties and spin-isospin structure of the one-body momentum distributions, related to to the spin-isospin structure of SRCs is presented in Section 4.; calculations of two-body momentum distributions are reviewed in Section 5., and one- and two-nucleon spectral functions are briefly discussed in Section 6.; a brief illustration of recent experimental data providing evidence of SRCs is given in Section 7.; the Conclusions are presented in Section 8..

# 2. Ab Initio Solutions of the Nuclear Many-Body Problem and Theoretical Predictions of SRCs in Configuration Space

## 2.1. The Standard Model of Nuclei

A description of nuclei in terms of quark and gluon d.o.f. implies the solution of non perturbative QCD problems, a very difficult and yet unsolved task. However, as in the case of various many-body systems composed of particles having their own structure, many-nucleon systems could be viewed as systems of point-like particles interacting via proper effective interactions that incorporate the leading d.o.f. of the system that, in case of nuclei, are those associated to nucleons and to the exchanged bosons. However, the reduction of a field theoretical problem to a non-relativistic potential description generates two-, three-,. . . , A-body interactions, so that the general potential energy operator assumes the following form

$$\widehat{V}(\mathbf{x}_1, \mathbf{x}_2, \mathbf{x}_3, \ldots, \mathbf{x}_A) = \sum_{n=2}^{A} \hat{v}_n(\mathbf{x}_1, \ldots \mathbf{x}_n), \tag{1}$$

where $\mathbf{x}_i \equiv \{\mathbf{r}_i, \mathbf{s}_i, \mathbf{t}_i,\}$ denotes the nucleon generalized coordinate, including spatial, spin and isospin coordinates. The relative weight of the various components in Eq. (1) has been estimated many years ago in Ref. [27], arguing that the relative strength between two- and $n$-body interactions should obey the following qualitative relation

$$(n - body\ potential) \simeq \left(\frac{v_N}{c}\right)^{(n-2)} \times (two - body\ potential), \tag{2}$$

where $v_N$ denotes the average nucleon velocity in a nucleus and $c$ the velocity of light. Taking $v_N \simeq 0.1c$, one is led to the conclusion that the two-nucleon interaction is the dominant one. Though such a statement is qualitatively correct, it is nowadays well established from a quantitative point of view that three-nucleon potentials have to be considered in order to explain the ground-state energy of light nuclei [28, 29, 30], with four-nucleon interactions playing only a minor role [28] (for a recent review on three- and many-nucleon forces see Ref. [31]). Therefore the non-relativistic Schrödinger equation assumes the following form

$$\left[\sum_i \frac{\hat{\mathbf{p}}_i^2}{2\,m_N} + \sum_{i<j} \hat{v}_2(\mathbf{x}_i,\mathbf{x}_j) + \sum_{i<j<k} \hat{v}_3(\mathbf{x}_i,\mathbf{x}_j,\mathbf{x}_k)\right] \Psi_f^A(\{\mathbf{x}\}_A) = E_f^A\,\Psi_f^A(\{\mathbf{x}\}_A), \quad (3)$$

where $\{\mathbf{x}\}_A \equiv \{\mathbf{x}_1,\mathbf{x}_2,\mathbf{x}_3,\ldots,\mathbf{x}_A\}$ denotes the set of A generalized coordinates (the spatial coordinates satisfying the condition $\sum_{i=1}^A \boldsymbol{r}_i = 0$) and $f$ denotes the complete set of quantum numbers of state $f$. Eq. (3) will be referred to as the *Standard Model* of nuclei and in what follows we will be mainly interested in the ground-state wave function $\Psi_{f=0}^A \equiv \Psi_0$. Once the interactions are fixed, Eq. (3) should be solved *ab initio*, i.e. without any significant approximation which could mask or distort the main features of $\Psi_f$. We will consider modern 2N bare interactions having the following general form

$$\hat{v}_2(\mathbf{x}_i,\mathbf{x}_j) = \sum_{p=1}^m v^{(p)}(r_{ij})\hat{\mathcal{O}}_{ij}^{(p)} \qquad r_{ij} \equiv |\boldsymbol{r}_i - \boldsymbol{r}_j|, \quad (4)$$

like, e.g., the AV18 [9] ($m = 18$) and AV8$'$ [10] ($m = 8$) interactions, whose main components are:

$$\mathcal{O}_{ij}^{(1)} = 1, \quad \mathcal{O}_{ij}^{(2)} = \boldsymbol{\sigma}_i\cdot\boldsymbol{\sigma}_j, \quad \mathcal{O}_{ij}^{(3)} = \boldsymbol{\tau}_i\cdot\boldsymbol{\tau}_j, \quad \mathcal{O}_{ij}^{(4)} = (\boldsymbol{\sigma}_i\cdot\boldsymbol{\sigma}_j)(\boldsymbol{\tau}_i\cdot\boldsymbol{\tau}_j),$$
$$\mathcal{O}_{ij}^{(5)} = S_{ij}, \quad \mathcal{O}_{ij}^{(6)} = S_{ij}\boldsymbol{\tau}_i\cdot\boldsymbol{\tau}_j, \quad \mathcal{O}_{ij}^{(7)} = (\boldsymbol{L}\cdot\boldsymbol{S})_{ij}, \quad \mathcal{O}_{ij}^{(8)} = (\boldsymbol{L}\cdot\boldsymbol{S})_{ij}\boldsymbol{\tau}_i\cdot\boldsymbol{\tau}_j, \quad (5)$$

Here $S_{ij} = 3(\hat{\boldsymbol{r}}_{ij}\cdot\boldsymbol{\sigma}_i)(\hat{\boldsymbol{r}}_{ij}\cdot\boldsymbol{\sigma}_j) - \boldsymbol{\sigma}_i\cdot\boldsymbol{\sigma}_j$ is the tensor operator, and $(\boldsymbol{L}\cdot\boldsymbol{S})_{ij}$ is the spin-orbit operator where $\boldsymbol{L} = \boldsymbol{r}_{ij}\times\boldsymbol{\pi}_{ij}$ with $\boldsymbol{\pi}_{ij} = -i(\partial/\partial\boldsymbol{r}_{ij})$ and $\boldsymbol{S} = \frac{1}{2}(\boldsymbol{\sigma}_i + \boldsymbol{\sigma}_j)$. As for 3N potentials, several models have been proposed in order to reproduce the binding energy of few-nucleon systems, that are underbound by about $0.2 - 0.3$ MeV per particle when only 2N interactions are used (see e.g.[32]). Within the MF approximation ($\sum_{i<j}\hat{v}_2(\mathbf{x}_i,\mathbf{x}_j) + \sum_{i<j<k}\hat{v}_3(\mathbf{x}_i,\mathbf{x}_j,\mathbf{x}_k) \Rightarrow \sum_i U(\mathbf{x}_i)$) the ground-state solution of Eq. (3) is an antisymmetrized product of single particle wave functions $\phi_{\alpha_i}$, i.e.

$$\Psi_0(\{\mathbf{x}\}_A) \Rightarrow \Phi_0(\{\mathbf{x}\}_A) = \hat{\mathcal{A}}\prod_i^A \phi_{\alpha_i}(\mathbf{x}_i) = \Phi_{0p0h}(\{\mathbf{x}\}_A), \quad (6)$$

where $\Phi_{0p0h}(\{\mathbf{x}\}_A)$ is a Slater determinant with zero particle, zero hole (0p-0h) excitations, i.e. with all states below the Fermi (F) level occupied and those above it empty ($\phi_{\alpha_i} = 0$ if $\alpha_i > \alpha_F$), and $\hat{\mathcal{A}}$ is the antisymmetrization operator. The general solution of Eq. (3) includes, on the opposite, a huge number of Slater determinants describing $np$-$nh$ excitations generated by SRCs

$$\Psi_0(\{\mathbf{x}\}_A) = c_0\,\phi_{0p0h}(\{\mathbf{x}\}_A) + c_1\,\Phi_{1p1h}(\{\mathbf{x}\}_A) + c_2\,\Phi_{2p2h}(\{\mathbf{x}\}_A) + \ldots. \quad (7)$$

*Ab initio* direct solutions of Eq. (3) in terms of bare realistic interactions are possible in the case of few-nucleon systems (A=3, 4) within several approaches (see e.g. Refs. [33, 34, 35, 36, 37, 38, 39, 40, 41]). For complex nuclei the fully *ab initio* solutions are still difficult to obtain, but for $A \leq 12$ ground-state energies and excitation spectra were obtained with the AV18 NN interaction plus 3N potentials, by means of the Green Function Monte Carlo (GFMC) method (see e.g. Ref.[42]); for $^{16}$O the Variational Monte Carlo (VMC) method has been used [43], and for $A \geq 16$ the *cluster expansion* approach has been adopted with success [44, 45]. The picture that emerges from these calculations is a structure of the ground-state wave function in the following form

$$\Psi_0(\{\mathbf{x}\}_A) = \hat{F}(\{\mathbf{x}\}_A)\,\Phi_0(\{\mathbf{x}\}_A), \tag{8}$$

where

$$\hat{F}(\{\mathbf{x}\}_A) = \hat{\mathcal{S}}\prod_{i<j}\hat{f}_{ij}(\mathbf{x}_i,\mathbf{x}_j) = \hat{\mathcal{S}}\prod_{i<j}\left[\sum_{p=1}^{m} f^{(p)}(r_{ij})\hat{\mathcal{O}}_{ij}^{(p)}\right] \tag{9}$$

is a correlation operator introducing SRCs into the MF wave functions $\Phi_0$, $\hat{\mathcal{S}}$ is a symmetrization operator, and $\hat{\mathcal{O}}_{ij}^{(p)}$ is the same operator appearing in Eq. (4). It can be seen that the many-body wave function exhibits a rich correlation structure, the dominant SRC effects arising from the short-range repulsion and the intermediate tensor attraction. Provided the many-body Schroedinger equation can be solved *ab initio* in terms of bare NN interactions, the comparison with proper experimental data could provide answer as to whether the short-range structure of nuclei can be described in terms of the free NN interaction. This is the central point of the problem that will be discussed in this contribution.

## 2.2. The One- and Two-Body Densities and SRCs

Once the many-body wave function $\Psi_0$ is at disposal, the relevant quantities of interest are the $n$-body densities, in particular:

1. the one-body non-diagonal spin-isospin independent density:

$$\rho(\boldsymbol{r}_1,\boldsymbol{r}_1') = A\int \Psi_0^*(\boldsymbol{r}_1,\{\boldsymbol{r}\}_{A-1})\Psi_0(\boldsymbol{r}_1',\{\boldsymbol{r}\}_{A-1})\prod_{i=2}^{A} d\boldsymbol{r}_i; \tag{10}$$

2. the two-body non-diagonal spin-isospin independent density

$$\rho(\boldsymbol{r}_1,\boldsymbol{r}_1';\boldsymbol{r}_2,\boldsymbol{r}_2') = \frac{A(A-1)}{2}\int \Psi_0^*(\boldsymbol{r}_1,\boldsymbol{r}_2,\{\boldsymbol{r}\}_{A-2})\Psi_0(\boldsymbol{r}_1',\boldsymbol{r}_2',\{\boldsymbol{r}\}_{A-2})\prod_{i=3}^{A} d\boldsymbol{r}_i; \tag{11}$$

3. the non-diagonal spin-isospin dependent two-body density

$$\rho_{(ST)}^{N_1N_2}(\mathbf{r}_1,\mathbf{r}_1';\mathbf{r}_2,\mathbf{r}_2') = \int \psi_0^{A*}(\{\widetilde{\mathbf{x}}\}_A)\sum_{i<j}\hat{P}_{ij}^{S}\,\hat{P}_{ij}^{T}\,\widehat{\rho}_{ij}(\mathbf{r}_1,\mathbf{r}_1';\mathbf{r}_2,\mathbf{r}_2')\psi_0^{A}(\{\widetilde{\mathbf{x}}'\}_A)d\mathbf{X}; \tag{12}$$

where $N_1$ and $N_2$ denote the two nucleons in state $(ST)$, $\hat{P}_{ij}^{S(T)}$ is a projection operator in the state with spin (isospin) S(T), $d\mathbf{X} \equiv \prod_{i=1}^{A} d\widetilde{\mathbf{x}}_i d\widetilde{\mathbf{x}}_i'$, and

$$\widehat{\rho}_{ij}(\mathbf{r}_1, \mathbf{r}_1'; \mathbf{r}_2, \mathbf{r}_2') = \delta(\widetilde{\mathbf{r}}_i - \mathbf{r}_1)\delta(\widetilde{\mathbf{r}}_j - \mathbf{r}_2)\delta(\widetilde{\mathbf{r}_i'} - \mathbf{r}_1')\delta(\widetilde{\mathbf{r}}_j' - \mathbf{r}_2') \prod_{k\neq\{i,j\}}^{A} \delta(\widetilde{\mathbf{r}}_k - \widetilde{\mathbf{r}}_k'). \quad (13)$$

is the non-diagonal two-body density operator. The one-body diagonal $\rho(\boldsymbol{r}_1)$, two-body diagonal $\rho(\boldsymbol{r}_1, \boldsymbol{r}_2)$, half-diagonal $\rho(\boldsymbol{r}_1, \boldsymbol{r}_2; \boldsymbol{r}_1')$ and $\rho_{(ST)}^{N_1N_2}(\mathbf{r}_1, \boldsymbol{r}_1'; \boldsymbol{r}_2)$ densities can easily be obtained from Eqs. (11), (12) and (13), by inserting proper $\delta$-functions into the integrals and properly generalizing the operator (13) (see Ref.[46]).

The diagonal spin-isospin independent two-body density matrix can be written as follows

$$\rho(\boldsymbol{r}_1; \boldsymbol{r}_2) = \sum_{ST} \rho_{(ST)}^{N_1N_2}(\mathbf{r}_1; \mathbf{r}_2) = \rho(\mathbf{r}_{rel}, \mathbf{R}_{c.m.}), \quad (14)$$

where the relative (*rel*) and center-of-mass (*c.m.*) coordinates are

$$\boldsymbol{r}_{rel} = \boldsymbol{r}_1 - \boldsymbol{r}_2 \equiv \boldsymbol{r} \qquad \boldsymbol{R}_{c.m.} = \frac{\boldsymbol{r}_1 + \boldsymbol{r}_2}{2} \equiv \boldsymbol{R}, \quad (15)$$

The integral of $\rho(\boldsymbol{r}_1; \boldsymbol{r}_2)$yields the total number of NN pairs

$$\int \rho(\boldsymbol{r}_1; \boldsymbol{r}_2) d\,\mathbf{r}_1\, d\,\mathbf{r}_2 = \sum_{ST} \int \rho_{(ST)}^{N_1N_2}(\mathbf{r}_1; \mathbf{r}_2) d\,\mathbf{r}_1\, d\,\mathbf{r}_2 = \sum_{ST} N_{(ST)}^{N_1N_2} = \frac{A(A-1)}{2}, \quad (16)$$

where $N_{(ST)}^{N_1N_2}$ is the number of NN pairs in state $(ST)$. The relative and *c.m.* two-nucleon densities can then be defined as follows

$$\rho_{rel}(\boldsymbol{r}) = \int \rho(\boldsymbol{r}, \boldsymbol{R})\, d\boldsymbol{R} \qquad \rho_{c.m.}(\boldsymbol{R}) = \int \rho(\boldsymbol{r}, \boldsymbol{R}) d\,\boldsymbol{r}. \quad (17)$$

As it will be shown in the next Section, these density matrices provide very clear definition of SRCs and their effects on NN densities in nuclei.

## 2.3. The Correlation Hole in Few-Nucleon Systems and Complex Nuclei

*Ab initio* calculations with bare realistic interactions show that, apart from an obvious normalization factor counting the

number of pairs in different nuclei, the relative two-body density $\rho_{rel}(r)$ and its spin-isospin components $\rho_{ST}^{N_1N_2}(r)$ exhibit at $r \lesssim 1.5\, fm$ a sharp damping with respect to the analogous MF densities. This is exactly the previously mentioned *correlation hole* ; it is illustrated in Fig. 1 for the nucleus of $^{16}O$. The correlation hole is generated by the cooperation of the short-range repulsion and the intermediate-range tensor attraction of the NN interaction, with the tensor force governing the overshooting at $r \simeq 1.0\, fm$ in the *np* distribution. Figs. 2 and 3 illustrate the universality of the correlation hole,

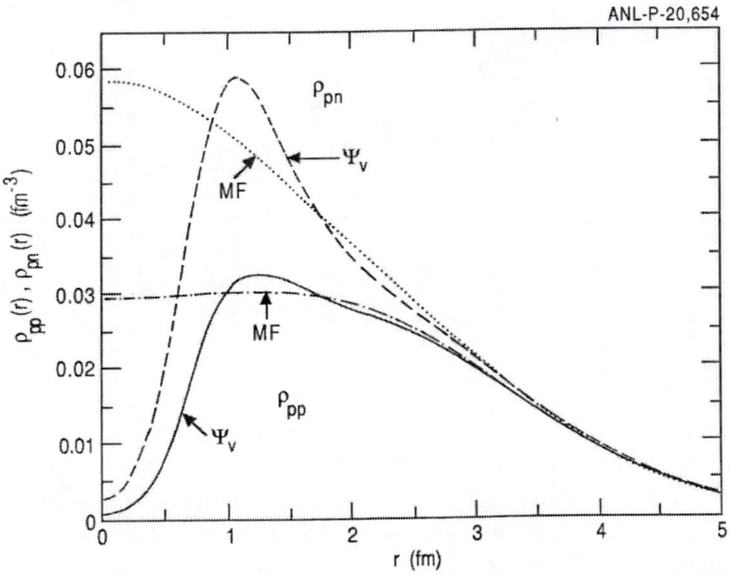


Figure 1. The two-body density distribution of $pn$ and $pp$ pairs in $^{16}$O corresponding to mean-field (MF) and correlated ($\Psi_V$) wave functions obtained within the Variational Monte Carlo approach with the AV14 NN interaction plus 3N forces (Figure reprinted from. [43]. Copyright (1992) by the American Physical Society).

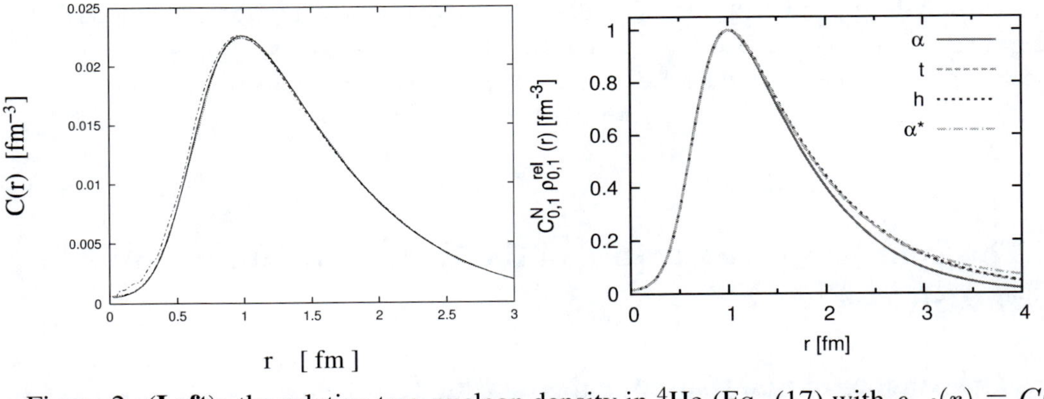


Figure 2. (**Left**): the relative two-nucleon density in $^4$He (Eq. (17) with $\rho_{rel}(r) \equiv C(r)$) calculated within six different *ab initio* many-body theories and the AV18 interaction leading practically to undistinguishable results. (Figure reprinted from. [47]. Copyright (1992) by the American Physical Society). (**Right**): the relative two-nucleon density (normalized at $r \simeq 1\ fm$) in $^2$H, $^3$H, $^4$He and $^4$He* for NN pairs in relative S=0 and T=1 state resulting from *ab initio* calculations within the method of Ref. [39] and the AV8′ interaction [10]. (Figure reprinted from. [48]. Copyright (2011) by the American Physical Society).

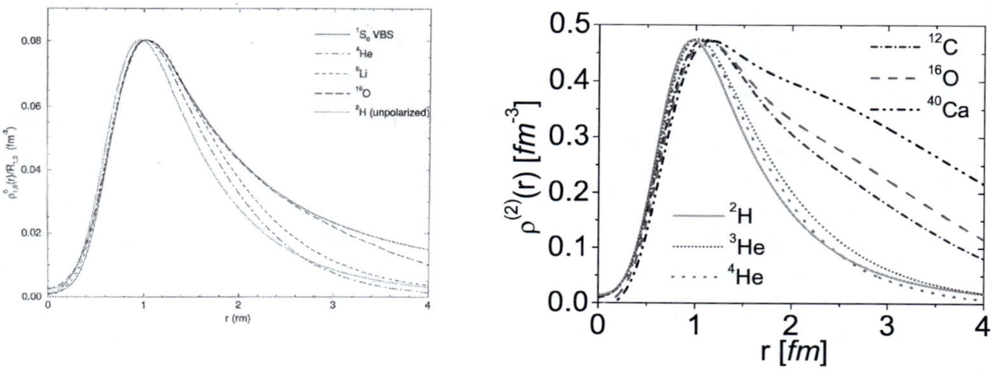


Figure 3. (**Left**): the relative two-nucleon density (Eq. (17) normalized at $r \simeq 1\,fm$) in $^2$H, $^4$He, $^6$Li and $^{16}$O obtained in Ref. [49] within the VMC method and AV18 interaction. (Figure reprinted from. [49]. Copyright (1996) by the American Physical Society). (**Right**): the two-nucleon density (Eq. (17) with $\rho_{rel}(r) \equiv \rho^{(2)}(r)$)) obtained with *ab initio* wave functions for $^3$He and $^4He$ and within the number-conserving linked-cluster expansion of Ref. [45] and the AV8$'$ interaction [10] for $^{12}$C, $^{16}$O and $^{40}$Ca [50].

i.e. its independence upon $A$. These Figures also demonstrate that different many-body approaches, ranging from the GFMC one to proper cluster expansion methods, which in principle may give different results for the ground-state energy, predict, practically, the same behavior of the correlation hole. In order to be able to obtain information about this important feature characterizing the relative NN motion in medium, we have first of all to shift to momentum space, expecting: (i) an increase of nucleon high-momentum components in the ground-state wave function, (ii) peculiar momentum configurations that are missing in a mean-field description, and, (iii) a variation of the spin-isospin structure of the ground-state wave function. Let us start by discussing the spin-isospin structure of nuclei and how it is affected by SRCs.

## 3. The Spin-Isospin Structure of the Nuclear Ground State and SRCs

### 3.1. The Number of Spin-Isospin Pairs in a Nucleus

The quantum numbers that characterize a two-nucleon pair in a nucleus are the relative orbital momentum L, the total spin S and the total isospin T. Pauli principle requires that L + S + T=*odd number*. In a pure shell-model picture and $A \leq 4$ L=0, so that (ST)=(10) and (01), whereas for $A > 4$ we can have both L even, with (ST)=(10) and (01), and L odd, with (ST)=(00) and (11). The deviations from the shell model originating from SRCs, are accompanied, in $A \leq 4$ nuclei, by the creation of $(00)$ and $(11)$ states, and in complex nuclei by a reduction of the number of $(01)$ and $(10)$ states in favor of $(11)$ and $(00)$ states.

The number of pairs in different $(ST)$ states in several nuclei given by

$$N_{(ST)} = \int d\boldsymbol{r}_1 \, d\boldsymbol{r}_2 \, \rho_{(ST)}(\boldsymbol{r}_1; \boldsymbol{r}_2) \tag{18}$$

and calculated by different groups, is reported in Table 1; it can be seen that: (i) SRCs do

**Table 1. The number of pairs $N_{(ST)}$, Eq. (18), in various spin-isospin states in the independent particle model (IPM) and taking into account SRCs within different many-body approaches (see text) with realistic interactions (AV18 and AV8'). (Table reprinted from Ref. [46]. Copyright (2013) by the American Physical Society)**

| Nucleus | | (ST) | | | |
|---|---|---|---|---|---|
| | | (10) | (01) | (00) | (11) |
| $^2$H | | 1 | - | - | - |
| $^3$He | IPM | 1.50 | 1.50 | - | - |
| | SRC [46] | 1.488 | 1.360 | 0.013 | 0.139 |
| | SRC [49] | 1.50 | 1.350 | 0.01 | 0.14 |
| | SRC [48] | 1.489 | 1.361 | 0.011 | 0.139 |
| $^4$He | IPM | 3 | 3 | - | - |
| | SRC [46] | 2.99 | 2.57 | 0.01 | 0.43 |
| | SRC [49] | 3.02 | 2.5 | 0.01 | 0.47 |
| | SRC [48] | 2.992 | 2.572 | 0.08 | 0.428 |
| $^{16}$O | IPM | 30 | 30 | 6 | 54 |
| | SRC[46] | 29.8 | 27.5 | 6.075 | 56.7 |
| | SRC [49] | 30.05 | 28.4 | 6.05 | 55.5 |
| $^{40}$Ca | IPM | 165 | 165 | 45 | 405 |
| | SRC[46] | 165.18 | 159.39 | 45.10 | 410.34 |

not practically affect the state $(10)$, but appreciably reduce the state $(01)$, in favor of the $(11)$ state; this is ascribed to a three-body-like mechanism originating from the tensor force [49, 48] illustrated in Fig. 4: tensor correlations between particles "2" and "3" generate a spin flip of particle "2", that gives rise to the state $(11)$ between particles "2" and "1" ; (ii) as in the case of the correlation hole, there is again a general agreement between the results by different groups using different many-body approaches, namely: the VMC with various Argonne interactions, in Ref.[49]; the correlated Gaussian basis approach [38] with the V8$'$ interaction, in Ref. [48]; the hyperspherical harmonic variational method with the $AV18$ interaction in Ref.[34]; the ATMS method of Ref. [35] with the AV8$'$ interaction and, eventually, the linked-cluster expansion of Ref. [45] with the AV8$'$ interaction, in Ref. [46].

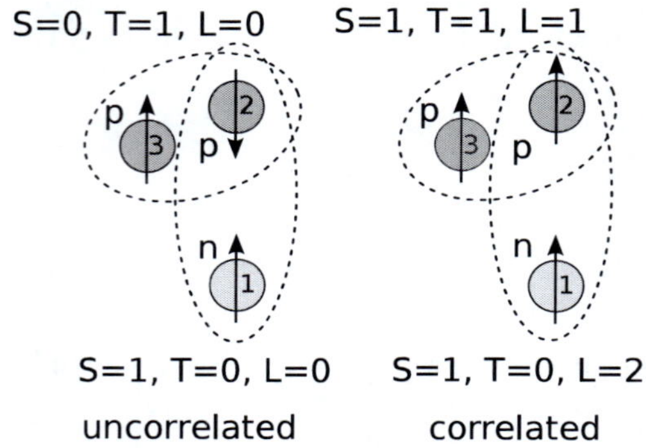


Figure 4. The three-body mechanism leading to the increase of the number of pairs in $(ST) = (11)$ state (After Ref. [48]). (Figure reprinted from Ref. [48]. Copyright (2011) by the American Physical Society).

## 4. One-Body Momentum Distributions and SRCs

Let us now discuss how and to what extent SRCs affect the one-body momentum distribution which is defined as the Fourier transform of the non diagonal one-body density matrix

$$n_A(\boldsymbol{k}_1) = \frac{1}{A(2\,\pi)^3}\int e^{-i\,\boldsymbol{k}_1\cdot(\boldsymbol{r}_1-\boldsymbol{r}'_1)}\rho(\boldsymbol{r}_1,\boldsymbol{r}'_1)d\boldsymbol{r}_1 d\boldsymbol{r}'_1 \tag{19}$$

Since $\rho(\boldsymbol{r}_1,\boldsymbol{r}'_1)$ represent density fluctuation, a large effect of SRCs on the high momentum components is expected to occur. As a matter of fact, recent systematic analysis of realistic calculations of $n_A(k)$ for A=2, 3, 4, 16, and 40 has been presented in Ref. [46]. The results for $^{16}$O, performed by different groups, is shown in Fig. 5(**a**), which is aimed at illustrating the convergence of different approaches that use similar NN interactions, whereas Fig. 5(**b**) shows that the high-momentum part of $n_A(k_1)$ of different nuclei exhibits a qualitative universal scaling behavior.

### 4.1. General Definitions and Two-Nucleon SRC (2N-SRC) Configurations

SRCs considerably increase the high-momentum content of the one-body momentum distributions through the term $\sum_{n=2}^{\infty} c_n \Phi_{npnh}$, in Eq. (7), i.e. via the population of $np$-$nh$ states with momentum much higher than the Fermi momentum $k_F \simeq 1.4 fm^{-1}$. SRCs, moreover, generate peculiar momentum configurations that are missing in a MF description [51]. As a matter of fact, since momentum conservation requires that

$$\sum_{i=1}^{A} \boldsymbol{k}_i = 0 \tag{20}$$

a nucleon with high momentum $\boldsymbol{k}_1$ in a MF configuration is expected to be balanced by the rest of the $(A-1)$ nucleons, i. e.

$$\boldsymbol{k}_1 \simeq -\sum_{2}^{A} \boldsymbol{k}_i \qquad \boldsymbol{k}_i \simeq \frac{\boldsymbol{k}_1}{A-1}, \tag{21}$$

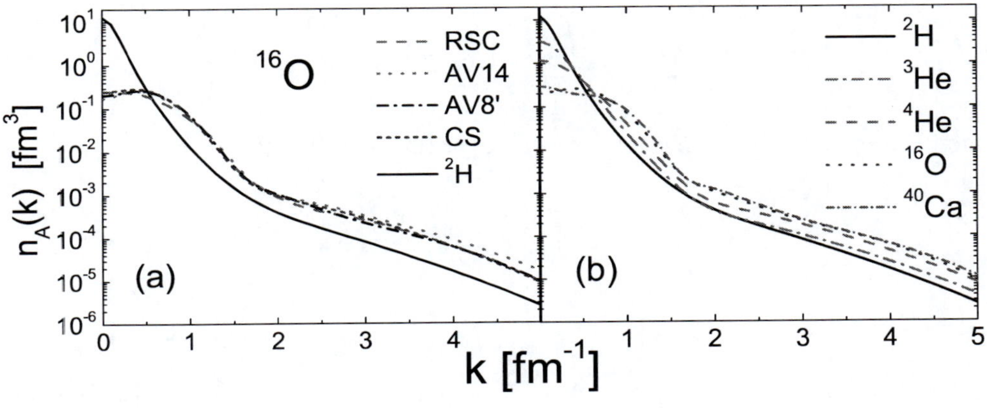


Figure 5. (**a**): the momentum distribution in $^{16}$O calculated with different NN interactions and theoretical approaches: RSC [52]; AV14 [43]; AV8' [45]. The phenomenological distribution of Ref. [53] is also shown (CS) and $^2$H denotes the deuteron momentum distribution. (**b**): the proton momentum distribution of different nuclei calculated within different many-body approaches with equivalent NN interactions, namely the AV18 one, in the case of $^2$H and $^3$He, and the AV8$'$ one, in the case of $^4$He, $^{16}$O, and $^{40}$Ca. Hereafter the notation $|\mathbf{k}_1| \equiv k$ will be adopted. (Figure reprinted from. [46]. Copyright (2013) by the American Physical Society).

whereas in a 2N-SRC configuration one has

$$\boldsymbol{k}_1 \simeq -\boldsymbol{k}_2 \qquad \boldsymbol{K}_{A-2} = \sum_{3}^{A} \boldsymbol{k}_i \simeq 0. \tag{22}$$

Therefore 2N-SRCs can be defined as those configurations of a pair of nucleons characterized by *high* relative and *small* c.m. momenta. The quantitative meaning of such a statement will be discussed later on.

## 4.2. The Probability of MF and SRC Configurations

The ground-state wave function $\Psi_0$, solution of Eq. (3) describes both MF and correlated nucleon motions. The latter, in turn, includes both long- and short-range correlations; long-range correlations (LRC) manifest themselves mostly in open shell nuclei, and are responsible for configuration mixing resulting in partial occupation of states which are empty in a simple independent particle model, with small effects, however, on high-momentum components; SRCs, on the contrary, generate high virtual particle-hole excitations even in closed-shell nuclei, and strongly affect the high-momentum content of the wave function. Therefore, assuming that the momentum distributions could be extracted from some experimental data, we have to figure out a clear-cut way to disentangle the momentum content generated by the MF and LRCs from the one arising from SRCs. Denoting by $\{|\psi_f^{A-1}>\}$ the complete set of eigenfunctions of nucleus $(A-1)$ described by the same Hamiltonian of nucleus A, and using the completeness relation

$$\sum_{f=0}^{\infty} |\Psi_f^{A-1}><\Psi_f^{A-1}| = 1, \tag{23}$$

the one-nucleon momentum distribution can be written as follows [54]

$$n_A(\boldsymbol{k}_1) = n_{gr}(\boldsymbol{k}_1) + n_{ex}(\boldsymbol{k}_1), \tag{24}$$

where

$$(2\pi)^3 n_{gr}(\boldsymbol{k}_1) = \\ = \sum_{f=0,\sigma_1} \left| \int e^{i\boldsymbol{k}_1\cdot\boldsymbol{r}_1} d\boldsymbol{r}_1 \int \chi^{\dagger}_{\frac{1}{2}\sigma_1} \Psi_{f=0}^{(A-1)*}(\{\boldsymbol{r}_i\}_{A-1}) \Psi_0(\boldsymbol{r}_1, \{\boldsymbol{r}_i\}_{A-1}) \prod_{i=2}^{A} d\boldsymbol{r}_i \right|^2 \tag{25}$$

and

$$(2\pi)^3 n_{ex}(\boldsymbol{k}_1) = \\ = \sum_{f\neq 0,\sigma_1} \left| \int e^{i\boldsymbol{k}_1\cdot\boldsymbol{r}_1} d\boldsymbol{r}_1 \int \chi^{\dagger}_{\frac{1}{2}\sigma_1} \Psi_{f}^{(A-1)*}(\{\boldsymbol{r}_i\}_{A-1}) \Psi_0(\boldsymbol{r}_1, \{\boldsymbol{r}_i\}_{A-1}) \prod_{i=2}^{A} d\boldsymbol{r}_i \right|^2. \tag{26}$$

Here the sum over $f$ stands also for an integral over the continuum final states that are present in Eq. (23). We see that the momentum distribution can be expressed through the overlap integrals between the ground-state wave function $\Psi_0$ of nucleus A and the wave function $\Psi_f^{(A-1)}$ of the state $f$ of nucleus $(A-1)$. The separation of the momentum distributions in $n_{gr}$ and $n_{ex}$ is particularly useful for $A = 3$, 4 nuclei, i.e. when the excited states of $(A-1)$ are in the continuum. For complex nuclei, where many discrete hole excited states are present, it is more convenient to use another representation where the particle-hole structure of the realistic solutions of Eq. (3) is explicitly exhibited by Eq. (7). Within such a representation, one has [55, 56]

$$n_A(\boldsymbol{k}_1) = n_0(\boldsymbol{k}_1) + n_1(\boldsymbol{k}_1), \tag{27}$$

where

$$(2\pi)^3 n_0(\boldsymbol{k}_1) = \\ = \sum_{f\leq F,\sigma_1} \left| \int e^{i\boldsymbol{k}_1\cdot\boldsymbol{r}_1} d\boldsymbol{r}_1 \int \chi^{\dagger}_{\frac{1}{2}\sigma_1} \Psi_{f}^{(A-1)*}(\{\boldsymbol{r}_i\}_{A-1}) \Psi_0(\boldsymbol{r}_1, \{\boldsymbol{r}_i\}_{A-1}) \prod_{i=2}^{A} d\boldsymbol{r}_i \right|^2 \tag{28}$$

$$(2\pi)^3 n_1(\boldsymbol{k}_1) = \\ = \sum_{f> F,\sigma_1} \left| \int e^{i\boldsymbol{k}_1\cdot\boldsymbol{r}_1} d\boldsymbol{r}_1 \int \chi^{\dagger}_{\frac{1}{2}\sigma_1} \psi_{f}^{(A-1)*}(\{\boldsymbol{r}_i\}_{A-1}) \psi_0(\boldsymbol{r}_1, \{\boldsymbol{r}_i\}_{A-1}) \prod_{i=2}^{A} d\boldsymbol{r}_i \right|^2. \tag{29}$$

The summation over $f$ in Eq. (28) includes all the discrete shell-model levels below the Fermi level in $(A-1)$ ("hole states" of $A$), and in Eq. (29) it includes all the discrete and continuum states above the Fermi level created by SRCs. In a fully uncorrelated MF approach, one has

$$n_A(\boldsymbol{k}_1) = n_0(\boldsymbol{k}_1) = \sum_{\alpha \leq F} |\phi_\alpha(\boldsymbol{k}_1)|^2\,; \qquad n_1(\boldsymbol{k}_1) = 0. \tag{30}$$

The modulus squared of the overlap integral represents the weight of the ground and excited

**Table 2. The proton MF, $\mathcal{P}^p_{gr(0)} = \int d\,\mathbf{k}_1\, n^p_{gr(0)}(\boldsymbol{k}_1)$, and SRC, $\mathcal{P}^p_{ex(1)} = \int d\,\mathbf{k}_1\, n^p_{ex(1)}(\boldsymbol{k}_1)$, probabilities, Eq. (31), in various nuclei obtained from AV18 and AV8' interactions. (Table reprinted from. [46]. Copyright (2013) by the American Physical Society)**

| MEAN FIELD AND SRC PROBABILITIES | | | |
|---|---|---|---|
| *Nucleus* | *Potential* | $\mathcal{P}_{gr}$ | $\mathcal{P}_{ex}$ |
| $^3$He [34] | AV18 [9] | 0.677 | 0.323 |
| $^4$He [35, 36] | RSC [7] AV8$'$[10] | 0.8 | 0.2 |
| *Nucleus* | *Potential* | $\mathcal{P}_0$ | $\mathcal{P}_1$ |
| $^{16}$O [45] | V8' [10] | 0.8 | 0.2 |
| $^{40}$Ca [45] | V8' [10] | 0.8 | 0.2 |

virtual states of $(A-1)$ in the ground state of $A$, so that the quantities

$$\mathcal{P}_{gr(0)} = \int_0^\infty n_{gr(0)}(\boldsymbol{k}_1)\, d\,\boldsymbol{k}_1 \quad \mathcal{P}_{ex(1)} = \int_0^\infty n_{ex(1)}(\boldsymbol{k}_1)\, d\,\boldsymbol{k}_1, \tag{31}$$

with

$$\mathcal{P}_{gr(0)} + \mathcal{P}_{ex(1)} = 1, \tag{32}$$

yield, respectively, the probability to find a MF and a correlated nucleon in the range $0 \leq k_1 \leq \infty$; they can therefore be assumed as the MF and SRC total probabilities. It is clear that both low- and high-momentum components contribute to mean-field and correlated momentum distributions but, as it should be expected, $n_{gr(0)}$ ($n_{ex(1)}$) should get contribution mainly from low (high) momentum components. This is clearly illustrated in Fig. 6, where the proton momentum distributions of A=3, 4, 16, and 40 nuclei are shown with the separation into the MF and correlation contributions: it can be seen that, starting from $k \gtrsim 2\, fm^{-1}$, the momentum distributions are dominated by the correlated part. The calculated values of $\mathcal{P}_{gr(0)}$ and $\mathcal{P}_{ex(1)}$ for several nuclei are listed in Table 2. Assuming that $n_0^{N_1}$ and $n_1^{N_1}$ could be obtained from some measurable cross section, it might well be that only a limited range of momenta is available experimentally, in which case it is useful to define the partial probabilities

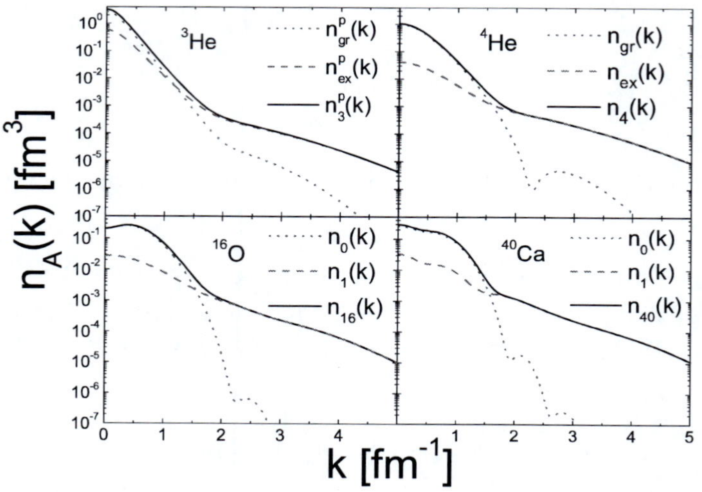


Figure 6. The proton momentum distribution $n^p_A(k_1) \equiv n_A(k)$ and its separation into the uncorrelated and correlated contributions, Eqs. (24-29), in A=3 (wave function from Ref. [34], AV18 interaction), A=4 (wave function from Ref. [35], AV8′ interaction), A= 16 (wave functions from Ref. [45], AV8′ interaction), and A=40 (wave function from Ref. [45], AV8′ interaction). The values of the probabilities $\mathcal{P}^p_{gr(0)} = 4\pi \int k^2\, dk\, n^p_{gr}(k)$ and $\mathcal{P}^p_{ex(1)} = 4\pi \int k^2\, dk\, n^p_{ex}$, Eq.(31), are listed in Table 2 and the partial probabilities, Eq. (33), in Table 3. (Figure reprinted from [46]. Copyright (2013) by the American Physical Society)

**Table 3. The values of the proton partial probability, Eq. (33), for $^3$He, $^4$He, $^{16}$O and $^{40}$Ca, calculated for different values of the momentum $k_1^-$ (in fm$^{-1}$) with $k_1^+ = \infty$. (Table reprinted from. [46]. Copyright (2013) by the American Physical Society)**

| | $^2$H | $^3$He(p) | | $^4$He | | $^{16}$O | | $^{40}$Ca | |
|---|---|---|---|---|---|---|---|---|---|
| $k_1^-$ | $\mathcal{P}$ | $\mathcal{P}_{gr}$ | $\mathcal{P}_{ex}$ | $\mathcal{P}_{gr}$ | $\mathcal{P}_{ex}$ | $\mathcal{P}_0$ | $\mathcal{P}_1$ | $\mathcal{P}_0$ | $\mathcal{P}_1$ |
| 0.0 | 1.0 | 0.7 | 0.3 | 0.8 | 0.2 | 0.8 | 0.2 | 0.8 | 0.2 |
| 0.5 | 0.3 | 0.3 | 0.2 | 0.5 | 0.1 | 0.7 | 0.2 | 0.7 | 0.2 |
| 1.0 | 0.08 | 0.03 | 0.07 | 0.1 | 0.1 | 0.2 | 0.1 | 0.2 | 0.1 |
| 1.5 | 0.06 | 0.005 | 0.04 | 0.008 | 0.08 | 0.008 | 0.1 | 0.01 | 0.1 |
| 2.0 | 0.04 | 0.002 | 0.02 | $7 \cdot 10^{-4}$ | 0.06 | $6 \cdot 10^{-4}$ | 0.06 | $3 \cdot 10^{-4}$ | 0.07 |

$$\mathcal{P}_{0(1)}(k_1^{\pm}) = 4\,\pi \int_{k_1^-}^{k_1^+} n_{0(1)}(\boldsymbol{k}_1) k_1^2\, d\, k_1 \tag{33}$$

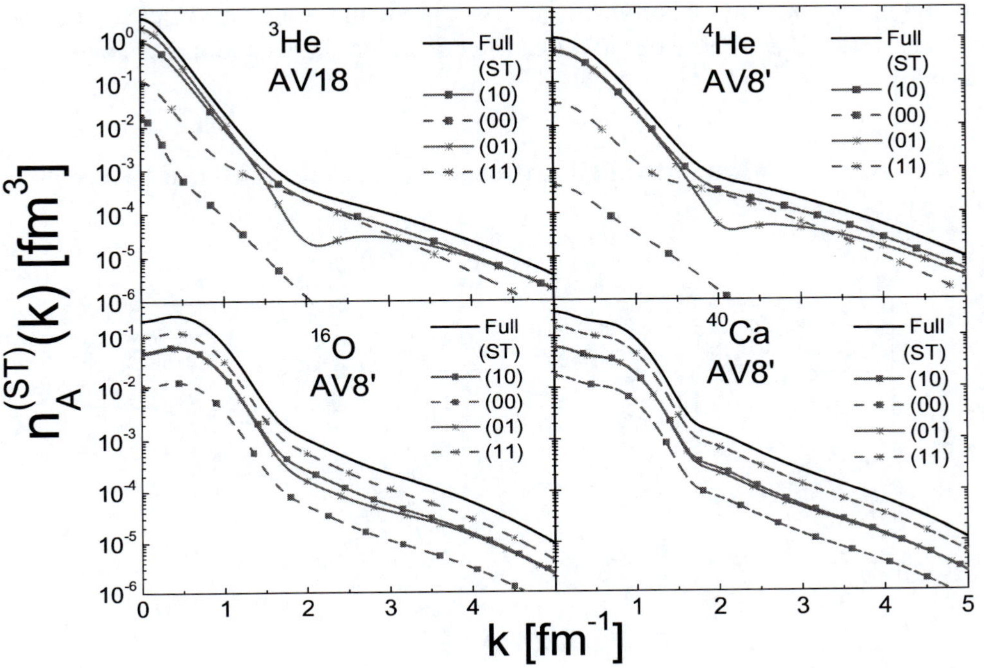


Figure 7. The various spin-isospin contributions to the proton distributions in $^3$He, $^4$He, $^{16}$O and $^{40}$Ca (Eq. (33)). Wave functions as in Fig. 6. (Figure reprinted from. [46]. Copyright (2013) by the American Physical Society)

i.e. the probability to observe a MF or a correlated nucleon with momentum in the range $k_1^- \leq k_1 \leq k_1^+$. The calculated values of $\mathcal{P}_{0(1)}^{N_1}(k_1^{\pm})$ are given in Table 3.

## 4.3. The Spin-Isospin Structure of the One-Body Momentum Distributions

By introducing the spin-isospin dependent half-diagonal density matrix $\rho_{(ST)}^{N_1N_2}(\boldsymbol{r}_1, \boldsymbol{r}_1'; \boldsymbol{r}_2)$, the one-body momentum distribution can be expressed in terms of its various spin-isospin components as follows [46]

$$n_A(\boldsymbol{k}_1) = \sum_{(ST)} n_A^{(ST)}(\boldsymbol{k}_1) = \int d\boldsymbol{r}_1\, d\boldsymbol{r}_1' e^{i\,\boldsymbol{k}_1\cdot(\boldsymbol{r}_1-\boldsymbol{r}_1')} \sum_{(ST)} \int d\boldsymbol{r}_2 \rho_{(ST)}^{N_1N_2}(\boldsymbol{r}_1, \boldsymbol{r}_1'; \boldsymbol{r}_2). \quad (34)$$

In Ref. [46] the spin-isospin dependent half-diagonal two-body density has been calculated for A=3, 4, 16 and 40, and the various spin-isospin contributions to $n_A(k)$ have been obtained as shown in Fig. 7. It appears that: (i) the contribution from the $(00)$ state is negligible, both in few-nucleon systems and complex nuclei; (ii) the contribution from the $(11)$ state in $^3$He and $^4$He is small, both at low and large values of $k$, but it plays a relevant role in the region $1.5 \lesssim k \lesssim 2.5\, fm^{-1}$; (iii) in the proton distribution of $^3$He the $(01)$ contribution is important everywhere except in the region $1.5 \lesssim k \lesssim 3\, fm^{-1}$; (iv) in complex nuclei, in agreement with the results shown in Table 1, the $(11)$ state (odd relative orbital momenta) plays a dominant role, both in the independent particle model and in many-body approaches. These observations are useful for understanding the content of the next Section.

### 4.4. The Nucleon Momentum Distributions in $^3$He and $^3$H and SRCs

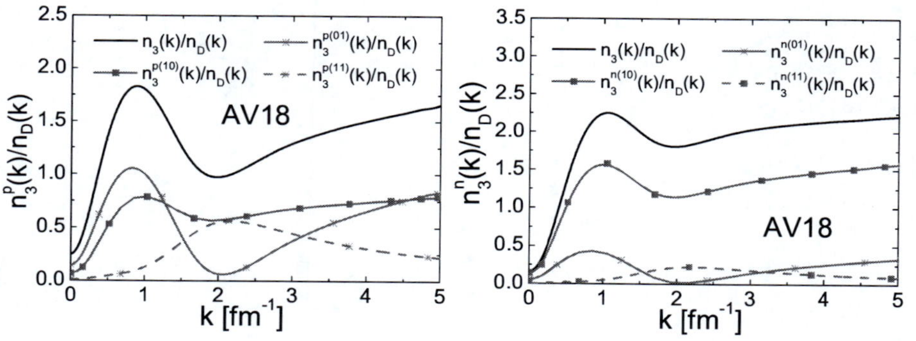


Figure 8. The spin-isospin components of the proton (**Left**) and neutron (**Right**) ratios $n_A^N(k)/n_D(k) = \sum_{ST} n_3^{N,(ST)}(k)/n_D(k)$ in $^3$He. Wave functions from Ref.[34]. (Figure reprinted from. [46]. Copyright (2013) by the American Physical Society).

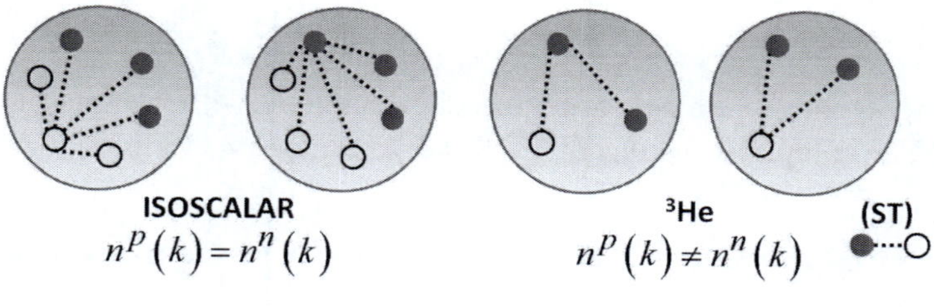


Figure 9. The number of $pn$ and $pp$ pairs affecting the high-momentum components of the nucleon momentum distributions. In isoscalar nuclei $n^p(k) = n^n(k)$, whereas in non isoscalar nuclei, e.g. in $^3$He, $n^p(k) \neq n^n(k)$ because the proton and the neutron are correlated with different nucleon pairs(full (open) dots denotes protons(neutrons)). (After Ref. [2]).

In order to better understand the role of SRCs in the one-body momentum distributions, let us analyze in details the spin-isospin structure in mirror nuclei $^3$He and $^3$H, in particular the ratio of proton and neutron distributions of $^3$He divided by the deuteron momentum distribution, as shown in Fig. 8. The different behavior of the proton and neutron momentum ratios there shown, can be understood in terms of SRC as follows [46]: a $pn$ pair can be either in deuteron-like $(10)$ state with probability $3/4$, or in $(01)$ state, with probability $1/4$;

a *pp* $(nn)$ pair can only be in $(01)$ state with probability one. [1] As illustrated in the cartoon in Fig 9, in $^3$He the proton momentum distribution is affected by SRCs acting in one $pn$ and one *pp* pairs; in the former pair the deuteron-like state $(10)$ is three times larger than the $(01)$ state, whereas in the latter pair the deuteron-like state is totally missing; on the contrary, the neutron distribution is affected by SRCs acting in two proton-neutron pairs, with a pronounced dominance of the deuteron-like state $(10)$; therefore, one expects that around $k \simeq 2\, fm^{-1}$, where $np$ SRCs dominate over *pp* SRCs [57, 58, 59], $n_3^n/n_D \simeq 2$ and $n_3^p/n_D \simeq 1$, which is indeed confirmed by the results presented in Fig. 8, where the various spin-isospin ratios $R_{A/D}^{N,(ST)}(k) = n_A^{N,(ST)}(k)/n_D(k)$ are presented.

# 5. Two-body Momentum Distributions

The most direct way to learn about SRCs would be through processes in which a pair of nucleons is knocked out from a nucleus A in kinematical conditions that would allow the reconstruction of the initial two-nucleon dynamics. The main nuclear quantities entering these processes is the two nucleon momentum distributions.

## 5.1. General Definitions

Introducing the relative and c.m. momenta,

$$\boldsymbol{k}_{rel} = \frac{1}{2}(\,\boldsymbol{k}_1 - \boldsymbol{k}_2\,) \equiv \boldsymbol{k} \qquad \boldsymbol{K}_{c.m.} = \boldsymbol{k}_1 + \boldsymbol{k}_2 \equiv \boldsymbol{K}, \tag{35}$$

the two-body momentum distribution is defined as follows

$$\begin{aligned} &n(\boldsymbol{k}_1, \boldsymbol{k}_2) = n(\boldsymbol{k}, \boldsymbol{K}) = n(k, K, \Theta) = \\ &= \frac{1}{(2\pi)^6} \int d\boldsymbol{r} d\boldsymbol{r}' d\boldsymbol{R} d\boldsymbol{R}'\, e^{-i\,\boldsymbol{K}\cdot(\boldsymbol{R}-\boldsymbol{R}')}\, e^{-i\,\boldsymbol{k}\cdot(\boldsymbol{r}-\boldsymbol{r}')} \rho(\boldsymbol{r}, \boldsymbol{r}'; \boldsymbol{R}, \boldsymbol{R}'), \end{aligned} \tag{36}$$

where $\rho(\boldsymbol{r}, \boldsymbol{r}'; \boldsymbol{R}, \boldsymbol{R}')$ is the non-diagonal two-body density (Eq. (11)), $k = |\boldsymbol{k}|$, $K = |\boldsymbol{K}|$ and $\Theta$ is the angle between $\mathbf{k}$ and $\mathbf{K}$. Three different types of two-body momentum distribution can be considered, namely:

1. the relative, $n_{rel}(k)$, and center-of-mass, $n_{c.m.}(K)$, momentum distributions, i.e. Eq. (36) integrated over the *c.m.* and relative momenta, respectively:

$$n_{rel}(\boldsymbol{k}) = \int n(\boldsymbol{k}, \boldsymbol{K})\, d\boldsymbol{K} \qquad n_{c.m.}(\boldsymbol{K}) = \int n(\boldsymbol{k}, \boldsymbol{K})\, d\boldsymbol{k}; \tag{37}$$

2. Eq. (36) in correspondence of $K_{c.m.} = 0$, describing back-to-back nucleons, as in the deuteron $(\boldsymbol{k}_2 = -\boldsymbol{k}_1)$ :

$$n(\boldsymbol{k}, 0) = \frac{1}{(2\pi)^6} \int d\boldsymbol{r} d\boldsymbol{r}'\; e^{-i\,\boldsymbol{k}\cdot(\boldsymbol{r}-\boldsymbol{r}')} \int d\boldsymbol{R} d\boldsymbol{R}'\, \rho(\boldsymbol{r}, \boldsymbol{r}'; \boldsymbol{R}, \boldsymbol{R}'); \tag{38}$$

[1] This is strictly true in the independent particle picture. SRCs change these probability according to the results presented in Table 1 without, however, affecting the correctness of our argument.

3. the full Eq. (36) as a function of $k$, $K$ and $\Theta$, a quantity that provides a three-dimensional picture of the two-body momentum distributions.

Hereafter, the two-body momentum distributions for a pair of nucleons $N_1N_2$ in spin-isospin state $(ST)$ will be denoted by $n_{(ST)}^{N_1N_2}(\boldsymbol{k}, \boldsymbol{K})$.

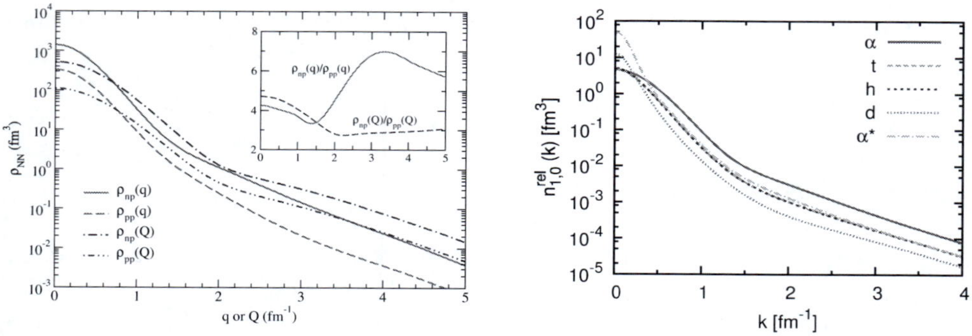


Figure 10. (Color online) (**Left**): the two-body momentum distribution of $np$ and $pp$ pairs in $^4$He, integrated over the the *c.m.* (relative) momentum $K \equiv Q$ $(k \equiv q)$ *vs.* the relative (c.m.) momentum (Eq. (37)), with $n_{rel}(k) \equiv \rho_{NN}(q)$ and $n_{c.m.}(K) \equiv \rho_{NN}(Q)$. The inset shows the ratios $\rho_{np}(q)/\rho_{pp}(q)$ and $\rho_{np}(Q)/\rho_{pp}(Q)$. (Figure reprinted from [58]. Copyright (2007) by the American Physical Society). (**Right**): the relative momentum distribution for $S = 1, T = 0$ pairs in $^2$H $\equiv$ d, $^3$He $\equiv$ h, $^3$H $\equiv$ t, $^4$He $\equiv \alpha$ and $^4$He* $\equiv \alpha^*$. (Figure reprinted from. [48]. Copyright (2011) by the American Physical Society).

## 5.2. The Momentum Distributions $n_{rel}(k)$ and $n_{c.m.}(K)$

Fig. 10 (**Left**) shows the relative and *c.m.* momentum distributions in $^4$He obtained in Ref.[58] whereas Fig. 10(**Right**) shows the the relative momentum distributions for $pn$ pairs in state $(ST) = (10)$ in $^2$H, $^3$He, $^3$H, $^4$He and$^4$He* from Ref.[48]. Both calculations are *ab initio* within the VMC method with the AV18 interaction (Ref. [58]) and the Gaussian correlated basis approach with the AV8$'$ interaction (Ref. [48]). The inset in Fig. 10 illustrates the dominance of the tensor force acting in $pn$ pairs: at low momenta, the ratio $n_{np}(k)/n_{pp}(k)$ is mostly governed by the ratio of the $pn$ to $pp$ pairs, $ZN/[Z(Z-1)/2] = 2$ but starting from $k \geq 1.5\, fm^{-1}$, the ratio sharply increases because of the action of the tensor force in the $(10)$ channel of the $np$ pair. The results exhibited in Fig. 10 demonstrate the universality of SRCs in few-nucleon system: at high values of $k$, the relative momentum distributions are very similar, thanks to the universality of the correlation hole previously discussed in Section 2.3.. The universality of the integrated momentum distributions is confirmed by the results for A=12, 16 and 40, obtained in Ref. [50] within the number-conserving linked-cluster expansion and the AV8$'$ $NN$ interaction.

## 5.3. The Momentum Distributions $n(k_{rel}, K_{c.m.} = 0)$

The momentum distribution $n(k_{rel}, K_{c.m.} = 0)$ is a very important quantity because, when compared with the deuteron momentum distribution, it can provide information on the short-range dynamics of a pair of nucleons in the medium and possible evidence of medium induced multi-nucleon correlations.

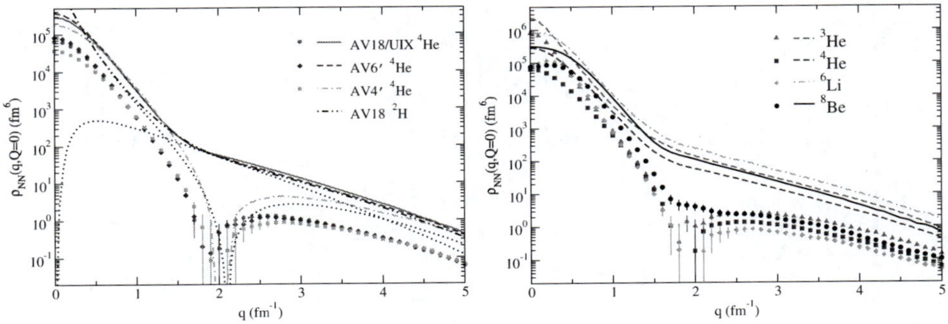


Figure 11. (Color online) (**Left**): the two-body momentum distributions of back-to-back nucleons (Eq. (38) with $K_{c.m.} \equiv Q = 0$ and $n(k,0) \equiv \rho(q, Q=0)$) for $np$ pairs (lines) and $pp$ pairs (symbols) in $^4$He, calculated with VMC wave functions and different NN interactions: AV18 plus UIX three nucleon interaction [32], AV6′ [11] and AV4′ [11] interactions. The dotted lines denote the S and D waves of the deuteron corresponding to the AV18 interaction. (**Right**): the same as in (**Left**) but for $^3$He, $^4$He, $^6$Li, and $^8$Be. AV18 interaction.(Figure reprinted from. [58]. Copyright (2007) by the American Physical Society)

The results for A=3, 4, 6 and 8 nuclei, obtained in Ref. [58] within the VMC method using different NN interactions plus 3N forces, are shown in Fig. 11, whereas the results for A=12, 16 and 40, obtained in Ref. [59] within the number-conserving linked-cluster expansion and the AV8′ NN interaction, are shown in Fig. 12.

The results for both few-nucleon systems and complex nuclei clearly show that: (i) the 3NF, which is essential to produce the correct binding energy of few-nucleon systems, appears to have tiny effects on the high-momentum components (Fig. 11(**Left**)), which is not surprising, in view of its long-range character; (ii) the universality of the relative momentum distributions, resulting from the universality of SRCs, is evident from Fig. 11(**Right**) and Fig. 12(**a**): in the range $3 \leq A \leq 40$ and $k \geq 2\,fm^{-1}$ a clear A-independence of the high relative-momentum behavior is exhibited; (iii) the results presented in Figs. 10(**a**) and 12(**b**) demonstrate the tensor dominance both in few-nucleon systems and complex nuclei; (iv) Fig. 11(**Left**) and 12(**b**) show that at high values of $k_{rel}$ the momentum distributions of deuteron and complex nucleus are very similar. This similarity is better illustrated in the next Section where the ratio $R_{A/D}(k, K_{c.m.} = 0) = n_A^{pn}(k_{rel}, K_{c.m.} = 0)/n_D(k)$ is presented for few-nucleon systems and complex nuclei.

### 5.4. The Momentum Distributions $n(k_{rel}, K_{c.m.}, \Theta)$

The knowledge of $n(k_{rel}, K_{c.m.}, \Theta)$ provides information on the three-dimensional picture of the two-nucleon momentum distribution. In this connection, it has to be stressed that the independence of $n(k_{rel}, K_{c.m.}, \Theta)$ upon the angle $\Theta$ is evidence of the factorization of the distributions in variables $k_{\rm rel}$ and $K_{\rm c.m.}$ [60, 61], i.e. $n^{NN}(k_{\rm rel}, K_{\rm c.m.}, \Theta) \Rightarrow \phi(k_{\rm rel})\chi(K_{\rm c.m.})$

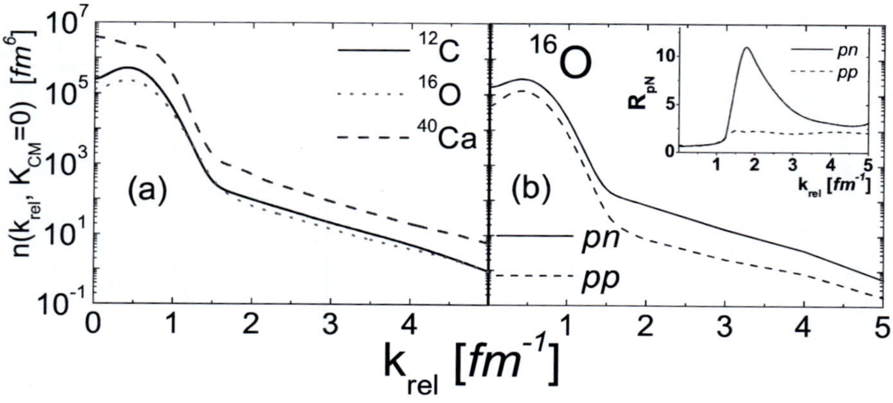


Figure 12. (Color online) (**a**): the two-nucleon momentum distribution in $^{12}$C, $^{16}$O and $^{40}$Ca for back-to-back nucleons (Eq. (38)) calculated within the number-conserving linked-cluster expansion of Ref. [45] with AV8' interaction. (**b**): the back-to-back *pn* and *pp* momentum distributions in $^{16}O$. The inset shows the ratio of the total momentum distributions to the distributions obtained by disregarding the tensor force, i.e. $R_{pN} = n_{pN}(k_{rel}, K_{c.m.} = 0)/n_{pN}^{central}(k_{rel}, K_{c.m.} = 0)$. (Figure reprinted from. [59]. Copyright (2008) by the American Physical Society)

where, for the time being, $\phi(k_{\rm rel})$ and $\chi(K_{\rm c.m.})$ denote two generic functions of $k_{rel}$ and $K_{c.m.}$ . The *pn* and *pp* two-body momentum distributions $n(k_{rel}, K_{c.m.}, \Theta)$ in few-nucleon systems [60] and complex nuclei [50] have been calculated with realistic wave functions. The results for $^3$He and $^4$He obtained with *ab initio* wave functions [34, 35] corresponding to the $AV18$ [9] and $AV8'$ [10] interactions are shown in Fig. 13 *vs.* $k_{\rm rel}$, in correspondence of several values of $K_{\rm c.m.}$ and two values of $\Theta$. The results for $^{16}$O are shown in Fig. 14. Apart from a different overall normalization, the results for few-nucleon systems at $K_{\rm c.m.} = 0$ fully agree with the ones of Ref. [58]. The peculiar and systematic results of these calculations can be summarized as follows: (i) with increasing values of the c.m. momentum, the high relative momentum part of the distributions strongly decreases; (ii) starting from a given value of $k_{\rm rel}$, being $k_{\rm rel} \simeq 1.5$ fm$^{-1}$ when $K_{\rm c.m.} = 0$ and assuming increasing values with increasing values of $K_{\rm c.m.}$, the *pn* distribution changes its slope and becomes close to the deuteron distribution. In particular, in the region ($k_{\rm rel} \gtrsim 2\,{\rm fm}^{-1}$, $K_{\rm c.m.} \lesssim 1$ fm$^{-1}$), $n^{pn}$ becomes $\Theta$-independent[2], assuming the form $n^{pn}(k_{\rm rel}, K_{\rm c.m.}, \Theta) \simeq n_D(k_{\rm rel}) n^{pn}_{\rm c.m.}(K_{\rm c.m.})$, where $n_D(k_{\rm rel})$ is the deuteron momentum distribution, and $n^{pn}_{\rm c.m.}(K_{\rm c.m.})$ describes the c.m. motion of the pair and provides the A-dependence of $n^{pn}(k_{\rm rel}, K_{\rm c.m.})$. The factorized property, that charaterizes also complex nuclei, as shown in the case of $^{16}$O in Fig. 14, represents a rigorous many-body demonstration that when the relative momentum of the *pn* pair is high, and, at the same time, the c.m. momentum is low, the two-body momentum distribution factorizes; (iii) when the *c.m.* momentum is of the same order as the (high) relative momentum, more than two particles can be locally correlated, with a resulting strong dependence upon the angle and the

[2]Such an independence has been checked in a wide range of angles

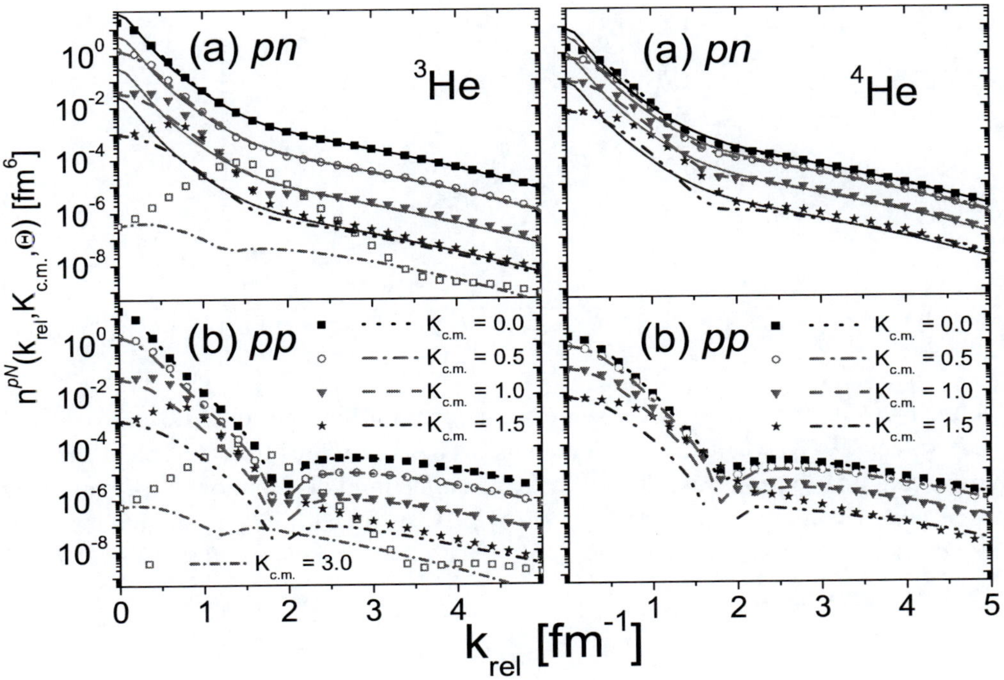


Figure 13. (Color online) (**Left**): the two-body momentum distributions of $pn$ (**a**) and $pp$ (**b**) pairs in $^3$He *vs.* the relative momentum $k_{rel}$, for fixed values of the c.m. momentum $K_{c.m.}$ and two orientations of the momenta, namely $\mathbf{k}_{rel}||\mathbf{K}_{c.m.}$ (*broken curves*) and $\mathbf{k}_{rel} \perp \mathbf{K}_{c.m.}$ (*symbols*). The continuous curves for the $pn$ pair represent the deuteron momentum distribution rescaled by the c.m. momentum distribution $n^{pn}_{c.m.}(K_{c.m.}) = \int n^{pn}(\boldsymbol{k}_{rel}, \boldsymbol{K}_{c.m.})\, d\boldsymbol{k}_{rel}$ (see Eq. (36)). $^3$He wave function from Ref. [34] and $AV18$ interaction [9]. (**Right**): the same as in (**Left**) but for $^4He$. Correlated variational wave function from [35] and $AV8'$ interaction [10].(Figure reprinted from. [60]. Copyright (2012) by the American Physical Society)

breaking down of factorization, as clearly appears in Fig. 13 for $K_{\rm c.m.} = 3$ fm$^{-1}$. These features are common to both few-nucleon systems and complex nuclei. A better evidence on the factorized behavior of the two-body momentum distributions for $pn$ pairs can be obtained by considering the ratio $R^{pn}(k) = n^{pn}(k_{\rm rel}, 0)/n_D(k_{\rm rel})$, which is presented in Fig. 15(**a**). The constant value exhibited by the ratio at $k_{\rm rel} \gtrsim 1.5$ fm$^{-1}$ is unquestionable evidence that in this region the dependence upon $k_{\rm rel}$ of the two-body momentum distribution $n^{pn}(k_{\rm rel}, 0)$ is the same as in the deuteron. As for the different magnitudes of the ratio for different nuclei, this is governed by the c.m. motion distribution of the pair, which is illustrated in Fig. 15(**b**). It can be seen that the difference in magnitude of the ratios in the region $k_{\rm rel} \gtrsim 1.5$ fm$^{-1}$ is governed by exactly the difference between the values of the *c.m.* momentum distributions at $K_{\rm c.m.} = 0$. The more rapid fall off of the *c.m.* momentum distributions of $^3$He is due to the weak binding of this nucleus, leading, with

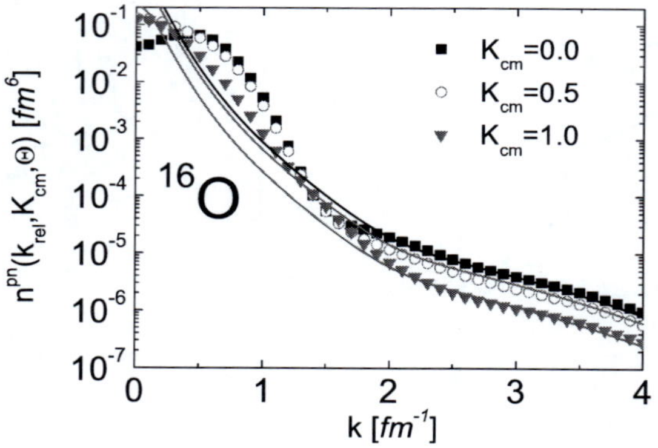


Figure 14. The two-body proton-neutron momentum distributions (Eq. (36)) in $^{16}$O for three values of $K_{c.m.}$ and $\Theta = 0$ (symbols). The continuous lines represent the deuteron momentum distributions rescaled by the c.m. momentum distribution of the pair calculated at the proper value of $K_{cm}$ (see Fig. 15(**b**)). Wave function from the number-conserved linked-cluster expansion of Ref. [45]. AV8′ NN interaction. (After Ref. [2]).

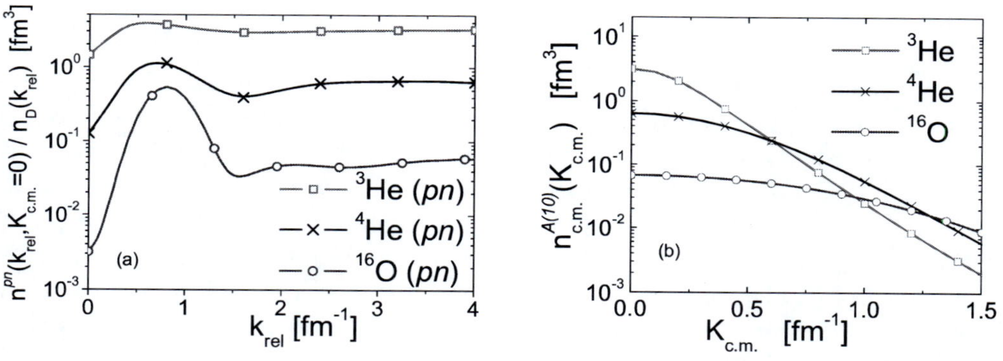


Figure 15. (**a**): the ratio of the $pn$ momentum distributions for back-to-back nucleons $n^{pn}(k_{\rm rel}, K_{\rm c.m.} = 0)$ in $^3$He,$^4$He, and $^{16}$O shown in Figs. 13 and 14, to the deuteron momentum distribution $n_D(k_{\rm rel})$ (full lines). The different magnitudes of the ratio for the three nuclei is due to the different values of the c.m. momentum distribution at $K_{\rm c.m.} = 0$ shown in Fig.15(**b**). (**b**): the *c.m.* momentum distribution in $^3$He, $^4He$, and $^{16}$O. (Figure reprinted from. [60, 46] Copyright (2012,2013) by the American Physical Society)

respect to the $^4$He and $^{16}$O, to the wider separation of the curves corresponding to various values of $K_{\rm c.m.}$ exhibited in Fig. 13. For nuclei with $A \geq 4$ and $K_{\rm c.m.} \lesssim 1.0 - 1.5$ fm$^{-1}$, the *c.m.* distribution can be associated to the average kinetic energy $< T >_{SM}$ of a pair moving in the mean field with a Gaussian distribution , $n_{c.m.}(K) \propto \exp\{-\alpha K^2_{c.m.}\}$ with $\alpha = 3/[2 < K^2_{c.m.} >] = [3(A-1)]/[4(A-2)m_N < T_{SM} >]$ as suggested in Ref. [53] and in agreement with the experimental finding for $^{12}$C (See Section VII).

# 6. The Convolution Model of SRCs and its Theoretical Many-Body Validation

Although the momentum distribution is not an observable, it is undisputable that it can play a role in particular scattering processes, that, at the same time, can also be influenced by other phenomena which could mask the effects generated by the momentum distributions. To clarify this point, let us consider the process $A(e, e'N)X$ in the Plane Wave Impulse Approximation (PWIA). This is the process when in the initial state an electron is impinging on nucleus A and, in the final state, the scattered electron and a nucleon N are detected in coincidence with the nucleus $X = (A-1)$ left in the energy state $E^f_{A-1}$; in the simplified assumption that the detected nucleon was knocked out by a direct interaction $\gamma^* N$ and left the nucleus with momentum $\mathbf{p}_N$ without interacting with the medium, the measurable missing momentum $\mathbf{p}_m = \mathbf{q} - \mathbf{p}_N$ and energy $E_m = \nu - T_N - T_{A-1}$ represent, respectively, the momentum of the nucleon before interaction $\mathbf{k}_1 = -\mathbf{p_m}$ and the intrinsic excitation energy $E^*_{A-1}$ of $(A-1)$. As is well known, even within such a severe approximation the cross section of the process is not proportional to the momentum distribution but to another quantity, the Spectral Function $S_A(\mathbf{k}_1, E)$ representing the joint probability that when a nucleon with momentum $\mathbf{k}_1 = -\mathbf{p}_m$ is removed instantaneously from the ground state of the nucleus A, the nucleus $(A-1)$ is left in the excited state $E^*_{A-1} = E - E_{min}$, where $E$ is the removal energy and $E_{min} = M_{A-1} + m_N - M_A$. The spectral function has the following form (from now-on spin indexes will be omitted for ease of presentation)

$$\begin{aligned}
S_A(\boldsymbol{k}_1, E) &= < \Psi_0^A | a^\dagger_{\mathbf{k}_1} \delta(E - \hat{H} + E_A) a_{\mathbf{k}_1} | \Psi_0^A >= \\
&= \sum_f \Big| \int e^{i\boldsymbol{k}_1 \cdot \boldsymbol{r}_1} d\boldsymbol{r}_1 \int \Psi_f^{(A-1)\,*}(\{\boldsymbol{r}\}_{A-1}) \Psi_0^A(\boldsymbol{r}_1, \{\boldsymbol{r}\}_{A-1}) \times \\
&\quad \times \prod_{i=2}^{A} d\boldsymbol{r}_i \Big|^2 \delta(E - E^f_{A-1} + E_A) \\
&= S_{gr(0)}(\boldsymbol{k}_1, E_{gr(0)}) + S_{ex(1)}(\boldsymbol{k}_1, E_{ex(1)})
\end{aligned} \tag{39}$$

where $E^f_{A-1} = E_{A-1} + E^*_{A-1}$, $E_A$ and $E_{A-1}$ denote the ground-state energies of initial and final nuclei, and $a^\dagger_{\boldsymbol{k}_1}$ ($a_{\boldsymbol{k}_1}$) is a creation (annihilation) operator. The two contributions, as in the case of the momentum distributions (cf. Eqs. (24) and (27)), arise from different final states of the system $(A-1)$, with $S_{ex(1)}(\boldsymbol{k}_1, E)$ governed by SRCs. Summing over the complete set of final states in Eq. (39) it is easy to obtain the *momentum sum rule*

$$n_A(\boldsymbol{k}_1) = \int_0^\infty S_A(\boldsymbol{k}_1, E)\, dE. \tag{40}$$

Eq. (40) clearly shows that the extraction of the momentum distribution from the experimental data implies a difficult integration over the full range of discrete and continuum excitation spectra of the residual nucleus $(A-1)$, up to very high values $E^*_{A-1}$, particularly in the interesting region of high values of $k$ (see Ref. [54, 55, 56]). Moreover, exact many-body spectral functions exist only for the three-nucleon system [62, 63, 64], (the complete set of final state is known), and for nuclear matter [65, 66], whereas for complex nuclei only model spectral functions have been developed, either within the local density approximation [67], or within the convolution model of Ref. [53]. The latter, which is aimed at describing the spectral function in the region of 2N SRCs, naturally arises from the behavior of the high-momentum part of the two-body momentum distributions described in the previous Sections. As a matter of fact, we have seen that at large values of $k_{rel}$ and small values of $K_{c.m.}$ the following relation holds

$$n^{pn}(\mathbf{k}_{rel}, \mathbf{K}_{c.m.}) \simeq n^{pn}(k_{rel}, K_{c.m.}) \simeq n_D(k_{rel}) n_{c.m.}(K_{c.m.}). \tag{41}$$

From momentum conservation, $\mathbf{k}_1 + \mathbf{k}_2 - \mathbf{K}_{c.m.} = 0$, $\mathbf{k}_{rel} = \mathbf{k}_1 - \mathbf{K}_{c.m.}/2$, one has

$$n_A(k_1) \simeq \int n_D(|\boldsymbol{k}_1 - \frac{\boldsymbol{K}_{c.m.}}{2}|) n_{c.m.}(K_{c.m.})\, d\,\boldsymbol{K}_{c.m.} = \int S_A(k_1, E)\, d\,E, \tag{42}$$

and assuming that the high values of the excitation energy $E^*_{A-1}$ are essentially given by the relative motion of nucleon "2" and nucleus $(A-2)$, one obtains

$$\begin{aligned} S_A(k_1, E) \;&\simeq\; \int n_D(|\boldsymbol{k}_1 - \frac{\boldsymbol{K}^N_{c.m.}}{2}|) n^N_{c.m.}(K_{c.m.}) d\,\boldsymbol{K}_{c.m.} \times \\ &\times\; \delta\left(E - E^{(2)}_{th} - \frac{A-2}{2m_N(A-1)}\left[\boldsymbol{k}_1 - \frac{A-1}{A-2}\boldsymbol{K}_{c.m.}\right]^2\right), \end{aligned} \tag{43}$$

where $E^{(2)}_{th}$ is the two-body threshold. Eq. (43) has been first obtained in Ref. [53], within several phenomenological assumptions, whose physical correctness are now justified by the many-body calculation of the momentum distributions. A convolution formula for the correlated part of the spectral function has also been shown to result from Brueckner-Bethe-Goldstone (BBG) theory of nuclear matter, where the spectral function corresponding to the nucleon self-energy $M(k, E) = V(k, S) + iW(k, E)$ is obtained from the single particle Green function $\mathcal{G}$ in the following form [68]

$$S_A(k, E) = -\frac{1}{\pi} Im\mathcal{G}(k, E) = \frac{1}{\pi} \frac{W(k, E)}{(-E - k^2/2m_N - V(k, E))^2 + W(k, E)^2} \tag{44}$$

which, at $E + \frac{k^2}{2m_N} >> |V(k, E)|; |W(k, E)|$, can be approximated by the following convolution integral[69]

$$\begin{aligned} S_A(k_1, E) \;&=\; \frac{\pi^2 \rho}{16} \int \frac{d^3\,\mathbf{K}_{c.m.}}{(2\pi)^3} n_{rel}(|\mathbf{k}_1 - \frac{1}{2}\mathbf{K}_{c.m.}|) n^{FG}_{c.m.}(K_{c.m.}) \times \\ &\times\; \delta\left(E - E^{(2)}_{thr} - \frac{1}{2\,m_N}(\mathbf{K}_{c.m.} - \mathbf{k}_1)^2\right). \end{aligned} \tag{45}$$

Here $\rho$ is the density of nuclear matter, $n^{FG}_{c.m.}$ the Fermi gas distribution and $n_{rel}$ the spin-isospin averaged two-body relative momentum distribution in nuclear matter. Eq. (45) in the region $E \simeq E^{(2)}_{thr} + k^2/(2m_N)$ agrees very well with the exact BBG spectral function, as shown in Fig. 16(**a**). Further confirmation of the convolution model, resulting from the factorization property of $n(\boldsymbol{k}_{rel}, \boldsymbol{K}_{c.m.})$, has been recently provided [61] by the analysis of the behavior of *ab initio* three-nucleon ground-state wave functions $\Psi_0$ in momentum space, by considering the following ratio

$$R = \frac{|\Psi_0(\mathbf{K}_{cm}, \mathbf{k}_{rel})|^2}{|\Psi_0(\mathbf{K}_{cm} = 0, \mathbf{k}_{rel})|^2}, \tag{46}$$

*vs.* $|\boldsymbol{k_{rel}}|$ at constant values of $|\boldsymbol{K_{c.m.}}|$. If factorization of $\Psi_0$ holds, i.e. $|\Psi_0(\mathbf{K}_{cm}, \mathbf{k}_{rel})|^2 \simeq n_{c.m.}(K_{c.m.})n_{rel}(k_{rel})$ the ratio becomes

$$R \simeq \frac{n_{c.m.}(K_{c.m.})}{n_{c.m.}(K_{c.m.} = 0)} = constant. \tag{47}$$

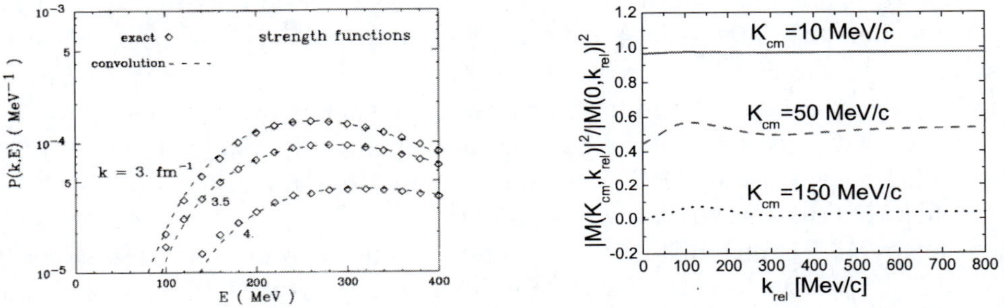


Figure 16. (**Left**): the exact BBG nuclear matter spectral function (*exact*) *vs* E in correspondence of three values of k compared with the BBG convolution model Eq.(45)(*convolution*). (Figure reprinted from [69]. Copyright (1996) by Elsevier). (**Right**): the ratio R (Eq. (46)) in $^3$He. Three-nucleon wave functions from Ref. [34]. AV18 interaction [9].(Figure reprinted from. [61]. Copyright (2011) by Springer & Verlag).

It can be seen from Fig. 16(**b**) that factorization indeed occurs starting, as expected, at a value of $k_{rel}$ which increases with increasing values of $K_{c.m.}$, in agreement with the results presented in Fig. 13. The magnitudes of the curves in Fig. 16(**b**) agree with the behavior of $n_{c.m.}(K_{c.m.})$ presented in Fig. 15. The most interesting quantity, as far as SRCs are concerned, is the two-nucleon momentum distributions, that in PWIA might in principle be extracted from the $A(e, e'2N)X$ process, when two nucleons are knocked out from the nucleus A and are detected with momenta $\mathbf{p}_1$ and $\mathbf{p}_2$ in coincidence with the scattered electron, with the nucleus $(A-2)$ left in the energy state $E^f_{A-2}$. The measurable missing momentum and energy are in this case the $\mathbf{p}_m = \mathbf{q} - \mathbf{p}_1 - \mathbf{p}_2$ and $E_m = \nu - T_{p1} - T_{p2} - T_{A-2} = E^*_{A-2}$. Assuming that the virtual photon has interacted with one nucleon (the fast one) of a correlated pair, with the second nucleon (the recoiling one) being emitted because of momentum conservation, the cross section will depend upon the two-nucleon spectral function

$$\begin{aligned} &S_A^{N_1N_2}(\boldsymbol{k}_1, \boldsymbol{k}_2, E) =< \Psi_0|a^\dagger_{\mathbf{k}_1} a^\dagger_{\mathbf{k}_2} \delta(E - \hat{H} + E_A) a_{\mathbf{k}_2} a_{\mathbf{k}_1}|\Psi_0 >= \\ &\sum_f \Big| \int e^{-i\boldsymbol{k}_1\cdot\boldsymbol{r}_1 - i\boldsymbol{k}_2\cdot\boldsymbol{r}_2} d\boldsymbol{r}_1 d\boldsymbol{r}_2 \langle \Psi_f^{(A-2)*}(\{\mathbf{r}\}_{A-2})|\Psi_0(\mathbf{r}_1, \mathbf{r}_2\{\mathbf{r}\}_{A-2})\rangle \times \\ &\times \delta(E - E^f_{A-2} + E_A). \end{aligned} \tag{48}$$

Summing over the final states of $(A-2)$ the *two-nucleon momentum sum rule*

$$n_A^{N_1N_2}(\boldsymbol{k}_1,\boldsymbol{k}_2)=\int d\,E\,S_A^{N_1N_2}(\boldsymbol{k}_1,\boldsymbol{k}_2,E)=n(k_{rel},K_{c.m.},\Theta) \tag{49}$$

is obtained. The two-nucleon Spectral Functions have been obtained within many body theories in Refs. [70] for finite nuclei, in Ref. [71] for nuclear matter, and in Ref. [76] for $^3$He.

## 7. Recent Experimental Evidence of SRCs

Let us eventually mention some recent evidence of SRCs. We discuss first of all the evidence of high-momentum components in $n_A(k)$. To this end, we show in Fig. 17 the one-body momentum distributions extracted from the exclusive, $A(e,e'p)X$, and inclusive, $A(e,e')X$, reactions, the latter analyzed in terms of y-scaling [72]. The y-scaling analysis produce large errors, but even in the worst case, it unambiguously demonstrates the dominant role played by SRCs in the high-momentum part of the one-body momentum distributions. Other evidence of SRCs from inclusive electron scattering is provided by the ratio of inclusive cross sections, e.g. $d\sigma_{eA}(x_{Bj},Q^2)/d\sigma_{eA'}(x_{Bj},Q^2)$. As a matter it has been suggested [51] that at sufficiently high values of the four momentum transfer $Q^2 \gtrsim 1.5 \div 2\,GeV^2$ when quasi elastic scattering from nucleons is the leading process, the ratio of the inclusive cross section per nucleon off nucleus A to nucleus A$'$

$$r(A,A')=\frac{d\sigma_{eA}(x_{Bj},Q^2)/A}{d\sigma_{eA'}(x_{Bj},Q^2)/A'} \tag{50}$$

plotted vs the Bjorken scaling variable $x_{Bj}$ should exhibits plateaux in the regions $1.5 \lesssim x_{Bj} \lesssim 2$ and $2.2 \lesssim x_{Bj} \lesssim 3$ that can be viewed as evidence that the virtual photon interacted with the same "quasi-free" two- and three-nucleon short-range correlated subsystems in nuclei A and A$'$. These plateaux were indeed observed at Jlab, as shown in Fig. 18 (for previous and recent references to this topic see e.g. [3, 74, 75]).

The best way to learn about SRCs are the processes in which two nucleons are knocked out from the nucleus at high momentum transfer, e.g. the process $A(e,e'N_1N_2)X$. In the past, the processes $A(p,p'N_1N_2)X$ and $A(e,e'N_1N_2)X$ have been intensively investigated theoretically (see e.g. [77, 78], and references therein quoted) and experimentally (see e.g. [79] and references therein quoted) but the experimental data were plagued by MEC and FSI and other competing effects and no conclusive quantitative information on SRCs could be obtained (for a critical discussion of this topic see [3]). Recently, however, high moment transfer experiments have been performed on $^{12}$C and $^4$He [80, 81, 82, 83, 84] that allowed one to detect a "fast" proton with momentum $\mathbf{p}_1$, identified as the member of a correlated pair kicked out by the high energy projectile, and a "slow" (or "recoil") nucleon (a proton or a neutron) with momentum $\mathbf{p}_2$, assumed to be the one emitted by momentum conservation in the correlated pair. By assuming the validity of the PWIA, which implies that $\mathbf{p}_1=\mathbf{k}_1+\mathbf{q}$, $\mathbf{p}_2=\mathbf{k}_2$ and $\mathbf{P}_{mis}=-(\mathbf{k}_1+\mathbf{k}_2)=\mathbf{K}_{c.m.}$, it is possible to reconstruct the momentum $\mathbf{k}_1$ that the struck nucleon had before interaction; by plotting the correlation between the value of $|\mathbf{p}_2|$ and the angle between $\mathbf{k}_1$ and $\mathbf{p}_2$ , it was found that

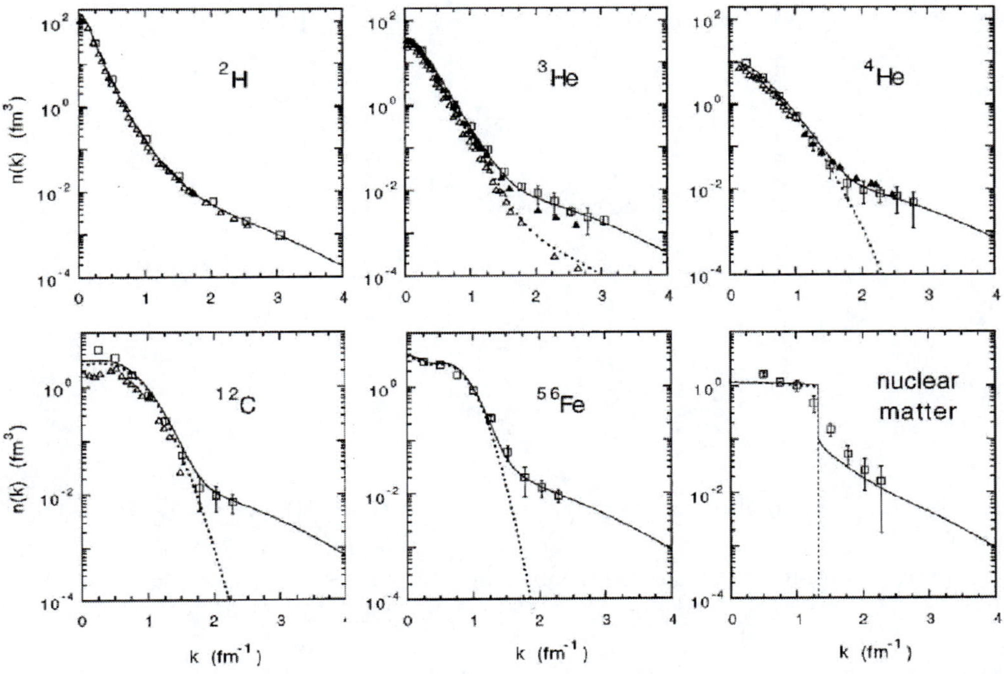


Figure 17. The momentum distributions of several nuclei and nuclear matter extracted from the analysis of inclusive, $A(e,e')X$ (open squares), and exclusive, $A(e,e'p)X$ (full and open triangles), cross sections. The full lines represent the results of many-body calculations, as in the previous Figures, and the dashed lines are MF predictions. For references to the original experimental and theoretical papers see Ref. [72].(Figure reprinted from. [72]. Copyright (1991) by Elsevier)

whereas recoiling nucleons with momentum of the order or less than the Fermi momentum were emitted isotropically, nucleons with momentum $p_2 \simeq 2 - 2.5\, fm^{-1}$ were emitted in a backward cone with respect to the direction of $\mathbf{k}_1$, in agreement with the picture of the absorption of the virtual photon by a correlated nucleon pair with a *c.m.* momentum distribution in $^{12}$C of the type $n_{c.m.}(K) \propto \exp[-\alpha K^2_{c.m.}]$ with $\alpha \simeq 1\, fm^2$ in agreement with the prediction of Ref. [53]. Furthermore by comparing with the same apparatus and kinematics the yield of $^{12}C(e,e'p)X$ with the yield of $^{12}C(e,e'pn)X$ it has been possible to obtain information about the ratio of $pn$ to $pp$ correlated pairs. A detailed discussion of these experiments and their interpretation is given in Ref. [3].

## 8. Conclusion

*Ab initio* many-body calculations performed in terms of realistic bare two-nucleon interactions show that two-nucleon short-range correlations (2NSRCs), characterized by the presence of a correlation hole in the two-nucleon density in nuclei, exhibit a universal character, manifesting itself in several A-independent features of nucleon momentum distributions. As a matter of fact, the calculated two-nucleon relative density displays a correlation

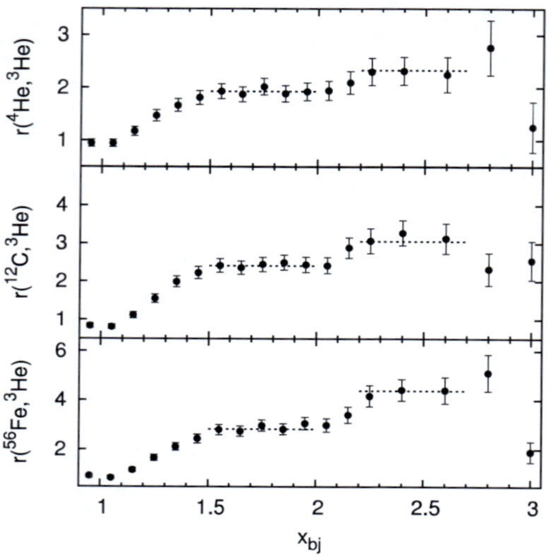


Figure 18. The experimental inclusive cross section ratio (Eq. (50)) vs the Bjorken scaling variable $x_{bj}$ for $A = [^4He, ^{12}C, ^{56}Fe]$ and $A' = [^3He]$ [73]. The horizontal dashed lines were drawn to indicate the regions of constant ratios interpreted as due to two- and three-nucleon SRCs. (Figure from Ref. [73]. Copyright (2008) by the American Physical Society)

hole which is essentially independent of the mass of the nucleus, a feature that demonstrates that two-nucleon motion at short relative distances is practically unaffected by the motion of nearby nucleons. This universal behavior in coordinate space reflects itself in peculiar universal features of one-nucleon, $n_A(|\mathbf{k}|)$, and two-nucleon, $n^{N_1N_2}(\mathbf{k}_{rel}, \mathbf{K}_{c.m.})$, momentum distributions. Concerning $n_A(|\mathbf{k}|)$, 2NSRCs increase the high-momentum part by orders of magnitude with respect to MF predictions; as for $n^{N_1N_2}(\mathbf{k}_{rel}, \mathbf{K}_{c.m.})$, particularly worth being stressed again is the following main feature characterizing the motion of a *pn* pair in medium: in the SRC region, where $2 \lesssim k_{\rm rel} \lesssim 5$ fm$^{-1}$ and, at the same time, $K_{\rm c.m.} \lesssim 1$ fm$^{-1}$, the relative and *c.m.* motions of the pair are decoupled, with the former described by a deuteron-like momentum distribution, and the latter, governing the A dependence of the motion, described by a momentum distribution linked to the average value of the MF kinetic energy. Such a decoupling of the relative and *c.m.* momenta have been theoretically justified by many-body calculations which predict factorization of the nuclear wave function at short inter-nucleon distances or, equivalently, at high values of $k_{rel}$ and low values of $K_{c.m.}$. Some aspects of this picture have already been partially confirmed by the experiments that provide evidence of the high-momentum content of the one-nucleon momentum distribution, by the experimental behavior of the inclusive electron-nucleus cross section ratios and, eventually, by the measurement of the percentage ratio of *pn* to *pp* correlated pairs in $^4$He and $^{12}$C. Much work however remains to be done in order to investigate the three-dimensional structure of the two-nucleon momentum distributions $n^{N_1N_2}(|\mathbf{k}_{rel}|, |\mathbf{K}_{c.m.}|, \Theta)$, with particular attention to their *c.m.* dependence in the SRCs region, as well as to its structure in the region where both $k_{rel}$ and $K_{c.m.}$ are large, characterized by the breaking down of c.m and relative momenta factorization due to the expected dominant role of many-nucleon SRCs.

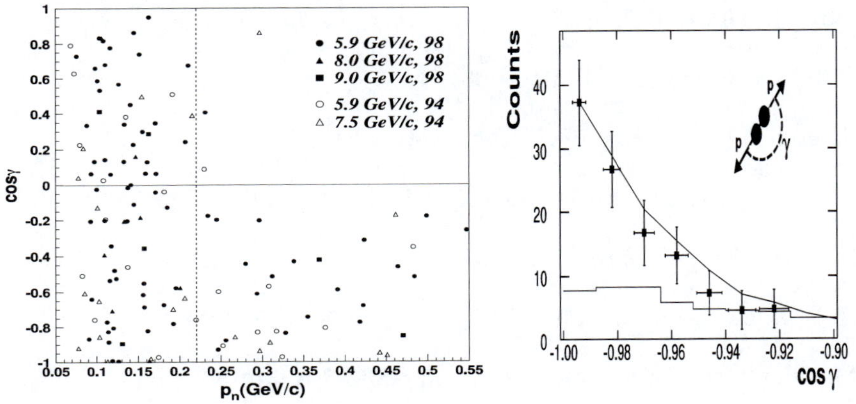


Figure 19. (LEFT)The correlation between the magnitude of the recoil neutron momentum $p_n$ and its direction $\gamma$ relative to $\vec{p_i}$ in the process $A(p,p'pn)X$. The legend gives the initial proton beam momentum and year of the run, with hollow points (labeled 94) from Ref. [80] and solid points (labeled 98) from Ref. [81]. In all cases, no angular correlation is observed for neutron momenta less than $k_{Fermi}$, while neutrons with larger momentum are produced preferably in the backward hemisphere of the reaction.(RIGHT) For the $^{12}$C(e,e′pp) reaction at $Q^2 > 1$ (GeV/c)$^2$ [83] the distribution of the cosine of the opening angle between the $\vec{p}_{miss}$ and $\vec{p}_{rec}$ for the $p_m = 0.55$ GeV/c setting. The curve is a simulation of the scattering off a moving pair with a width of 0.136 GeV/c for the pair center-of-mass momentum and the histogram shows the distribution of random events.(Figure from Ref.[81]. Copyright (2002) by the American Physical Society)

Unveiling the correlation structure of nuclei is a fundamental task of nuclear physics, because by this way information on the basic in-medium NN interaction can be obtained. Moreover, it should also be stressed that, recently, a non trivial impact of 2NSRCs on different fields, such as high-energy hadron-nucleus [85, 86, 87] and nucleus-nucleus scattering [88], deep inelastic scattering [4] and the equation of state of unconventional nuclear matter [89, 5], has been demonstrated.

# References

[1] "Nuclear Science: A Long Range Plan". The DOE//NSF Nuclear Science Advisory Committee, uSA, February 1996

[2] M. Alvioli, C. Ciofi degli Atti, L. P. Kaptari, C. B. Mezzetti and H. Morita, *Int. J. Mod. Phys.E* **22** (2013) 1330021.

[3] J. Arrington, D. W. Higinbotham, G. Rosner, and M. Sargsian, *Prog. in Part. and Nucl. Phys.* **67** (2012) 898.

[4] O. Hen, D. W. Higinbotham, G. A. Miller, E. Piasetzky and L. B. Weinstein, *Int. J. Mod. Phys.E* **22**(2013) 1330017.

[5] L. Frankfurt, M. Sargsian and M. Strikman, *Int. J. Mod. Phys.* **A23** (2008) 2991.

[6] A. Bohr, B. N. Mottelson, *Nuclear Structure*, World Scientific (1969).

[7] R.V. Reid, *Ann. Phys. N.Y.* **50** (1968) 411.

[8] M. Lacombe *et al.*, *Phys. Rev.* **C 21** (1980), 861.

[9] R. B. Wiringa, V. G. J. Stoks and R. Schiavilla, *Phys. Rev.* **C51** (1995) 38.

[10] B. S. Pudliner *et al*, *Phys. Rev.* **C56** (1997) 1720.

[11] R. B. Wiringa and S. C. Pieper, *Phys. Rev. Lett.* **8**9 (2002) 182501.

[12] R. Machleidt, and D. R. Entem, *Phys. Rep* **503**(2011)1.

[13] E. Epelbaum, H.-W. Hammer and U.-G Meissner, *Rev. Mod. Phys.* **81** (2009) 1773.

[14] W. N. Polizou and W. Glöckle, *Few-Body Systems* **9** (1990) 97.

[15] J. P. Vary, *Phys. Rev.* C**7**, (1973) 521.

[16] R. Roth, T. Neff and H. Feldmeier, *Prog. Part. Nucl. Phys.* **65** (2010) 50.

[17] B. R. Barrett, P. Navrtil and J. P. Vary, *Progr. in Part. and Nucl. Phys.* **69** (2013) 131.

[18] K. Suzuki and S. Y. Lee, *Progr. Theor. Phys.* **64** (1980) 2091.

[19] S. K. Bogner, R. J. Furnstahl and A. Schwenk, *Prog. Part. Nucl. Phys.* **65** (2010) 94.

[20] S. K. Bogner, T. T. S. Kuo and A. Schwenk, *Phys. Rep.* **386** (2003) 1.

[21] E. R. Anderson, S. K. Bogner, R. J. Furnstahl and R. J. Perry, *Phys. Rev.* **C82** (2010) 054001.

[22] S. K. Bogner and D. Roscher, *Phys. Rev.* **C86** (2012) 064304.

[23] R. J. Furnstahal and K. Hebeler, arXiv:1305.3800v1 [nucl-th] (2013).

[24] J. W. Holt, N. Kaiser and W. Weise, arXiv:1304.6350v1 [nucl-th] (2013).

[25] R. Jastrow, *Phys. Rev.* **98** (1955) 1479.

[26] K. Gottfried, *Annals of Physics* **21** (1963) 29.

[27] H. Primakoff, T. Holstein, *Phys. Rev.* **55** (1939) 1218.

[28] A. Nogga, H. Kamada, W. Gloeckle, *Phys. Rev. Lett.* **785** (2000) 944.

[29] A. Kievsky, M. Viviani, L. Girlanda and L.E. Marcucci, *Phys. Rev.* **C81** (2010) 044003.

[30] S. C. Pieper, and R. B. Wiringa, *Annu. Rev. Nucl. Part. Sci.* **51** (2011) 53.

[31] H.-W. Hammer, A. Nogga, and A. Schwenk, *Rev. Mod. Phys.* **85** (2013) 197.

[32] B. S. Pudliner *et al*, *Phys. Rev. Lett.* 73 (1995) 4396.

[33] W. Gloeckle, H. Witala, D. Huber, H. Kamada and J. Golak, *Phys. Rept.* **274** (1996) 107.

[34] A. Kievsky, S. Rosati, M. Viviani, *Nucl. Phys.* **A551** (1993) 241.
L.E. Marcucci, L. Girlanda, A. Kievsky, S. Rosati and M. Viviani, *Few-Body Syst.* **44** (2008) 227.

[35] H. Morita, Y. Akaishi, H. Tanaka, *Prog. Theor. Phys.* **79** (1988) 863.
H. Morita *et al*, to be published

[36] R. Schiavilla, V. R. Pandharipande, R. B. Wiringa, *Nucl. Phys.* **A449** (1986) 219.

[37] K. Varga and Y Suzuki, *Stochastic Variational Approach to Quantum-Mechanical Few-Body Problems*, Lecture Notes in Physics **54** (Springer, 1998)

[38] K. Varga and Y. Suzuki, *Phys. Rev.* **C52** (1995) 2885.

[39] Y.Suzuki and H. Horiuchi, *Nucl. Phys.* **A81** (2009) 188.

[40] V. Efros, W. Leidemann and G. Orlandini, *Few-Body. Syst.* **14** (1993) 151.

[41] W. Leidemann and G. Orlandini, *Prog. in Part. and Nucl. Phys.* **68** (2013) 158.

[42] S. C. Pieper, *Rivista Nuovo Cimento* 31 (2008) 709.

[43] S. C. Pieper, R. B. Wiringa and V. R. Pandharipande, *Phys. Rev.* **C46** (1992) 1741.

[44] F. Arias de Saavedra, C. Bisconti, G. Co' and A. Fabrocini, *Phys. Rep.* **450** (2007) 1.

[45] M. Alvioli, C. Ciofi degli Atti and H. Morita, *Phys. Rev.* **C72** (2005) 054310.

[46] M. Alvioli, C. Ciofi degli Atti, L. P. Kaptari, C. B. Mezzetti and H. Morita, *Phys. Rev.* **C87** (2013) 034603.

[47] H. Kamada, A. Nogga, W. Gloeckle *et al*, *Phys. Rev.* **C64** (2001) 044001.

[48] H. Feldmeier, W. Horiuchi, T. Neff and Y. Suzuki, *Phys. Rev.* **C84** (2011) 054003 .

[49] J. L. Forest, V. R. Pandharipande, S. C. Pieper, R. B. Wiringa, R. Schiavilla and A. Arriaga, *Phys. Rev.* **C54** (1996) 646.

[50] M. Alvioli, C. Ciofi degli Atti and H.Morita, ArXiv: 0709:3989 (2007).

[51] L. L. Frankfurt, M. I. Strikman, *Phys. Rept.* **160** (1988) 235.

[52] J. G. Zabolitzki and W. Ey, *Phys. Lett.* **B76** (1978) 527.

[53] C. Ciofi degli Atti, S. Simula, *Phys. Rev.* **C53** (1996) 1689.

[54] C. Ciofi degli Atti, E. Pace and G. Salmè, *Phys. Lett.* **B141** (1984) 14.

[55] C. Ciofi degli Atti and S. Liuti, *Phys. Lett.* **B225** (1989) 215.

[56] C. Ciofi degli Atti, S. Liuti and S. Simula, *Phys. Rev.* **C41** (1990) R2474.

[57] M. M. Sargsian, T. V. Abrahamyan, M. I. Strikman and L. L. Frankfurt, *Pys. Rev.* **C71** (2005) 044615.

[58] R. Schiavilla, R. B. Wiringa, S. C. Pieper and J. Carlson, *Phys. Rev. Lett.* **98** (2007) 132501.

[59] M. Alvioli, C. Ciofi degli Atti and H. Morita, *Phys. Rev. Lett.* **100**(2008) 162503.

[60] M. Alvioli, C. Ciofi degli Atti, L. P. Kaptari, C. B. Mezzetti, H. Morita and S. Scopetta, *Phys. Rev.* **C85** (2012) 021001.

[61] C. Ciofi degli Atti, L. P Kaptari, H. Morita and S. Scopetta, *Few-Body Systems* **50**(2011) 243.

[62] A. Dieperink, T. de Forest, I. Sick, and R. Brandenburg, *Phys. Lett.* **B63** (1976) 261.

[63] C. Ciofi degli Atti, E. Pace, and G. Salmè, *Phys. Rev.* **C21** (1980) 805.

[64] H. Meier-Hajduk, C. Hajduk, P. Sauer and W. Theis, *Nucl. Phys.* **A395**, (1983) 332.

[65] O. Benhar, A. Fabrocini and S. Fantoni, *Nucl. Phys.* **A505**, (1989) 267.

[66] A. Ramos , A. Polls and W. Dickhoff, *Nucl. Phys.* **A503** (1989) 1.

[67] O. Benhar, A. Fabrocini, S. Fantoni, and I. Sick, *Nucl. Phys.* **A579** (1994) 493.

[68] W. H. Dickhoff and D. Van Neck, *Many-Body Theory Exposed!*, World Scientific, 2007.

[69] M. Baldo, M. Borromeo and C. Ciofi degli Atti, *Nucl. Phys.* **A604** (1996) 429.

[70] W. J. W. Geurts, K. Allart, W. H. Dickhoff and H. Müther, *Phys. Rev.* **C54** (1996) 1144.

[71] O. Benhar and A. Fabrocini, *Phys. Rev.* **C62** (2000) 034304.

[72] C. Ciofi degli Atti, E. Pace and G. Salme, *Phys. Rev.* **C43** (1991) 1155.

[73] K. S. Egiyan *et al*, *Phys. Rev. Lett.* **96** (2006) 082501.

[74] N. Fomin *et al.*, *Phys. Rev. Lett.* **108** (2012) 092502.

[75] C. Ciofi degli Atti and C. B. Mezzetti, *Phys. Rev.* **C79** (2009) R051302.

[76] C. Ciofi degli Atti, L. P. Kaptari, *Phys. Rev.* **C66** (2002) 044004.

[77] J. Ryckebusch, M. Vanderhaegen, K. Heyde and M. Varoquier, *Phys. Lett.* **B350** (1995) 1.

[78] C. Giusti, F. D. Pacati, M. Schwamb and S. Boffi, *Eur. Phys. J.* **A33** (2007) 29.

[79] D. L. Groep *et al.*, *Phys. Rev.* **C63** (2001) 014005.

[80] J. L. Aclander *et al.*, *Phys. Lett.* **B453** (1999) 211.

[81] A. Tang, J. W. Watson, J. L. S. Aclander, J. Alster, G. Asryan, *et al.*, *Phys. Rev. Lett.* **90** (2003) 042301.

[82] E. Piasetzky *et al*, *Phys. Rev. Lett.* **97** (2006) 162504.

[83] R. Shneor, *et al.*, *Phys. Rev. Lett.* **99** (2007) 072501.

[84] R. Subedi *et al.*, *Science* **320** (2008) 1476.

[85] M. Alvioli, C. Ciofi degli Atti, H. Morita, I. Marchino and V. Palli, *Phys. Rev.* **C78** (2008) 031601(R).

[86] M. Alvioli, C. Ciofi degli Atti, B. Z. Kopeliovich, I. K. Potashnikova and I. Schmidt, *Phys. Rev.* **C81** (2010) 025204.

[87] C. Ciofi degli Atti, B. Z. Kopeliovich, C. B. Mezzetti, I. K. Potashnikova and I. Schmidt, *Phys. Rev.* **C84** (2011) 025205.

[88] M. Alvioli, M. Strikman, *Phys. Rev.* **C83** (2011) 044905.

[89] T. Frick, H. Muther, A. Rios, A. Polls and A. Ramos, *Phys. Rev.* **C71**(2005) 014313.

In: NN and 3N Interactions
Editors: Leonid Blokhintsev and Igor Strakovsky
ISBN: 978-1-63321-053-0


*Chapter 6*

# NUCLEON - NUCLEON INTERACTIONS AT SHORT DISTANCES

***Misak M. Sargsian*** *
Florida International University,
Miami, Florida, US

**Abstract**

Despite the progress made in understanding the NN interactions at long distances based on effective field theories, the understanding of the dynamics of short range NN interactions remains as elusive as ever. One of the most fascinating properties of short range interaction is its repulsive nature which is responsible for the stability of strongly interacting matter. The relevant distances, $\leq 0.5$ fm, in this case are such that one expects the onset of quark-gluon degrees of freedom with interaction being dominated by QCD dynamics. We review the current status of the understanding of the QCD dynamics of NN interactions at short distances, highlight outstanding questions and outline the theoretical foundation of QCD description of hard NN processes. We present examples of how the study of the hard elastic NN interaction can reveal the symmetry structure of valence quark component of the nucleon wave function and how the onset of pQCD regime is correlated with the onset of color transparency phenomena in hard *pp* scattering in the nuclear medium. The discussions show how the new experimental facilities can help to advance the knowledge about the QCD nature of nuclear forces at short distances.

**PACS** 12.38.−t, 13.75.Cs, 21.30.−x
**Keywords:** Nuclear Forces, QCD, NN Interaction, Hard Processes

## 1. Introduction

After the discovery of the Atomic Nuclei in 1911 [1] and the observation of strong nuclear forces in 1919 [2, 3], the understanding of Nuclear Physics on a fundamental level is still an enduring intellectual challenge. We know why atomic nuclei are bound but we don't know why they are stable in the way they are, especially heavy nuclei. The Stability Theorem [4]

*E-mail address: sargsian@fiu.edu

requires a repulsive interaction in order to heavy nuclei not to collapse to the configurations with the radius of $\approx 0.5$ fm and per nucleon binding energy of 1500 MeV!

It is remarkable that such a repulsive interaction was eventually found in the short range part of the Nucleon-Nucleon (NN) interaction, when the analysis of the NN scattering data by Jastrow in 1951 [5] unambiguously demonstrated the existence of the repulsive interaction in the $NN$ systems at $\leq 1$ fm.

The short-range nature of nuclear forces indicates that the bulk of the dynamics of atomic nuclei can be understood from the dynamics of the NN interaction. Therefore it is not surprising that special emphasis is given to the studies of NN interaction throughout the history of nuclear physics.

If one studies the NN interaction starting from large distances (say $\geq$ 2fm), one first observes a clear dominance of the long range one-pion exchange interaction. By decreasing the separation of nucleons to $1.2 - 1.5$ fm one will observe the onset of the strong contribution due to two-pion exchange forces resulting in the tensor interaction. The observed magnitude of the tensor as well as strong spin-orbit interactions, however, requires an addition of the vector component to the exchanged forces whose contribution gradually increases with the decrease of NN separations and dominates the overall interaction in the region of the repulsive core at $\leq 0.5$ fm. The above picture allows us to identify three main regions (similar to Ref. [6]) for NN interactions which are characterized by their unique dynamics and therefore by their specific theoretical approaches in describing the NN forces.

**Long to Intermediate Range:** The biggest success in understanding the nuclear forces is achieved in describing long range and long to intermediate range pion exchange interactions based on the effective theories (e.g. [7, 8, 9, 10]). The advantage of these theories is that they satisfy the symmetries of QCD Lagrangian and the interaction vertex is fixed by the mechanism of the Chiral symmetry breaking. Effective theories can be extended to the intermediate-short range distances by introducing contact interactions. Such an approach is successful in describing the intermediate - long range part of the NN interaction, however no dynamical insight can be achieved on the structure of the short-range interaction.

**Intermediate to Short Range:** Intermediate to short range interactions are currently based fully on phenomenological approaches. In One Boson Exchange Potential (OBEP) models the vector meson exchanges are introduced with phenomenological coupling constants, with the assumption that they account for the field theoretical effects of multiple scattering, self interaction of exchanged mesons, and other higher order corrections (for review see Ref. [11]). The another approach is in phenomenological parameterization of the NN interaction potential in the form of the sum of Yukawa type interactions (e.g. Ref. [12]). In both cases the parameters entering the approximations are fixed by fitting the calculations to the phase shifts of interactions extracted from the analysis of the NN scattering data. In fact the major success of these models are in part due to the significant improvements in quality and quantity of NN phase-shift data by SAID [13, 14] and Nijmegen [15] groups.

Despite the progress in quantitative description of NN interactions up to the lab kinetic energies of 350 MeV the difficulty is however in the fact the there is no clear cut constraints that QCD can impose on the dynamics of the interaction in these approaches.

**Short Range:** Concerning the short range ($< 0.5 - 0.7$ fm) interaction, the most fascinating part of it is the repulsive core which we know is responsible for the stability of atomic nuclei. This is the largely unknown part of the nuclear forces. It is worth mentioning that the most modern phenomenological parameterizations [12] of NN potentials still use the Wood-Saxon type function for the repulsive part of the interaction parameterized in 60's. What concerns OBEP models, here the biggest inconsistency is that one considers the exchange of composite particles (bosons) which have sizes comparable to or larger than the internucleon distances being considered. In the language of Feynman diagrams this situation is reflected in the fact that exchanged mesons are highly virtual (their virtuality being comparable with their masses). In this respect it is worth to quote Richard Feynman who noticed that "In fact a "pion far off its mass shell" may be a meaningless - or at least highly complicated idea" [16].

**Short Ranges and QCD:** The interesting aspect of the short range repulsive interaction is that it operates at distances of $\leq 0.5$ fm at which one expects full onset of the QCD degrees of freedom. Therefore our expectation is that the NN interaction should be strongly related to the QCD dynamics at these distances.

The current experimental and theoretical situation in understanding the QCD dynamics of NN interaction at short distances is very incomplete. The available experimental data indicate the rich dynamics at short-range NN interaction. However, no new data were taken since mid 1990's after re-profiling the AGS at Brookhaven National Lab as an injector for Relativistic Heavy Ion Collider as well as completion of the Saturne II project at Saclay [17]. The halt of the flow of experimental data significantly restricted the possibility of further theoretical development. This situation is expected to change in the near future with the advent of high energy hadronic machines at J-PARC (Japan) [18], PANDA (Germany) [19] as well as the currently discussed NICA(Russia) [20] and HIAF(China) [21] projects.

**Outline:** In this work we review briefly the history of QCD studies of NN interactions (Sec.2.) and highlight the outstanding theoretical and experimental issues in understanding the QCD structure of high energy and momentum transfer (hard) NN interactions (Sec.3.). We then present the theoretical foundation of hard QCD processes relevant to NN interaction (Sections 5.-6.). First we consider the amplitudes of NN interaction in helicity basis, most appropriate for high energy scattering. Then, anticipating the helicity conservation in the perturbative regime of QCD in Sec.4. we discuss in detail the polarization observables in elastic NN interaction and derive several useful sum rules and equalities for polarization asymmetries that can serve as a tool for identifying the onset of helicity conservation in the hard QCD regime. In Sec.5. we discuss the other signature of the QCD regime of hard NN Scattering - the quark counting rule. Sec. 6. elaborates the light-front dynamics of hard NN scatterings, the definition of light-front wave function of nucleons as well as quark helicity spinors in the light-front, giving us the main theoretical tools for studying hard NN interactions. In Sec. 7. we consider a few examples of the application of the above mentioned theoretical approaches in the quark-interchange model of NN interaction, comparing the results with the limited experimental data currently available. Section 8. summarizes the results and gives some outlook on other possibilities of studying NN systems at short distances.

## 2. Brief History of QCD approaches to Short Range NN Interaction

The QCD approach in studies of high momentum and energy transfer NN interactions revealed a very rich picture of strong interactions, introducing completely new subjects into the discussion of nuclear forces such as dimensional counting rules, helicity conservations, minimal Fock components, hidden color, etc.

**Quark-Gluon Interactions and NN Scattering:** First attempts to understand short range properties of $NN$ interaction within QCD were made in the 1970's [22, 23, 24], in which the phenomenology of $NN$ interaction was confronted with the quark-interchange and three gluon exchange mechanisms of hadronic interactions. The real boost in studies of hard hadronic processes in general and $NN$ interaction in particular came after the discovery of dimensional counting rules [25, 26, 27] for fixed large center of mass angle high energy processes. The counting rules gave new observables to explore in hard processes such as the exponents of the energy dependences of scattering cross sections. The most intriguing part of it was that such dependences were naturally predicted within QCD in the picture of hard gluon exchanges between current quarks in hadrons.

The another feature of the processes proceeding through the hard gluon exchanges was the prediction of helicity conservation and new relations between different spin observables in the hard hadronic processes [28, 29]. Naturally, the most accessible hadronic processes were the $NN$ scatterings and there were several dedicated experiments on measuring different spin observables in hard NN processes. However, the experimental results were at best controversial [30, 31] with the most "damaging" being the measurement of the $A_{nn}$ polarization [32, 33], which observed that for $90^0$ center of mass elastic $pp$ scattering at $p_{Lab} = 11.75$ GeV/c, the cross section of protons scattering with parallel spins normal to the scattering plane is four times the one with the antiparallel spins.

**The Quark-Gluon Wave Function of the Nucleon:** To calculate the NN scattering cross section within QCD, one needs, as an input, the quark-wave function of the nucleon. However, this wave function is poorly known due to the immense complexity of the QCD picture of the nucleon in which the number of quark and gluon constituents are not conserved (for the recent review on bounds states in QCD see Ref.[34]).

On the other hand, the advantage of high energy and momentum transfer reactions is that they allow a successful application of Fock-component approach to describing the wave function of a nucleon. This approach allows us to start with the minimal Fock component of the nucleon consisting of only three valence quarks and treat the higher order Fock-components as corrections since they are suppressed by additional inverse orders of the invariant momentum transfer.

Concerning the minimal-Fock component of the nucleon, the next question is, what symmetry of valence quarks one needs to consider in the wave function of the nucleon. Here the most natural is to consider the SU(6) symmetry based on the fact that the very same symmetry seems to be realized in constituent quark model which describes the baryonic spectrum reasonably well. Indeed the SU(6) symmetry was one of the the first approaches in describing hard NN processes [28, 29]. Also the constituent quark formalism described reasonably well several properties of intermediate energy NN scattering within the SU(6) symmetry [35].

**QCD and Short-Range Structure of the NN-Bound System:** The third aspect which is worth to emphasize in relation to studies of NN interaction at short distances is the dynamics of the NN-bound system probed at short distances. The deuteron studies opened up a new realm in studies of QCD dynamics of strong forces. It was realized in Refs. [36, 37, 38, 39, 40], that the fact, that the deuteron is a colorless 6-valence-quark system, creates an additional possibility for existence of color-octet three-quark (3q) states that combine into color-singlet 6-quark combinations (referred as hidden color states). The calculations indicate that there is a substantial hidden color component in the NN bound system at short distances in which the 6q system becomes a relevant degree of freedom.

The notion of the hidden-color component gave a new possible meaning to the NN repulsion which can be in part due to the orthogonality between the initial two color octet and final two color singlet nucleons. The other implication of the hidden color component is the prediction of the large contribution from the $\Delta - \Delta$ component in the NN interaction following from the decomposition of the color-singlet 6-quark system.

## 3. Outstanding Questions of NN Interactions at Short Distances

There are many unresolved questions which are related to the nucleon-nucleon interactions at short distances with implications ranging from particle physics to astrophysics. With the expectation of a new generation of experimental studies of hard NN processes in the near future, it is worth enumerating several of the problems which can be addressed in the experiments at facilities mentioned earlier (Sec.1.).

**Persistence of the Nucleonic Degrees of Freedom:** Recent observations of large ($\sim 2M_{Sun}$) neutron star masses [41] indicates existence of rather unreasonably stiff equation of state of the nuclear matter, which is related to the persistence of the nucleonic degrees of freedom [42] at such high densities in which one expects plenty of inelastic transitions and strong overlap of nucleon wave functions. Such persistence is also observed in probing short-range proton-neutron correlations in the nuclei [43, 44] in which the theoretical analysis [45] shows that for up to $\leq$ 1 fm separations, the NN system has no apparent non-nucleonic component, consisting almost entirely from proton and neutron.

It is interesting that this observation also has its reflection in the modification of partonic distributions of bound nucleons in the nuclear medium (EMC effect). Here, the recent analysis [46] indicates a rather small modification of nucleons in the medium of heavy nuclei, which seems puzzling.

Such a persistence of the nucleons in the high density nuclear environment can be due to the short range repulsion, since the attractive interaction will make the composite system very responsive to medium modifications.

**Oscillatory Energy Dependence of the Scaled Cross Section of Hard NN Interaction:** Even though the fixed angle high energy $NN$ scattering demonstrated rather unambiguous $s^{-10}$ invariant energy dependence, in agreement with quark counting rules [26, 25, 27], the more careful analysis revealed a violation of this dependence in a rather peculiar form: while the quark-counting rule predicts energy independence of $s^{10}\frac{d\sigma}{dt}$, the data clearly demonstrate an oscillatory energy dependence of this combination (Fig.1). These regularities observed in the residual energy dependence are striking and despite several intriguing explanations [47, 48], their origin is not understood. In Ref. [47] it is assumed that

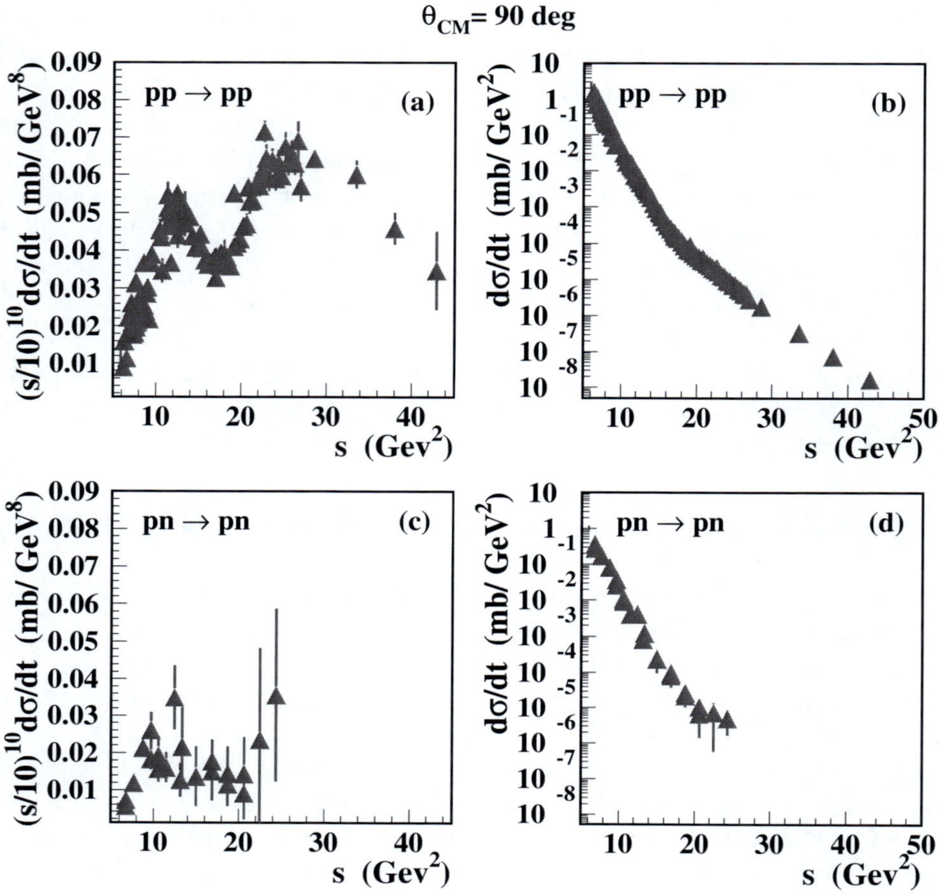


Figure 1. The invariant energy dependence of elastic $pp$ and $pn$ differential cross sections unweighted (b) (d) and weighted by $s^{10}$ factor (a), (c). The experimental data are from Ref. [49, 50, 51, 52, 53].

in addition to the quark-interchange mechanism of hard NN scattering which is sensitive to the quark-gluon interaction at very short distances, the three-gluon exchange mechanism contributes to the overall scattering amplitude. The latter (often referred to as Landshoff mechanism [24]) does not require a quark-interchange, thus, the short distances are not necessarily involved in the NN interaction. Because of the possible Chromo-Coulomb phase shift, the two amplitudes enter with the relative phase that can be tuned to describe the observed energy oscillations.

In Ref. [48] it is assumed that the oscillations are due to the interference of the quark-interchange pQCD amplitudes and the amplitudes of near-threshold $s$-channel strangeness and charm productions with possible resonating configurations having quantum numbers of $J = 1$, $L = 1$ and $S = 1$. With the relative phases of the pQCD and resonating amplitudes taken as a free parameter, it was possible to describe the observed energy dependences reasonably well.

Overall, in both approximations the assumption is that the pQCD short-range quark-interchange amplitude is interfering with another amplitude which represents a spatially extended system.

Concluding this discussion, it is worth noting that the oscillations in the energy dependence is on a level of 60% in the region where the magnitude of the hard elastic *pp* cross section changes by *eight* orders of magnitude (Fig.1(b)). As Figs.1(b) and (c) show, the same is apparently true for *pn* hard elastic scattering.

**Anomalous Polarization Asymmetries in Hard NN Scattering:** One of the biggest challenges in understanding the QCD dynamics of hard $NN$ interactions came from the observation of large out-of-plane polarization effects at $90^0$ cm scattering for incoming proton momenta as large as $p_{lab} = 11.75$ GeV/c (Fig.2). The surprise in these results is that the experiment found that the elastic cross section of the protons that scatter with their spins parallel and normal to the scattering plane is almost *four* times larger than the cross section when the spins are antiparallel.

The $A_{nn}$ experiment resulted in a flurry of theoretical activity (see e.g. [28, 29, 54, 55, 56, 57, 58, 59, 60]. However the origin of large asymmetry and its energy dependence are still unresolved.

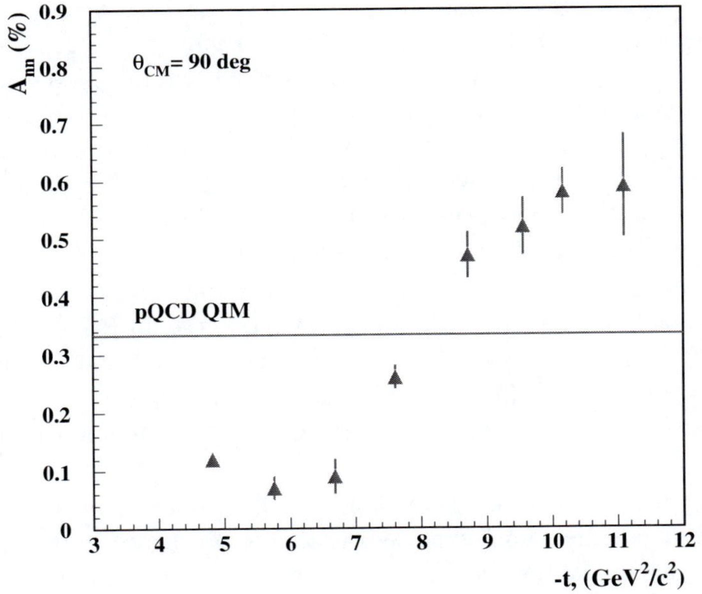


Figure 2. The $-t$ dependence of the analyzing power parameter ($A_{nn}$) of hard elastic pp scattering. Data are from Ref. [33]. The curve labelled as "pQCD QIM" is the prediction within pQCD quark interchange model (Sec.7).

The implication of this result is that, in order to generate such a large polarization effects, one needs to have a large contribution from the double-spin flip amplitude $\phi_2$ or negligible contribution from the helicity conserving $\phi_1$ amplitude (see Sec.7.). However, the challenge here is that in the hard QCD regime, the $\phi_2$ is the most suppressed and the $\phi_1$ is the largest. Another challenge that the experiment posed was (again) the "oscillatory" energy dependence of the asymmetry (Fig.2). Its difference from the basically energy independence of the pQCD quark-interchange prediction is rather striking (Fig.2),

The above situation indicates that, whatever the origin of this large asymmetry is, it is related to the contribution of a spatially extended configuration into the process of the $pp$ scattering. The oscillatory energy dependence indicates that there is an interference between short and long range phenomena. For example, the earlier mentioned charm threshold enhancement effect with the intermediate $J = 1$, $L = 1$, $S = 1$ quantum state results in $\phi_1 = 0$ and its interference with the background pQCD QIM process can in principle generate the energy dependence of Fig.2 (see Ref. [48]).

**Color Transparency in Hard $NN$ Scattering:** One of the most remarkable predictions of QCD is the existence of color transparency phenomena in hard processes taking place in the nuclear environment. The possibility of producing small-sized color-neutral object in hard processes will be evidenced by the reduced absorption of hadrons in propagating through the nuclear medium. The interesting QCD feature is that such a neutral object can be produced both for mesons and baryons. In large momentum transfer, the small distances involved in quark-gluon interaction will require that the quarks in the minimal Fock component of hadron wave function will populate distances $\leq 1/\sqrt{-t}$. Such configurations will have small color dipole moments which will result in reduced strong interaction in the nuclear medium [61, 62]. These expectation resulted in tremendous theoretical and experimental activities in mid 1990's (for review of the subject see e.g. Refs. [63, 64]). The realization that these small-sized configurations are not eigenstates of QCD Hamiltonian of free hadrons and once produced at finite energies they will evolve to normal size hadrons, suggested that the experimental verification of color transparency phenomena is more complex than first expected(see e.g. Ref. [65, 66]).

While the Color Transparency phenomenon or the reduction of the absorption in the nuclear medium is observed for the hard production of $q\bar{q}$ systems [67, 68, 69], the similar effect is still elusive for a $qqq$ system. The first experiments based on the hard $pp \to pp$ elastic scattering from nuclei [70, 71, 72] demonstrated another "oscillatiory" feature - this time in the proton beam momentum dependence of the nuclear absorption parameter (Fig.3).

The interesting feature of these data is that initially the transparency grows with the proton momentum above the values expected from Glauber theory, and then after $p_{lab} =$ 9 GeV, it falls back leading to transparency values apparently below the Glauber prediction at $p_{Lab} \approx 14$ GeV. The latter indicates that whatever propagates at these energies in the nuclear medium has total cross section larger than that of the $NN$ scattering.

**Oscillations Superimposed:** If now one superimposes Figs.(1),(2) and (3) recalculating them for invariant energy dependences at given $\theta_{cm} = 90^0$ one observes apparent correlations between these three different measurements.

First, the data suggest an inverse correlation between the $s^{10}$ weighted $pp$ cross section (Fig.4(a)) and nuclear transparency $T_{pp}$(Fig.4(c)). This correlation strongly indicates that the broad peak in the $pp$ cross section at $s \approx 25$ GeV$^2$ is related to the spatially large

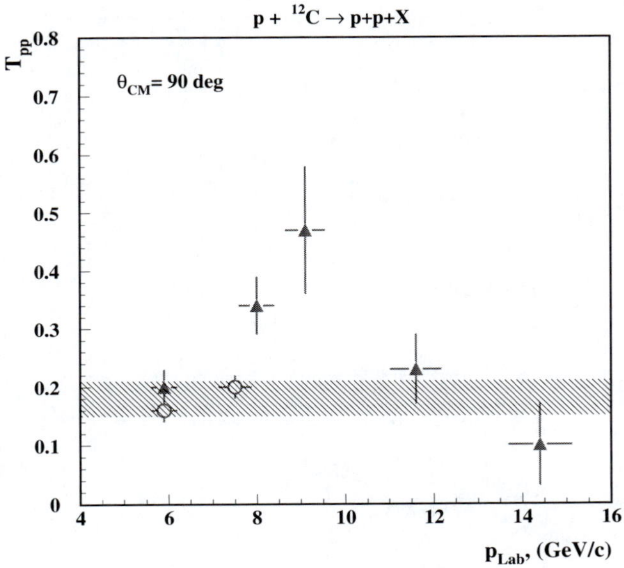


Figure 3. Dependence of the nuclear transparency parameter on the momentum of the incoming proton. Data are from Ref. [71, 72]. Shaded area represents the calculation based on the Glauber theory [73, 74] in which the transparency is defined based on the total cross section of $NN$ scattering.

configuration which is strongly absorbed in the nuclear medium as indicated by the drop of the $T_{pp}$ at the same $s$. Thus, the observed correlation suggests that the nuclei filter the large-size component from the amplitude of hard *pp* scattering [64]. This picture is reinforced when one compares the $s$-dependences of the same $s^{10}$ weighted *pp* cross section (Fig.4(a)) and the polarization asymmetry, $A_{nn}$ (Fig.4(b)). As Fig.4(b)) shows, at $s \approx 25$ Gev$^2$ the $A_{nn}$ produces the largest asymmetry well above the pQCD prediction [48] (see Sec.7.), while the pQCD prediction is closest to the $A_{nn}$ data at $s \sim 18 - 20$ GeV$^2$. The interesting result here is that, at the latter energies ($s \sim 18 - 20$ GeV$^2$) the onset of Color Transparency (less attenuation) is observed in Fig.4(c)). Such a correlation between the agreement of the pQCD prediction of $A_{nn}$ with the data and the increase of the nuclear transparency, $T_{pp}$, seems to validate the main premise of the Color Transparency prediction, that the small-sized configurations are produced in the pQCD regime which propagate in the nuclear medium with less absorption than ordinary hadrons.

**Diquark Signatures of Nucleons:** There is a rather rich phenomenology about the quark-diquark structure of the nucleon. This phenomenology follows from the hadron spectroscopy, deep-inelastic scattering, baryon form-factors as well as exclusive hadron interactions (e.g. [75, 76]). The diquark structure reveals itself both as a specific symmetry of quark wave function of the nucleon as well as dynamical component that contributes to the scattering process (for example through the diquark exchange). However, the impossibility of formulation of diquark-structures from the first principles of QCD and its inclusion into

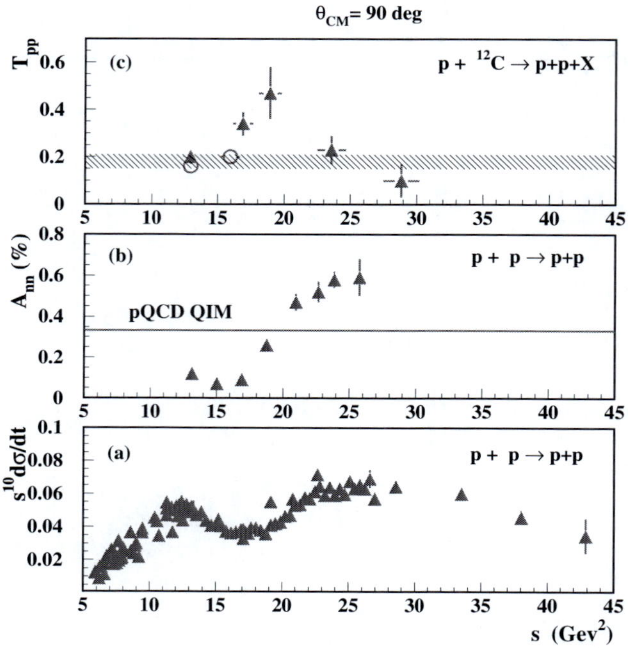


Figure 4. Dependence of the nuclear transparency (c), asymmetry, $A_{nn}$ (b) and $s^{10}$ scaled elastic *pp* cross section (a) on the invariant energy $s$.

the self-consistent evolution equation scheme makes the issue of diquarks an open question in NN scattering physics.

**Hidden Color Component in the NN-System:** The hidden color component is one of the unique QCD effects in NN interaction that can not be imitated by any baryons degrees of freedom. The hypothesis of the color-neutrality of the observed strongly interacting composite systems introduces the possibilities for a new reality in which baryonic systems with high degree of compositeness (one example is NN system) contain explicitly colored three-quark clusters that combine into a color neutral object. One of the examples is the contribution of two colored baryons, $N_c$, into the colorless NN system [36, 38, 39]. While existence of such hidden-color components are accepted within QCD there is no clear experimental evidence yet for such components.

## 4. Helicity Amplitudes and Observables of Elastic NN Scattering

In the following three sections an outline of the theoretical approach is given, which is most suitable for the description of the hard $NN$ scattering in QCD.

The natural framework in describing $NN$ scattering amplitudes at high energies is the helicity description of the quantum states of the nucleon. This representation has the obvious advantage of satisfying the boost invariance of the observable quantities as well as allowing us to define their asymptotic limits in the case of helicity conservation, which is the case in the pQCD regime of the scattering.

For exclusive, $1 + 2 \to 3 + 4$ scattering of spin-half particles, in general, one has a total of 16 helicity amplitudes $\langle \lambda_3\lambda_4 \mid M \mid \lambda_1\lambda_2\rangle$, where $\lambda_i$ describes the helicity state of the $i$'th nucleon having $+\frac{1}{2}$ and $-\frac{1}{2}$ eigenvalues for which we also use the notations $+$ and $-$ respectively[1]. Parity conservation and time reversal invariance of strong interaction as well as the application of generalized Pauli principle (or Isospin symmetry) introduces the following relations between helicity amplitudes of the elastic NN scattering (see e.g. [77, 78]):

$$\begin{aligned}\langle -\lambda_3 - \lambda_4 \mid M \mid -\lambda_1 - \lambda_2\rangle &= (-1)^{(\lambda_1-\lambda_2-(\lambda_3-\lambda_4))}\langle \lambda_3\lambda_4 \mid M \mid \lambda_1\lambda_2\rangle \\ \langle \lambda_1\lambda_2 \mid M \mid \lambda_3\lambda_4\rangle &= (-1)^{(\lambda_1-\lambda_2-(\lambda_3-\lambda_4))}\langle \lambda_3\lambda_4 \mid M \mid \lambda_1\lambda_2\rangle \\ \langle \lambda_4\lambda_3 \mid M \mid \lambda_2\lambda_1\rangle &= (-1)^{(\lambda_1-\lambda_2-(\lambda_3-\lambda_4))}\langle \lambda_3\lambda_4 \mid M \mid \lambda_1\lambda_2\rangle. \end{aligned} \quad (1)$$

These relations reduce the number of independent elastic $NN$ helicity amplitudes to five which we define as follows:

$$\begin{aligned}&\phi_1 \equiv \langle ++ \mid M \mid ++\rangle = \langle -- \mid M \mid --\rangle \\ &\phi_2 \equiv \langle ++ \mid M \mid --\rangle = \langle -- \mid M \mid ++\rangle \\ &\phi_3 \equiv \langle +- \mid M \mid +-\rangle = \langle -+ \mid M \mid -+\rangle \\ &\phi_4 \equiv \langle +- \mid M \mid -+\rangle = \langle -+ \mid M \mid +-\rangle \\ &\phi_5 \equiv \langle ++ \mid M \mid +-\rangle = \langle -+ \mid M \mid --\rangle = \langle -- \mid M \mid +-\rangle = \langle -+ \mid M \mid ++\rangle = \\ &\quad -\langle -- \mid M \mid -+\rangle = -\langle +- \mid M \mid ++\rangle = -\langle ++ \mid M \mid -+\rangle = -\langle +- \mid M \mid --\rangle, \end{aligned} \quad (2)$$

where $\phi_i$'s are functions of invariants, $s$ and $t$, or can be alternately presented as a function of $s$ and the scattering angle, $\theta_{cm}$ in the center of mass reference frame. From the above relations it follows that for isotriplet states, particularly for elastic $pp$ scattering at $\theta_{cm} = 90^0$, one has $\phi_5^{pp}(\theta_{cm} = 90^0) = 0$.

## 4.1. Practical Observables for Hard Elastic $NN$ Scattering

There is one unpolarized observable which is the cross section of elastic $NN \to NN$ scattering defined as:

$$\frac{d\sigma}{dt} = \frac{1}{16\pi}\frac{1}{s(s-4m_N^2)}\sigma_0 \quad \text{with} \quad \sigma_0 = \frac{1}{2}(|\phi_1|^2 + |\phi_2|^2 + |\phi_3|^2 + |\phi_4|^2 + 4|\phi_5|^2) \quad (3)$$

and a total of 256 possible polarization observables in the center of mass reference frame. However, the application of the above mentioned parity conservation, time reversal invariance and generalized Pauli principle reduces this number to 24 independent polarization observables (see e.g. [78]).

Considering the fact that the most of the possible $NN$ scattering experiments in the near future will be performed at fixed-target kinematics, we will consider the polarization observables for the lab frame kinematics. For this purpose we first define the momenta of

[1]Subscripts "1", "2", "3" and "4" hereafter represent the incoming, target, scattered and recoil nucleons respectively.

incoming, target, scattered and recoil nucleons as $\mathbf{p_1}$, $\mathbf{p_2}(=0)$, $\mathbf{p_3}$ and $\mathbf{p_4}$ respectively. Then in addition to three directions defined by above vectors:

$$\hat{\mathbf{p}}_1 = \frac{\mathbf{p_1}}{|p_1|}, \quad \hat{\mathbf{p}}_3 = \frac{\mathbf{p_3}}{|p_3|}, \quad \hat{\mathbf{p}}_4 = \frac{\mathbf{p_4}}{|p_4|} \tag{4}$$

one has four following vectors: The one, $n$, is transverse to the scattering plane:

$$\mathbf{n} = \frac{\mathbf{p_1} \times \mathbf{p_3}}{|\mathbf{p_1} \times \mathbf{p_3}|} \tag{5}$$

and three remaining define in-scattering-plane directions:

$$\mathbf{s_1} = \frac{\mathbf{n} \times \mathbf{p_1}}{|\mathbf{n} \times \mathbf{p_1}|}, \quad \mathbf{s_3} = \frac{\mathbf{n} \times \mathbf{p_3}}{|\mathbf{n} \times \mathbf{p_3}|}, \quad \mathbf{s_4} = \frac{\mathbf{n} \times \mathbf{p_4}}{|\mathbf{n} \times \mathbf{p_4}|}. \tag{6}$$

Based on these directions we define a polarization observable

$$X_{3,4,1,2}, \tag{7}$$

where subscripts $1, 2, 3, 4$ correspond to the directions of the polarization of the initial, target, scattered and recoil nucleons. Each of these subscripts can have polarizations along all the above defined directions.

Furthermore, we will present only those polarization asymmetry observables which are experimentally most feasible (or practical) categorizing them by their *degree* of polarization. Also at the right hand side of the each polarization asymmetry observable we give the corresponding expression in the limit of helicity conservation.

**Single Polarization:**
There is only one nonzero single polarization observable defined by the direction of $\mathbf{n}$:

$$X_{n,0,0,0} = X_{0,n,0,0} = X_{0,0,n,0} = X_{0,0,0,n} = -\frac{1}{\sigma_0} Im\left[\phi_5^\dagger(\phi_1 + \phi_2 + \phi_3 - \phi_4)\right] \to 0. \tag{8}$$

**Double Polarizations:**
(a) Out-of-plane initial polarization asymmetry and polarization transfer observables:

$$\begin{aligned}
X_{0,0,n,n} &= X_{n,n,0,0} = \frac{1}{\sigma_0}\left(Re\left[\phi_1^\dagger\phi_2 - \phi_3^\dagger\phi_4\right] + 2|\phi_5|^2\right) \to -\frac{1}{\sigma_0} Re\left[\phi_3^\dagger\phi_4\right] \\
X_{n,0,n,0} &= X_{0,n,0,n} = \frac{1}{\sigma_0}\left(Re\left[\phi_1^\dagger\phi_3 - \phi_2^\dagger\phi_4\right] + 2|\phi_5|^2\right) \to \frac{1}{\sigma_0} Re\left[\phi_1^\dagger\phi_3\right] \\
X_{0,n,n,0} &= X_{n,0,0,n} = -\frac{1}{\sigma_0}\left(Re\left[\phi_1^\dagger\phi_4 - \phi_2^\dagger\phi_3\right] + 2|\phi_5|^2\right) \to -\frac{1}{\sigma_0} Re\left[\phi_1^\dagger\phi_4\right]
\end{aligned} \tag{9}$$

(b) Scattering plane initial state polarization asymmetry observables:

$$\begin{aligned}
X_{0,0,s_1,s_1} &= \frac{1}{\sigma_0} Re\left[\phi_1^\dagger\phi_2 + \phi_3^\dagger\phi_4\right] \to \frac{1}{\sigma_0} Re\left[\phi_3^\dagger\phi_4\right] \\
X_{0,0,s_1,\hat{p}_1} &= X_{0,0,\hat{p}_1,s_1} = \frac{1}{\sigma_0} Re\left[\phi_5^\dagger(\phi_1 + \phi_2 - \phi_3 + \phi_4)\right] \to 0 \\
X_{0,0,\hat{p}_1,\hat{p}_1} &= -\frac{1}{2\sigma_0}\left(|\phi_1|^2 + |\phi_2|^2 - |\phi_3|^2 - |\phi_4|^2\right) \to -\frac{1}{2\sigma_0}\left(|\phi_1|^2 - |\phi_3|^2 - |\phi_4|^2\right)
\end{aligned} \tag{10}$$

(c) Helicity transfer and double helicity final state observables:

$$X_{\hat{p}3,0,\hat{p}1,0} = \frac{1}{\sigma_0}\left(-Re\left[\phi_5^\dagger(\phi_1 - \phi_2 + \phi_3 + \phi_4)\right]\sin\theta_3\right.$$
$$\left.+\frac{1}{2}\left[|\phi_1|^2 - |\phi_2|^2 + |\phi_3|^2 - |\phi_4|^2\right]\cos\theta_3\right) \to \frac{1}{2\sigma_0}\left[|\phi_1|^2 + |\phi_3|^2 - |\phi_4|^2\right]\cos\theta_3$$
$$X_{0,\hat{p}4,\hat{p}1,0} = \frac{1}{\sigma_0}\left(-Re\left[\phi_5^\dagger(-\phi_1 + \phi_2 + \phi_3 + \phi_4)\right]\sin\theta_4\right.$$
$$\left.-\frac{1}{2}\left[-|\phi_1|^2 + |\phi_2|^2 + |\phi_3|^2 - |\phi_4|^2\right]\cos\theta_4\right) \to \frac{1}{2\sigma}\left[|\phi_1|^2 - |\phi_3|^2 + |\phi_4|^2\right]\cos\theta_4$$
$$X_{\hat{p}3,\hat{p}4,0,0} = \frac{1}{\sigma_0}\left(\frac{1}{2}\left[|\phi_1|^2 + |\phi_2|^2 - |\phi_3|^2 - |\phi_4|^2\right]\cos\theta_3\cos\theta_4\right.$$
$$\left.+Re\left[\phi_1^\dagger\phi_2 + \phi_3^\dagger\phi_4\right]\sin\theta_3\sin\theta_4 - Re\left[\phi_5^\dagger(\phi_1 + \phi_2 - \phi_3 + \phi_4)\right]\sin(\theta_3 - \theta_4)\right)$$
$$\to \frac{1}{\sigma_0}\left(\frac{1}{2}\left[|\phi_1|^2 - |\phi_3|^2 - |\phi_4|^2\right]\cos\theta_3\cos\theta_4 + Re\left[\phi_3^\dagger\phi_4\right]\sin\theta_3\sin\theta_4\right), \quad (11)$$

where $\theta_3$ and $\theta_4$ are the lab scattering angles of scattered "3" and recoil "4" nucleons.

**Triple Polarizations:** Here we consider the situations in which the transverse polarization of only one of the final nucleons are measured in addition to the polarizations of initial nucleons:

$$X_{n,0,s_1,s_1} = X_{0,n,s_1,s_1} = -X_{n,0,\hat{p}_1,\hat{p}_1} = -X_{0,n,\hat{p}_1,\hat{p}_1} =$$
$$= -\frac{1}{\sigma_0}Im\left[\phi_5^\dagger(\phi_1 + \phi_2 - \phi_3 + \phi_4)\right] \to 0$$
$$X_{n,0,\hat{p}_1,s_1} = X_{0,n,\hat{p}_1,s_1} = \frac{1}{\sigma_0}Im\left[\phi_1^\dagger\phi_4 - \phi_2^\dagger\phi_3\right] \to 0$$
$$X_{n,0,s_1,\hat{p}_1} = X_{0,n,s_1,\hat{p}_1} = -\frac{1}{\sigma_0}Im\left[\phi_1^\dagger\phi_3 - \phi_2^\dagger\phi_4\right] \to 0 \quad (12)$$

## 4.2. Sum Rules and Equalities for Scattering of Identical Particles

The polarization asymmetry observables, defined above, allow us to extract all the individual helicity amplitudes of $NN$ elastic scattering. In practice it is less likely that all of these observables will be experimentally available at some point. However, several sum rules and equalities between different observables (which have better chance of being experimentally measured) will allow us to ascertain the extent of the polarization in the hard scattering regime.

First, it is worth emphasizing that, without invoking the assumption of helicity conservation, for identical particles one can demonstrate the existence of large asymmetries for initial state polarizations at any given high energy scattering at $\theta_{cm} = 90^0$. For this, one observes that for identical particles at $\theta_{cm} = 90^0$ elastic scattering, (i.e. $pp \to pp$ scattering) due to the generalized Pauli principle, the amplitudes with a single helicity flip are zero ($\phi_5 = 0$). From the same principle it follows that $\phi_4(\theta_{cm} = 90^0) = -\phi_3(\theta_{cm} = 90^0)$.

These properties of helicity amplitudes allow us to derive the following relations between the above defined polarization observables at $\theta_{cm} = 90^0$ elastic scattering of protons: The first relation is for double polarized initial states [28, 29]:

$$X^{pp}_{0,0,n,n} - X^{pp}_{0,0,s_1,s_1} - X^{pp}_{0,0,\hat{p}_1,\hat{p}_1} = 1, \tag{13}$$

where for these polarization observables in the literature often the following notations are used: $A_{nn} \equiv X_{0,0,n,n}$, $A_{ss} \equiv X_{0,0,s_1,s_1}$ and $A_{ll} \equiv X_{0,0,\hat{p}_1,\hat{p}_1}$.

The second relation is for the out-of-plane polarization and helicity transfer scatterings:

$$\begin{aligned} X^{pp}_{n,0,n,0} &= X^{pp}_{0,n,n,0} = \frac{1}{\sigma_0} Re\left[\phi_1^{pp,\dagger}\phi_3^{pp} - \phi_2^{pp,\dagger}\phi_4^{pp}\right] \\ X^{pp}_{\hat{p}_3,0,\hat{p}_1,0} &= X^{pp}_{0,\hat{p}_4,\hat{p}_1,0} = \frac{1}{2\sigma_0}\left(|\phi_1^{pp}|^2 - |\phi_2^{pp}|^2\right)\cos\theta_3, \end{aligned} \tag{14}$$

where $\theta_3 = \theta_4$ for $\theta_{cm} = 90^0$.

And the third relation is for the triple polarization case:

$$X^{pp}_{n,0,\hat{p}_1,s_1} = X^{pp}_{n,0,s_1,\hat{p}_1} = \frac{1}{\sigma_0} Im\left[\phi_1^\dagger\phi_4 - \phi_2^\dagger\phi_3\right]. \tag{15}$$

It follows from Eqs.(13) and (14) that the polarization correlations do not vanish with an increase of energy and they could be quite large. For Eq.(14) the large polarization follows from the expected dominance of the real parts of the amplitudes and observed hierarchy [60] in the hard scattering regime (for any large and fixed $\theta_{cm}$):

$$|\phi_1| \geq |\phi_3| \sim |\phi_4| \gg |\phi_5| \gg |\phi_2|. \tag{16}$$

## 4.3. Signatures of Helicity Conservation

The onset of the helicity conservation in QCD is associated with the two following properties of helicity amplitudes: *first* the relation of Eq.(16) is established with $\phi_5$ and $\phi_2$ being negligible and *second* the amplitudes are predominantly real.

For the particular case of the elastic *pp* scattering at $\theta_{cm} = 90^0$ the onset of the helicity conservation will result in the following relations between double initial state polarization asymmetries:

$$X^{pp}_{0,0,n,n} \approx -X^{pp}_{0,0,s_1,s_1} \approx \frac{1 + X^{pp}_{0,0,\hat{p}_1,\hat{p}_1}}{2} \approx |\phi_3^{pp}|^2 = |\phi_4^{pp}|^2 \tag{17}$$

and most interestingly between initial state polarizations and helicity transfer asymmetries:

$$X^{pp}_{0,0,n,n} - X^{pp}_{0,0,\hat{p}_1,\hat{p}_1} \approx -(X^{pp}_{0,0,s_1,s_1} + X^{pp}_{0,0,\hat{p}_1,\hat{p}_1}) \approx \frac{X^{pp}_{\hat{p}_3,0,\hat{p}_1,0} + X^{pp}_{0,\hat{p}_4,\hat{p}_1,0}}{2\cos\theta_3} \approx \frac{|\phi_1^{pp}|^2}{2}. \tag{18}$$

For the more general case of hard elastic $NN$ scattering in the helicity conservation regime at large and fixed $\theta_{cm} \sim 90^0$ a rather non-trivial relation exists:

$$X_{\hat{p}_3,0,\hat{p}_1,0} \to -X_{0,0,\hat{p}_1,\hat{p}_1} \tag{19}$$

which indicates that the helicity transfer asymmetry approaches to the asymmetry of double initial helicity scattering.

For all other observables in the helicity conservation regime one obtains expressions which are given in the RHS parts of Eqs.(8,9,10,11,12). It is worth mentioning that the measurement of the triple-polarizations in Eq.(12) are important for verifying that the helicity amplitudes become real, especially in relation to $\phi_1$, $\phi_3$ and $\phi_4$ amplitudes.

## 5. Quark Counting Rules of QCD

About 40 years ago in two seminal papers [25, 26], predictions were made that at fixed $\theta_{cm}$ the high energy hadronic processes probe the constituents of hadrons in their minimal Fock component states. This observation stimulated extensive research on hard exclusive hadronic processes.

Within automodelism (scale invariance) assumption the invariant energy dependence of the Feynman amplitude, $\mid \langle cd \mid M \mid ab \rangle \mid^2$, of hard exclusive $a + b \to c + d$ scattering is such that it compensates the combined dimensionality of the Fock states of participating hadrons, $[N_a N_b N_c N_d]^{\frac{1}{2}} = m^{n_a+n_b+n_c+n_d-4}$, where $N_j$'s are the normalization factors and $n_j$'s are the number of constituents of the Fock component of the $j$ hadron:

$$|j>= N_j^{\frac{1}{2}} |n_j, constituents\rangle. \tag{20}$$

This results in the simple relation for the energy dependence of the differential cross sections of hard exclusive reactions:

$$\frac{d\sigma^{ab\to cd}}{dt} = F(\theta_{cm})_{ab\to cd} s^{-(n_a+n_b+n_c+n_d-2)}, \tag{21}$$

where $F(\theta_{cm})_{ab\to cd}$ is a function of $\theta_{cm}$ only.

As it was observed in Refs. [79, 80] the automodelism assumption is justified for local field theories with vector interaction and thus it could be extended to electromagnetic processes in which case leptons and bare photons would be counted as a one unit of constituent.

One of the most beautiful observations was [26, 27] that the energy counting rule naturally follows from perturbative QCD in which the gluon exchange between massless quarks will provide the scale invariance of the scattering amplitude. Within pQCD one can prove the two main asymptotic rules presented in Fig.5: (a) the sub-process of hard elastic $quark - quark$ interaction is energy independent (depends only on the scattering angle) and (b) each energetic propagator of intermediate quark line enters with factor, $\frac{1}{s}$. These two rules in addition to the above mentioned counting of leptons and bare photons define all the necessary rules for estimating the energy dependence of differential cross sections of hard exclusive scattering of hadrons, leptons as well as photons in the fixed $\theta_{cm}$ kinematics in the $s \to \infty$ limit.

The important consequence of quark-counting rules is that it explains why one should expect the dominance of the minimal-Fock component in the wave function of the interacting hadrons. The simple estimate shows that higher order Fock components will result in larger negative exponent in the $s$-dependence of the scattering amplitude and therefore will decrease much faster than the contribution due to minimal-Fock component in the high energy limit.

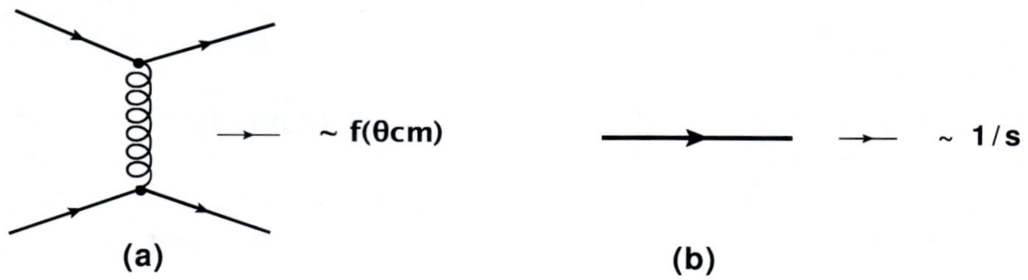


Figure 5. Quark counting rules.

# 6. Basics of Light-Front Dynamics in QCD

**Bound State Problem:** One of the fundamental problems in field theory is the description of the quantum bound states in a relativistically invariant form. The problem inevitably arises with the proper treatment of the time at which the bound state is observed as well as the unambiguous identification of the compositeness of a bound state due to the non-trivial structure of the vacuum in field theories. The latter is related to the differentiation of the constituents of the bound system from the particles arising from the vacuum fluctuations.

The problem of the treatment of the time in the relativistic dynamics of the bound states can be seen in the following consideration of the wave function of the system consisting of $n$ particles. To "measure" the wave function of the system, the observer $X$ in Fig.6 (a) must instantaneously measure the positions and times of all $n$ particles. In nonrelativistic dynamics due to Galilean relativity, the $n$ positions can be measured instantaneously at the given absolute time $t$, allowing one to talk about the wave function of the system defined as $\Psi(z_1, z_2, z_3, ., z_n, t)$.

In the relativistic case, the measurement will assume sending light signals from the position of the observer $X$ (Fig.6 (a)) to all $n$ particles resulting in the measurement of the function $\Psi(z_1, t_1, z_2, t_2, z_3, t_3, ..z_n, t_n)$. Even if the latter function can be arranged to be relativistically invariant it does not have a meaning of the bound state of the system at the given time, $t$.

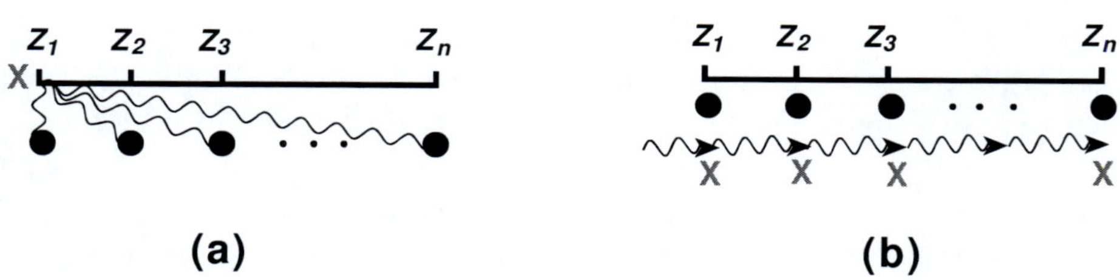


Figure 6. Relativistic measurement of the wave function.

This problem can be solved if the observer "rides" on the front of the light beam that passes through the all $n$ particles. In this case the observer will find that all constituents are measured at the same "light front" time $\tau = t_1 - z_1 = t_2 - z_2 = t_3 - z_3 = \cdots = t_n - z_n$ and can reconstruct the wave function of the system with constituents being measured at light-cone positions of $\mathcal{Z}_i = t_i + z_i$, $i = 1, \cdots n$ at the same light-cone time $\tau$: $\Psi_{LF}(\mathcal{Z}_1, \mathcal{Z}_2, \mathcal{Z}_3, \cdots \mathcal{Z}_n, \tau)$.

That the relativistic dynamics simplifies in the light-front due to the maximum number of kinematic generators of the Lorentz group, was first observed by P. A. M. Dirac [81]. In relation to relativistic bound states, S. Weinberg [82] demonstrated how the simple structure of the vacuum allows the separation of the constituent structure of the bound state from the vacuum fluctuation contributions.

The light-front dynamics is a natural approach in QCD, in which due to light masses of quarks, one naturally deals with relativistic bound states [83]. In this case, choosing quantization axis by the direction of the momentum of the hadron, its wave function is quantized at fixed $z^+ = t + z$ time with $p^-$ component representing the light-front energy of the hadron. In this approach the Weinberg type equation specifies the quark-gluon wave function of the hadron which depends on the transverse momenta $\mathbf{k}_{\perp,\mathbf{i}}$ and the light-front longitudinal momentum fractions (defined as $x_i \equiv \frac{k_i^+}{p^+}$) of the constituent quarks and gluons. The wave function defined in such way is a Lorentz boost invariant and addresses the main issues of the description of the relativistic bound states - Lorentz invariance and definite light-cone time of the quantization.

In practical applications, the quark-gluon wave function of hadrons is expanded into the Fock component representations that allows to account fully the compositeness of the hadrons as a bound system of quarks. For example the Fock-state expansion of the light cone wave function of the nucleon can be presented as [84, 85]

$$\begin{aligned} \mid \Psi_N^{S_z}(P^+, \mathbf{P}_\perp)\rangle &= \sum_n \prod_{i=1}^{n} \frac{dx_i d^2 k_{\perp,i}}{\sqrt{x_i} 16\pi^3} (2\pi)^3 2\delta(1 - \sum_{j=1}^{n} x_j) \delta^{(2)}(\sum_j^{n} \mathbf{k}_{\perp,\mathbf{j}}) \\ &\times \ \psi_{n/N}(x_i, \mathbf{k}_{\perp,\mathbf{i}}, \lambda_i) \mid n; x_i p^+, x_i \mathbf{P}_\perp + \mathbf{k}_{\perp,\mathbf{i}}, \lambda_i\rangle \end{aligned} \tag{22}$$

where $\mid n; x_i p^+, x_i \mathbf{P}_\perp + \mathbf{k}_{\perp,\mathbf{i}}, \lambda_i\rangle$ represents the eigenstates of Fock-components with $n$ constituents. The coefficients: $\psi_{n/N}(x_i, \mathbf{k}_{\perp,\mathbf{i}}, \lambda_i)$ are the probability amplitudes (wave functions) for given $n$ constituents with relative transverse momenta, $\mathbf{k}_{\perp,\mathbf{i}}$ and "longitudinal" momentum fractions $x_i$.

**Hard pQCD Interaction in the Light-Front:** Using the light cone - helicity spinors in the form [86]:

$$u_\lambda(p^+, p_t) = (p^+ + \beta m + \alpha_t \cdot p_t)\chi_\lambda, \tag{23}$$

where

$$\chi_{\pm\frac{1}{2}} = \frac{1}{\sqrt{2}}(1 + \alpha_3)\chi_\pm, \quad \text{with} \quad \chi_+ = \begin{pmatrix} 1 \\ 0 \\ 0 \\ 0 \end{pmatrix} \quad \text{and} \ \chi_- = \begin{pmatrix} 0 \\ 1 \\ 0 \\ 0 \end{pmatrix}, \tag{24}$$

for the fermion-vector field vertex one obtains:

$$\begin{aligned} \bar{u}_{\lambda'}(p')\gamma^+ u_\lambda(p) &= 2\sqrt{p'_+ p_+}\delta^{\lambda,\lambda'} \\ \bar{u}_{\lambda'}(p')\gamma^- u_\lambda(p) &= \frac{1}{\sqrt{p'_+ p_+}} \left[(1 + S_{\lambda,\lambda'}) p_R p'_L + (1 - S_{\lambda,\lambda'}) p_L p'_R + m^2\right] \delta^{\lambda,\lambda'} \end{aligned}$$

$$
\begin{aligned}
&+ \frac{\Delta_{\lambda,\lambda'} m}{\sqrt{p'_+ p_+}} [(p_R - p'_R)(1 + \Delta_{\lambda,\lambda'}) + (p_L - p'_L)(1 - \Delta_{\lambda,\lambda'}] \\
\bar{u}_{\lambda'}(p')\gamma^{R/L} u_\lambda(p) &= \left[\sqrt{\frac{p_+}{p'_+}} p'_{R/L}(1 \mp S_{\lambda\lambda'}) + \sqrt{\frac{p'_+}{p_+}} p_{R/L}(1 \pm S_{\lambda\lambda'})\right] \delta^{\lambda,\lambda'} \\
&+ m\Delta_{\lambda,\lambda'} \left[\sqrt{\frac{p_+}{p'_+}} - \sqrt{\frac{p'_+}{p_+}}\right](1 - \Delta_{\lambda,\lambda'}), \qquad (25)
\end{aligned}
$$

where "$R$" and "$L$" are indices of the given vector $A$ correspond to $A_{R/L} = A_x \pm iA_y$. Also, $S_{\lambda,\lambda'} \equiv \lambda + \lambda'$ and $\Delta_{\lambda,\lambda'} \equiv \lambda - \lambda'$.

By using the light cone gauge for the exchanged gluon propagators, one avoids the necessity of inclusion of the ghost fields to cancel the nonphysical contributions of gluonic fields and, for the interaction "cell" of Fig.5(a) one obtains:

$$
\bar{u}\gamma^\mu u d_{\mu\nu} \bar{u}\gamma^\nu u = \bar{u}\left[\frac{\gamma^+}{r^+} r_\perp - \gamma_\perp\right] u\bar{u}\left[\frac{\gamma^+}{r^+} r_\perp - \gamma_\perp\right] u, \qquad (26)
$$

where $r$ is the four-momentum of the exchanged gluon.

Applying the above defined vertex rules, it is now easy to show, (a) the validity of counting rules of Fig.5 in perturbative QCD and (b) that the counting rules are accompanied by the helicity conservation of interacting quarks. Thus the signature of the quark-counting rules together with the helicity conservation will represent a powerful indicator for the onset of the pQCD regime in strong interactions.

### 6.1. Non-Perturbative QCD Contribution and Factorization Properties of Hard Scattering Amplitudes

Even though in probing short distance properties of $NN$ interaction our focus is on the pQCD aspects of hard quark-gluon interactions, we cannot avoid the non-perturbative contributions. The latter will appear prominently in the wave functions of nucleons. The necessary assumption that allows us to proceed with the calculation of hard $NN$ scattering amplitudes is the assumption of the factorization of nonperturbative and perturbative QCD contributions [86]. This assumption is based on the very plausible expectation that the dominant contribution from the wave functions of incoming and outgoing nucleons come from the collinear states of their minimal Fock components. In this case, each three quarks of initial nucleon "arrive" at hard scattering "point" with momenta almost parallel to their parent nucleons and after the hard scattering they emerge parallel to the respective momenta of outgoing nucleons. Clearly the wave functions of nucleons containing almost collinear quarks are dominated by the nonperturbative QCD contribution.

Even if such factorization is justified and applied, another problem is the number of possible diagrams (which is in the order of tens of thousands) which take into account all possible combinations of five hard gluon exchanges between incoming and outgoing collinear quarks. This and the issue of nonperturbative wave functions can be addressed if the calculated quantities are expressed through the experimentally measured quantities such as nucleon-form factors [87] or if we consider the observables such as cross section ratios, polarization and angular distribution asymmetries which are not sensitive to the absolute normalization of the amplitudes and are sensitive either to the structure of hard scattering or symmetry of the nucleon wave functions.

Below we will consider a few examples in which the above approach is applied in describing the ratios of the hard elastic $pn$ and $pp$ scatterings as well as the angular asymmetry of the hard $pn \to pn$ scattering around $\theta_{cm} = 90^0$.

# 7. Quark Interchange Mechanism of Hard Elastic $NN$ Scattering

As it follows from Eq.(21), the quark counting rule applied to the hard elastic $NN$ scattering predicts

$$\frac{d\sigma}{dt} = F(\theta_{cm})_{NN \to NN} \cdot s^{-10}, \tag{27}$$

which, as it follows from the discussion of Sec.3., describes the bulk of the energy dependence of hard $NN$ scattering at large $\theta_{cm}$. According to the application of the quark counting rules, the $s^{-10}$ energy dependence requires the hard exchange of five gluons between six quarks representing the minimal Fock components of two nucleons. One such mechanism that satisfies the above requirement is the quark-interchange mechanism of the $NN$ interactions.

## 7.1. Dominance of Quark-Interchange Diagrams

Considering the topology of hard QCD scatterings in elastic NN processes one can identify a few possibilities in which the collinear quarks from minimal Fock components of the nucleon wave function can scatter into the final state (Fig.7). In fact the mechanism of independent three-gluon exchanges (Landshoff mechanism [24]) (Fig.7 (c) predicts softer ($s^{-8}$) energy dependence since it requires less number of gluon exchanges. However this diagram (as well as the one in Fig.7(b)) does not exchange quarks and will contribute equally to $NN$ and $\bar{N}N$ scattering. Thus, comparison of the experimental data of these two reactions will allow the estimation of the contributions of such diagrams. Such a comparison was made in the experiment of Ref. [88] which found the upper limit of the ratio of elastic differential cross sections of $\bar{p}p$ to $pp$ at $p_{Lab} = 9.9$ GeV/c at $\theta_{cm} = 90^0$ to be less than $4\%$. The measurement of the same ratio at $p_{Lab} = 5.9$ GeV/c yielded $\sim 2.5\%$. These results, together with the comparison of the cross sections of twenty different hard exclusive hadronic reactions unambiguously confirmed the dominance of the hard scattering amplitudes of hadrons containing quarks of the same flavor with respect to the scattering amplitudes of hadrons that share no common flavors of quarks.

Thus, one can start with a rather solid experimental justification of the dominance of the quark-interchange (QIM) diagrams of Fig.7(a) in the hard elastic NN scattering amplitude. In fact the theoretical analysis demonstrates that the QIM diagrams represent the dominant mechanism of hard elastic scattering for up to ISR energies (see discussion in [29]). It is worth mentioning that the quark-interchange mechanism also describes reasonably well the 90 c.m. hard break-up of two nucleons from the deuteron [89, 90, 91, 92] and $^3He$ [93].

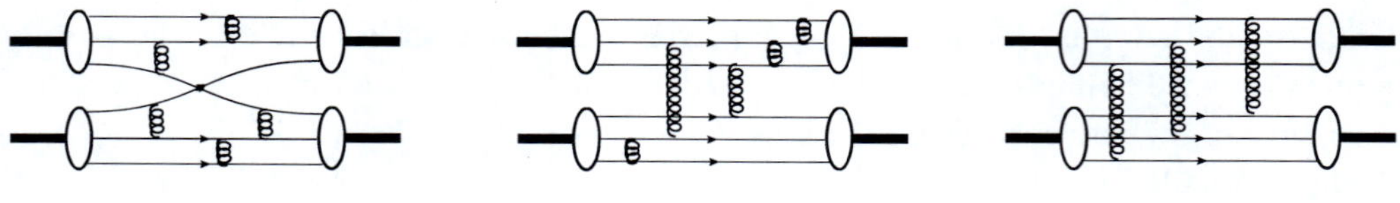


Figure 7. Possible scattering diagrams for elastic NN processes.

Based on the above arguments we now focus on the quark-interchange diagrams which contain five gluon exchanges as shown in Fig.8. Furthermore, we assume the validity of earlier discussed factorization of the minimal (3q) Fock components of initial and final state wave functions of nucleons and the hard interaction kernel, which represents the quark interchange scattering through the five hard gluon exchanges. These assumption allows to

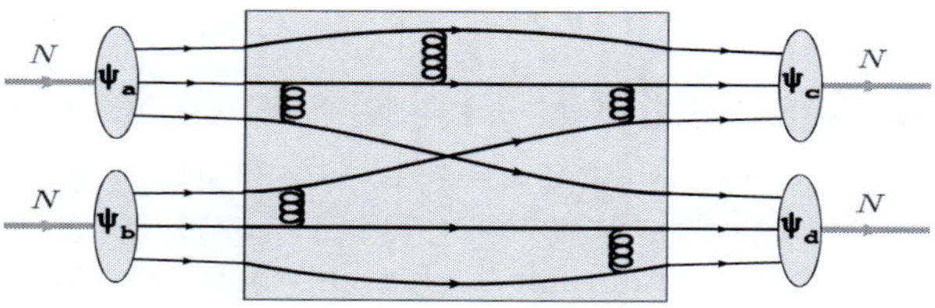


Figure 8. Typical diagram for quark-interchange mechanism of $NN \to NN$ scattering.

represent the amplitude of hard $N(a) + N(b) \to N(c) + N(d)$ scattering of Fig.8, within quark-interchange approximation in the following form [94]:

$$\langle cd \mid T \mid ab\rangle = \sum_{\alpha,\beta,\gamma} \langle \psi_c^\dagger \mid \alpha_2', \beta_1', \gamma_1'\rangle \langle \psi_d^\dagger \mid \alpha_1', \beta_2', \gamma_2'\rangle \times$$
$$\langle \alpha_2', \beta_2', \gamma_2', \alpha_1'\beta_1'\gamma_1' \mid H \mid \alpha_1, \beta_1, \gamma_1, \alpha_2\beta_2\gamma_2\rangle \cdot \langle \alpha_1, \beta_1, \gamma_1 \mid \psi_a\rangle \langle \alpha_2, \beta_2, \gamma_2 \mid \psi_b\rangle, \tag{28}$$

where $(\alpha_i, \alpha_i')$, $(\beta_i, \beta_i')$ and $(\gamma_i, \gamma_i')$ describe the spin-flavor quark states of minimal-Fock component of nucleon wave function before and after the hard scattering, $H$, and

$$C^j_{\alpha,\beta,\gamma} \equiv \langle \alpha, \beta, \gamma \mid \psi_j\rangle \tag{29}$$

describes the probability amplitude of finding the $\alpha, \beta, \gamma$ helicity-flavor combination of three valence quarks in the nucleon $j$ [28, 94]. Note that in Eq.(28) the factorization of nucleon wave functions from the hard scattering kernel is justified by the energies, characteristic to the $C^j_{\alpha,\beta,\gamma}$ factors, being of the order of the nucleon mass, while the $H$-kernel is characterized by the transfered momenta $-t, -u \gg m_N^2$.

To be able to calculate the $C^j_{\alpha,\beta,\gamma}$ factors, one represents the nucleon wave function through the helicity-flavor basis of the valence quarks. We use a rather general form, separating the nucleon wave function into two parts characterized by two (e.g. second and third) quarks being in spin zero - isosinglet and spin one - isotriplet states as follows:

$$\psi^{i_N^3, h_N} = \frac{1}{\sqrt{2}} \left\{ \Phi_{0,0}(k_1, k_2, k_3)(\chi^{(23)}_{0,0}\chi^{(1)}_{\frac{1}{2},h_N}) \cdot (\tau^{(23)}_{0,0}\tau^{(1)}_{\frac{1}{2},i_N^3}) + \Phi_{1,1}(k_1, k_2, k_3) \times \right.$$

$$\sum_{i^3_{23}=-1}^{1} \sum_{h^3_{23}=-1}^{1} \langle 1, h_{23}; \frac{1}{2}, h_N - h_{23} \mid \frac{1}{2}, h_N \rangle \langle 1, i^3_{23}; \frac{1}{2}, i^3_N - i^3_{23} \mid \frac{1}{2}, i^3_N \rangle \times$$
$$(\chi^{(23)}_{1,h_{23}} \chi^{(1)}_{\frac{1}{2},h_N-h_{23}}) \cdot (\tau^{(23)}_{1,i^3_{23}} \tau^{(1)}_{\frac{1}{2},i^3_N-i^3_{23}}) \Big\}, \tag{30}$$

where $j^3_N$ and $h_N$ are the isospin component and the helicity of the nucleon. Here $k_i$'s are the light cone momenta of quarks which should be understood as $(x_i, k_{i\perp})$, where $x_i$ is a light cone momentum fraction of the nucleon carried by the $i$-quark. We define $\chi_{j,h}$ and $\tau_{I,i^3}$ as helicity and isospin wave functions, where $j$ is the spin, $h$ is the helicity, $I$ is the isospin and $i^3$ its third component. The Clebsch-Gordan coefficients are defined as $\langle j_1, m_1; j_2, m_2 \mid j, m\rangle$. Here, $\Phi_{I,J}$ represents the momentum dependent part of the wave function for ($I = 0, J = 0$) and ($I = 1, J = 1$) two-quark spectator states respectively. We also introduce a parameter, $\rho$:

$$\rho = \frac{\langle \Phi_{1,1} \rangle}{\langle \Phi_{0,0} \rangle}, \tag{31}$$

which characterizes an average relative magnitude of the wave function components corresponding to ($I = 0, J = 0$) and ($I = 1, J = 1$) quantum numbers of two-quark "spectator" states. Note that the two extreme values of $\rho$ define two well known approximations: $\rho = 1$ corresponds to the exact SU(6) symmetric picture of the nucleon wave function and $\rho = 0$ will correspond to the contribution of only good-scalar diquark configuration in the nucleon wave function (see e.g. Ref. [75, 96, 82] where this component is referred to as a scalar or good diquark configuration ($[qq]$) as opposed to a vector or bad diquark configuration denoted by $(qq)$). In further discussions we will keep $\rho$ as a free parameter. Note that in our approach diqaurks represent a $qq$-component of the nucleon wave function and quarks from the diquark participate in the hard scattering through the quark-interchange.

To calculate the scattering amplitude of Eq.(28) we take into account the conservation of the helicities of quarks participating in the hard scattering. This allows us to approximate the hard scattering part of the amplitude, $H$, in the following form:

$$H \approx \delta_{\alpha_1\alpha_1'} \delta_{\alpha_2\alpha_2'} \delta_{\beta_1,\beta_1'} \delta_{\gamma_1,\gamma_1'} \delta_{\beta_2,\beta_2'} \delta_{\gamma_2,\gamma_2'} \frac{F(\theta)}{s^4}. \tag{32}$$

Inserting this expression into Eq.(28) for the QIM amplitude one obtains [28, 94]:

$$\langle cd \mid T \mid ab \rangle = Tr(M^{ac} M^{bd}) \tag{33}$$

with:

$$M^{i,j}_{\alpha,\alpha'} = C^i_{\alpha,\beta\gamma} C^j_{\alpha',\beta\gamma} + C^i_{\beta\alpha,\beta} C^j_{\beta\alpha',\beta} + C^i_{\beta\gamma\alpha} C^j_{\beta\gamma\alpha'}, \tag{34}$$

where we sum over all the possible values of $\beta$ and $\gamma$.

Furthermore, separating the energy dependence and angular parts for nonvanishing helicity amplitudes of Eq.(2) one obtains:
for $pp \to pp$:

$$\begin{aligned}
\phi_1 &= C(s)\,[(3+y)F(\theta) + (3+y)F(\pi-\theta)] \\
\phi_3 &= C(s)\,[(2-y)F(\theta) + (1+2y)F(\pi-\theta)] \\
\phi_4 &= -C(s)\,[(1+2y)F(\theta) + (2-y)F(\pi-\theta)]
\end{aligned} \tag{35}$$

and for $pn \to pn$:

$$\begin{aligned}\phi_1 &= C(s)\left[(2-y)F(\theta)+(1+2y)F(\pi-\theta)\right]\\ \phi_3 &= C(s)\left[(2+y)F(\theta)+(1+4y)F(\pi-\theta)\right]\\ \phi_4 &= C(s)\left[2yF(\theta)+2yF(\pi-\theta)\right]\end{aligned} \tag{36}$$

with $\phi_2 = \phi_5 = 0$ due to helicity conservation. Here the function $C(s) = \frac{N}{s^4}$ is due to quark-counting rule with the normalization factor, N, and $F(\theta)$ accounts for the angular dependence generated by hard scattering kernel. In the above equation the parameter:

$$y = x(x+1) \text{ with } x = \frac{2\rho}{3(1+\rho^2)} \tag{37}$$

defines the symmetry of the minimal-Fock (3q) component of nucleon wave function. For example, $\rho = 1$ case reproduces the SU(6) result of Refs. [28] and [29].

As Eqs.(35) and (36) show the helicity amplitudes reveal strong sensitivity to the underlaying symmetry of $3q$ nucleon wave function. The hard kernel, $H$, as it follows from Eq.(32) defines mainly the overall $s$- dependence of the amplitudes, thus verifying the dominance of the minimal-Fock component of the quark wave function of nucleon and the QCD mechanism of hard interaction. What concerns the overall normalization of the helicity amplitudes they are defined both by the normalization of the nucleon wave function and the sum of the multitude of hard scattering amplitudes.

In the following we will consider three examples in which the specific measurements can verify the different aspects of the above described model of NN interaction.

**Ratio of Differential Cross Sections of Elastic $pn$ and $pp$ Scatterings at $\theta_{cm} = 90^0$:** As it can be checked from Eqs.(35) and (36), at $\theta_{cm} = 90^0$ the helicity amplitudes of $pp$ scattering depend identically on the helicity-flavor symmetry factor, $y$ of the $3q$ wave function of the proton, while no such identity exists for $pn$ amplitudes. Therefore one should expect rather strong sensitivity of the ratio of $pn$ to $pp$ elastic cross sections on the underlying spin-flavor symmetry of nucleon wave functions [95]. These can be seen in Fig.9, where the ratio of the differential cross sections of elastic $pn$ to $pp$ scatterings is presented as a function of $s$ at $\theta_{cm} = 90^0$. The calculations are performed for the case of $SU(6)$ symmetry ($\rho = 1$) and scalar diquark model, for which $\rho = 0$. The results are compared with the available experimental data. Despite the large errors present in the data, one can clearly see the sensitivity of the considered ratio to the symmetry structure of nucleon wave function with data favoring the scalar diquark model of nucleon.

**Angular Asymmetry of Elastic $pn$ Scattering:** In this example we demonstrate that an observable such as the asymmetry of a hard elastic proton-neutron scattering with respect to $\theta_{cm} = 90^0$ may provide another insight into the helicity-flavor symmetry of the 3q wave function of the nucleon. The measurable quantity we consider is:

$$A_{90^0}(\theta) = \frac{\sigma(\theta)-\sigma(\pi-\theta)}{\sigma(\theta)+\sigma(\pi-\theta)}, \tag{38}$$

where $\sigma(\theta)$ is the differential cross section of the elastic $pn$ scattering.

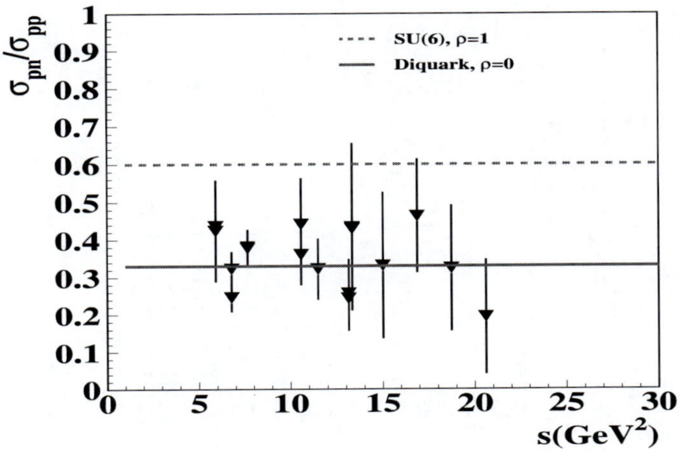


Figure 9. Ratio of the $pn \to pn$ to $pp \to pp$ elastic differential cross sections as a function of $s$ at $\theta^N_{c.m.} = 90^0$. The data are from Refs. [50, 51, 52, 53].

In Fig.10 the asymmetries of $pn$ scattering calculated with SU(6) ($\rho = 1$) and pure scalar-diquark ($\rho = 0$) models are compared with the data [94]. The comparisons show that the nucleon wave function (30) with the good-scalar diquark component ($\rho = 0$) produces the right sign for the angular asymmetry. On the other the data are in qualitative disagreement with the prediction of exact SU(6) symmetry ($\rho = 1$) of the quark wave function of nucleon.

In Ref. [94] we used $\rho$ as a free parameter to obtain the best fit of the data, which was found for

$$\rho \approx -0.3 \pm 0.2. \tag{39}$$

The nonzero magnitude of $\rho$ indicates the small but finite relative strength of a bad/vector diquark configuration in the nucleon wave function. It is intriguing that the obtained magnitude of $\rho$ is consistent with the $10\%$ probability of "bad" diquark configuration discussed in Ref. [97].

Another interesting property of Eq.(39) is the negative sign of the parameter $\rho$. Within qualitative quantum-mechanical picture, the negative sign of $\rho$ may indicate for example the existence of a repulsion in the quark-(vector- diquark) channel as opposed to the attraction in the quark - (scalar-diquark) channel. It is rather surprising that both the magnitude and sign agree with the result of the phenomenological interaction derived in the one-gluon exchange quark model discussed in Ref. [96].

**The $A_{nn}$ Asymmetry of Elastic $pn$ Scattering:** In this last example we consider the $A_{nn} \equiv X_{0,0,n,n}$ asymmetry defined according to Eq.(9). This asymmetry is especially interesting to consider for $\theta_{cm} = 90^0$ scattering, since in this case as it was mentioned before the all non-zero helicity amplitudes depend identically on the helicity-flavor factor of $y$. As a result, by using helicity amplitudes of Eq.(35) in the RHS part of Eq.(9) one

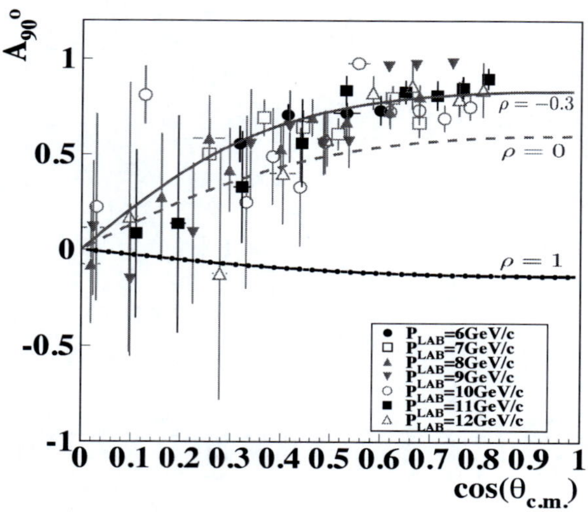


Figure 10. Asymmetry of $pn$ elastic cross section. Solid dotted line - SU(6), with $\rho = 1$, dashed line - diquark model with $\rho = 0$, solid line - fit with $\rho = -0.3$. The data are from Refs. [52, 53].

obtains a constant value, $A_{nn} = 0.333$, at $\theta_{cm} = 90^0$ independent on the specific helicity-flavor symmetry of 3q wave functions.

This result is compared with the data in Figs.2 and 4, which demonstrates that agreement with the data happens at $s \approx 18 - 20$ GeV$^2$ at the very same energy range where transparency studies demonstrate the drop of the absorption of propagating "protons" in the nuclear medium (Fig.(3) and (4)).

## 8. Conclusion and Outlook

In this work we reviewed the current status of QCD studies of hard $NN$ processes that probe quark-gluon interactions at short distances. We gave brief history of these studies and highlighted the outstanding questions that may be resolved in the future experiments at the emerging facilities worldwide. We outlined the theoretical foundation of QCD hard processes and demonstrated that in high energy regime the hard processes probe the minimal (3q) Fock component of the nucleon wave function which is factorized from the kernel that defines the dynamics of the hard scattering.

We considered few examples of application of the outlined approach and demonstrated that even with the limited quantity and quality of the data one can conclude that most probably SU(6) symmetry is breaking down at large $x$ limit of the valence quark component of nucleon wave function. Another observation we made is that, apparently, the phenomenology of hard $pp$ and $pn$ scattering favors the diquark picture of the nucleon with small and finite mixture from the vector diquarks which enter with the negative phase relative to the scalar diquark contribution. Finally, our estimate of the $A_{nn}$ asymmetry of elastic $pp$ scattering within pQCD indicates that it agrees with the data at same energies at which color transparency signature is observed in the hard $A(p, 2p)X$ reactions.

This review covered only the hard elastic NN scatterings. However, it is worth noting that there is a growing activity in studies of NN bound systems at short distances by probing short range correlations in the nuclear wave functions (see e.g. [45, 98, 99, 100, 101]). Currently these studies unambiguously identified the tensor component of short-range $NN$ interactions in the nuclear medium. The planned experiments at the 12 GeV energy upgraded Jefferson Lab [102, 103] will be able to probe the bound $NN$ systems at distances relevant to the nuclear core, where one may expect the onset of QCD degrees of freedom in the similar way as in the hard NN interactions.

Overall new experiments in studies of both hard $NN$ scattering processes and $NN$ short-range correlations in nuclei will provide the necessary ground for advancing the understanding of QCD dynamics of strong forces at short distances.

The author is thankful to Drs. W. Boeglin, S. J. Brodsky, R. Gilman, L. Frankfurt, G. Miller, E. Piasetzky and M. Strikman for illuminating discussions and many years of collaboration on the physics of hard processes. Special thanks to Dr. Igor Strakovsky for the suggestion and opportunity to write this review. This work is supported by the U.S. DOE grant under contract DE-FG02-01ER41172.

## References

[1] Rutherford, E. *Phil. Mag.* 1911, *21*, 669-688.

[2] Rutherford, E. *Phil. Mag.* 1919, *37*, 571-580.

[3] Rutherford, E. *Phil. Mag.* 1919, *37*, 581-587.

[4] Blatt, J.M.; Weisskopf, V.F. *Theoretical Nuclear Physics*; Dover Books on Physics; Dover Publications, 2010, pp 1-896.

[5] Jastrow, R. *Phys. Rev.* 1951, *81*, 165-170.

[6] Taketani, M,; S. Nakamura, S.; Sasaki, M. *Proc. Theor, Phys.* 1951, *6*, 581-586.

[7] Kaplan, D.B.; M. J. Savage, M.J.; Wise, M.B. *Nucl. Phys. B* 1998, *534*, 329-355.

[8] Epelbaum, E. (2010). Nuclear Forces from Chiral Effective Field Theory: A Primer. [arXiv:1001.3229] [nucl-th], 1-60.

[9] Epelbaum, E. *Prog. Part. Nucl. Phys.* 2006, *57*, 654-741.

[10] Yang, C. J.; Elster, C.; Phillips, D. R. *Phys. Rev. C* 2009, *80*, 044002 (1-27).

[11] Machleidt, R. *Adv. Nucl. Phys.* 1989, *19*, 189-376.

[12] Wiringa, R. B.; Stoks, V. G. J.; Schiavilla, R. *Phys. Rev. C* 1995, *51*, 38-51.

[13] Arndt, R. A.; Strakovsky, I. I.; Workman, R. L. *Phys. Rev. C* 2000, *62*, 034005 (1-16).

[14] Arndt, R. A.; Briscoe, W. J.; Strakovsky, I. I.; Workman, R. L. *Phys. Rev. C* 2007, *76*, 025209 (1-10).

[15] Bergervoet, J. R. et al., *Phys. Rev. C* 1990, *41*, 1435-1452.

[16] Feynman, R. P. *Photon-Hadron Interactions*; Advanced Book Classics; Addison-Wesley Reading: Massachusetts, 1998, pp 1-282.

[17] Gaillard, G. et al., *Helv. Phys. Acta* 1991, *64*, 201-202.

[18] Kumano, S. *Nucl. Phys. A* 2007, *782*, 442-449.

[19] Lutz, M. F. M. et al., [PANDA Collaboration], (2009). Physics Performance Report for PANDA: Strong Interaction Studies with Antiprotons [arXiv:0903.3905] [hep-ex], 1-216.

[20] Sissakian, A. N. et al., [NICA Collaboration], *J. Phys. G* 2009, *36*, 064069 (1-5).

[21] Yang, J. C. et al., *Nucl. Instrum. Meth. B* 2013, *317*, 263-265.

[22] Gunion, J. F.; Brodsky, S. J.; Blankenbecler, R. *Phys. Rev. D* 1973, *8*, 287-312.

[23] Sivers, D. W.; Brodsky, S. J.; Blankenbecler, R. Phys. Rept. 1976, *23*, 1-121.

[24] Landshoff, P. V. *Phys. Rev. D* 1974, *10*, 1024-1030.

[25] Matveev, V. A.; Muradian, R. M.; Tavkhelidze, A. N. *Lett. Nuovo Cim.* 1973, *7*, 719-723.

[26] Brodsky, S. J.; Farrar, G. R. *Phys. Rev. Lett.* 1973, *31*, 1153-1156.

[27] Brodsky, S. J.; Farrar, G. R. *Phys. Rev. D* 1975, *11*, 1309-1330.

[28] Farrar, G. R.; Gottlieb, S. A.; Sivers, D. W.; Thomas, G. H. *Phys. Rev. D* 1979, *20*, 202-210.

[29] Brodsky, S. J.; Carlson, C. E.; Lipkin, H. J. *Phys. Rev. D* 1979, *20*, 2278-2289.

[30] Auer, I. P. et al., *Phys. Rev. Lett.* 1976, *37*, 1727-1730.

[31] Auer, I. P. et al., *Phys. Lett. B* 1977, *70*, 475-478.

[32] Crabb, D. G. et al., *Phys. Rev. Lett.* 1978, *41*, 1257-1259.

[33] Crosbie, E. A. et al., *Phys. Rev. D* 1981, *23*, 600-603.

[34] Hoyer, P. (2014). Bound states – from QED to QCD. [arXiv:1402.5005] [hep-ph], 1-45.

[35] Maltman, K.; Isgur, N. *Phys. Rev. D* 1984, *29*, 952-977.

[36] Harvey, M. *Nucl. Phys. A* 1981, *352*, 326-342.

[37] Obukhovsky, T. I.; Smirnov, Y. .F.; Chuvilsky, Y. M. *J. Phys. A* 1982, *15*, 7-23.

[38] Brodsky, S. J.; Ji, C. -R. *Phys. Rev. D*, 1986, *33*, 1951-1960.

[39] Ji, C. -R.; Brodsky, S. J. *Phys. Rev. D* 1986, *34*, 1460-1473.

[40] Kusainov, A. M.; Neudatchin, V. G.; Obukhovsky, I. T. *Phys. Rev. C* 1991, *44*, 2343-2357.

[41] Demorest, P.; Pennucci, T.; Ransom, S.; Roberts, M.; Hessels, J. *Nature* 2010, *467*, 1081-1083.

[42] Heiselberg, H.; Pandharipande, V. *Ann. Rev. Nucl. Part. Sci.* 2000, *50*, 481-524.

[43] Piasetzky, E.; Sargsian, M.; Frankfurt, L.; Strikman, M.; Watson, J. W. *Phys. Rev. Lett.* 2006, *97*, 162504 (1-4).

[44] Subedi, R. et al., *Science* 2008, *320*, 1476-1478.

[45] Frankfurt, L.; Sargsian, M.; Strikman, M. *Int. J. Mod. Phys. A* 2008, *23*, 2991-3055.

[46] Frankfurt, L.; Strikman, M. *Int. J. Mod. Phys. E* 2012, *21*, 1230002 (1-29).

[47] Ralston, J. P.; Pire, B. *Phys. Rev. Lett.* 1982, *49*, 1605-1608.

[48] Brodsky, S. J.; de Teramond, G. F. *Phys. Rev. Lett.* 1988, *60*, 1924-1927.

[49] HEPDATA, (2013). The Durham HEP database http://durpdg.dur.ac.uk/hepdata/.

[50] Allaby, J. V. et al., *Phys. Lett. B* 1968, *28*, 67-71.

[51] Akerlof, C. W. et al., *Phys. Rev.* 1967, *159*, 1138-1149.

[52] Perl, M. L. et al., *Phys. Rev. D* 1970, *1*, 1857-1882.

[53] Stone, J. L. et al., *Nucl. Phys. B* 1978, *143*, 1-31.

[54] Avilez, C.; G. Cocho, G.; Moreno, M. *Phys. Rev. D* 1981, *24*, 634-646.

[55] Lipkin, H. J. *Nature* 1986, *324*, 14-16.

[56] Bourrely, C.; Soffer, J. *Phys. Rev. D* 1987, *35*, 145-151.

[57] Preparata, G.; Soffer, J. *Phys. Lett. B* 1986, *180*, 281-284.

[58] Troshin, S. M.; Tyurin, N. E. *JETP Lett.* 1986, *44*, 149-152.

[59] Anselmino, M.; P. Kroll, P.; Pire, B. *Z. Phys. C* 1987, *36*, 89-103.

[60] Gordon, L. E.; Ramsey, G. P. *Phys. Rev. D* 1999, *59*, 074018 (1-12).

[61] Brodsky, S. J. In *Proceedings of the Thirteenth International Symposium on Multi-particle Dynamics*; Editors Kittel, W.; Metzeger, W.; Stergiou, A.; World Scientific, Singapore, 1981; pp 963-966.

[62] Mueller, A. H. In *Proceedings of the Seventeenth Rencontre de Moriond*; Editor Tran Thanh Van, J.; Editions Frontiers, Gif-sur-Yvette, France, 1982; pp 13-16.

[63] Frankfurt, L. L.; Miller, G. A.; Strikman, M. I. *Ann. Rev. Nucl. Part. Sci.* 1994, *44*, 501-560.

[64] Jain, P.; Pire, B.; Ralston, J. P. *Phys. Rept.* 1996, *271*, 67-179.

[65] Farrar, G. R.; Liu, H.; Frankfurt, L. L.; Strikman, M. I. *Phys. Rev. Lett.* 1988, *61*, 686-689.

[66] Kopeliovich, B. Z.; Nemchik, J.; Schmidt, I. *Phys. Rev. C* 2007, *76*, 015205 (1-8).

[67] Frankfurt, L.; Miller, G. A.; Strikman. M. *Phys. Lett. B* 1993, *304*, 1-7.

[68] El Fassi, L. et al., *Phys. Lett. B* 2012, *712*, 326-330.

[69] Clasie, B. et al., *Phys. Rev. Lett.* 2007, *99*, 242502 (1-5).

[70] Carroll, A. S. et al., *Phys. Rev. Lett.* 1988, *61*, 1698-1701.

[71] Mardor, I. et al., *Phys. Rev. Lett.* 1998, *81*, 5085-5088.

[72] Aclander, J. et al., *Phys. Rev. C* 2004, *70*, 015208 (1-21).

[73] Frankfurt, L.; Strikman, M.; Zhalov, M. *Phys. Lett. B* 2001, *503*, 73-80.

[74] Frankfurt, L. L.; Moniz, E. J.; Sargsian, M. M.; Strikman, M. I. *Phys. Rev. C* 1995, *51*, 3435-3444.

[75] Anselmino, M.; Predazzi, E.; Ekelin, S.; Fredriksson, S.; Lichtenberg, D. B. *Rev. Mod. Phys.* 1993, *65*, 1199-1234.

[76] Wilczek, F. In *From fields to strings*; Editors, Shifman, M. et al.; World Scientific Pub Co Inc, 2005; Vol. 1, pp 77-93.

[77] Perl, M. L. *High Energy Hadrons*; John Wiley & Sons Inc, 1974; pp 1-582.

[78] Bystricky, J.; Lehar, F.; Winternitz, P. *J. Phys.* 1978, *39*, 1-32.

[79] Bogolyubov, N. N.; Vladimorov, V. S.; Tavkhelidze, A. N. *Teor. Mat. Fiz.* 1972, *12*, 3-17.

[80] Bogolyubov, N. N.; Tavkhelidze, A. N.; Vladimirov, V. S. *Teor. Mat. Fiz.* 1972, *12*, 305-330.

[81] Dirac, P. A. M. *Rev. Mod. Phys.* 1949, *21*, 392-399.

[82] Weinberg, S. *Phys. Rev.* 1966, *150*, 1313-1318.

[83] S.J. Brodsky, S. J. In *Proceedings, Particle physics at the Fermi scale*; Editors, Qiu, J.; Pang, Y.; GRC Press, 1994; pp 271-374.

[84] Brodsky, S. J.; Pauli, H. -C.; Pinsky, S. S. *Phys. Rept.* 1998, *301*, 299-486.

[85] Brodsky, S. J. In *Lectures at the 58th Scottish University Summer School in Physics*; St. Andrews, Schotland; hep-ph/0412101; 2004; pp 1-66.

[86] Lepage, G. P.; Brodsky, S. J. *Phys. Rev. D* 1980, *22*, 2157-2198.

[87] Brodsky, S. J.; Hiller, J. R. *Phys. Rev. C* 1983, *28*, 475-482.

[88] White, C. et al., *Phys. Rev. D* 1994, *49*, 58-78.

[89] Frankfurt, L. L.; Miller, G. A.; Sargsian, M.M.; Strikman, M. I. *Phys. Rev. Lett.* 2000, *84*, 3045-3048.

[90] Sargsian, M. M. *Phys. Lett. B* 2004, *587*, 41-51.

[91] Brodsky, S. J.; Frankfurt, L.; Gilman, R. A.; Hiller, J. R.; Miller, G. A.; Piasetzky, E.; Sargsian, M.; Strikman, M. *Phys. Lett. B* 2004, *578*, 69-77.

[92] Sargsian, M. M. *AIP Conf. Proc.* 2008, *1056*, 287-294.

[93] Sargsian, M. M.; Granados, C. *Phys. Rev. C* 2009, *80*, 014612 (1-18).

[94] Granados, C. G.; Sargsian, M. M. *Phys. Rev. Lett.* 2009, *103*, 212001 (1-4).

[95] Granados, C. G.; Sargsian, M. M. *Phys. Rev. C* 2011 *83*, 054606 (1-9).

[96] Jaffe, R. L. *Phys. Rept.* 2005, *409*, 1-45.

[97] Selem. A,; Wilczek, F. In the proceedings of *Ringberg Workshop on New Trends in HERA Physics*; Editors Grindhammer, G.; World Scientific; hep-ph/0602128; 2005; pp 337-356.

[98] Frankfurt, L. L.; Strikman, M. I.; Day, D. B.; Sargsian, M. *Phys. Rev. C* 1993, *48*, 2451-1461.

[99] Arrington, J.; Higinbotham, D. W.; Rosner, G.; Sargsian, M. *Prog. Part. Nucl. Phys.* 2012, *67*, 898-938.

[100] Sargsian, M. M. *Phys. Rev. C* 2010, *82*, 014612 (1-12).

[101] Boeglin, W. U. et al., [Hall A Collaboration], *Phys. Rev. Lett.* 2011, *107*, 262501 (1-6).

[102] Sargsian, M. M. et al., *J. Phys. G* 2003, *29*, 1-52.

[103] White Paper: *The Science Driving the 12 GeV Upgrade of CEBAF* Jefferson Lab, Newport News, VA, 2000; pp 1-278.

In: NN and 3N Interactions
Editors: Leonid Blokhintsev and Igor Strakovsky
ISBN: 978-1-63321-053-0


*Chapter 7*

# SHORT-RANGE NN INTERACTIONS AND Δ-ISOBAR EXCITATION IN THE $pd \to \{pp\}_s n$ REACTION WITH A FAST PP PAIR

***Yu.N. Uzikov*** *
Joint Institute for Nuclear Researches,
Dubna, Moscow Region, Russia

**Abstract**

The deuteron breakup reaction $pd \to \{pp\}_s n$, where $\{pp\}_s$ is a fast proton pair in forward direction at small excitation energy $E_{pp} \leq 3$ MeV, is discussed at beam energies in the GeV region. The spin-singlet *pp*-state $^1S_0$ dominates at these conditions. Dynamics of this reaction is considered on the basis of the main mechanisms of the backward elastic *pd*-scattering. High sensitivity of the $pd \to \{pp\}_s n$ cross section to the short-range NN-interaction and to the three-body forces of the $N - \Delta - N$-type is emphasized.

**PACS** 24.70.+s; 24.50.+g; 21.45.+v

**Keywords:** Nuclear Forces, Short-Range Correlations, Nucleon Isobars

## 1. Introduction

The structure of the lightest nuclei at short $NN$ distances, and specifically in the nucleon overlap region $r_{NN} \leq 0.5$ fm, is related to fundamental problems of the physics of strong interactions. It is expected that these distances correspond to the transition region from meson-nucleon to quark-gluon degrees of freedom. One of the signals of this transition could be a failure of the conventional picture, based on meson-exchange models of $NN$-interaction. Electromagnetic probes are generally considered the best instrument for these investigations. Most of our knowledge about short-range structure of the deuteron was obtained from elastic electron-deuteron scattering [1, 2].

*E-mail address: uzikov@jinr.ru

A consistent picture of different electromagnetic process on the deuteron $ed \to ed$ [3], $d(e, ep')n$ [4], $\gamma d \to pn$ [5] at high transferred momenta $Q > 1$ GeV/c has not yet been established within the conventional approach mainly due to the unknown strength of the *meson-exchange currents* (MEC). Nevertheless, it is worth to mention two remarkable features, related to the phenomenological repulsive core in the NN-interaction. (i) One of them is the node in the deuteron charge form factor $G_C(Q^2)$ observed at transferred 3-momentum $Q \approx 0.8$ GeV/c [6]. In the impulse approximation this node is directly related to the node in the s-wave component of the deuteron wave function in momentum space $u(q)$, which occurs in all realistic models of the NN($^3S_1$)-interaction [7, 8, 9] at internal momentum of the nucleon in the deuteron $q = q_c \approx 0.4$ GeV/c. Due to the zero isospin of the deuteron only isoscalar MEC contribute to the elastic electron-deuteron scattering. The node observed in the form factor $G_C(Q^2)$ implies that the contribution of the isoscalar MEC is very small. (ii) Another important feature is found in the deuteron electro-disintegration near the threshold $ed \to e\{pn\}_s$ at low excitation energy of the final $pn$ pair, $E_{pn} = 0 - 3$ MeV, and large transferred momentum. In this case the isosinglet and isotriplet states of the final pn-pair are allowed. In the impulse approximation the isovector M1-transition $(^3S_1 - {}^3D_1) \to {}^1S_0$ predicts a dip in the breakup cross section at $Q \approx 2q_c \approx 0.7$ GeV/c, which is related to the repulsive core in the $pn(^1S_0)$-interaction (for review see Ref.[10]. However, the predicted node was not observed in the data [11]. This fact was interpreted as a first direct indication of the existence of the (isovector) MEC, which contribution is supposed to be significant [10].

In this connection, one should note that in proton-deuteron scattering a similar feature occurs. So, considering energy dependence of the cross section of the reaction

$$p + d \to \{pp\}_s + n \tag{1}$$

at backward scattered neutron at the cms angle $\theta_n^{cm} = 180°$ and forward emitted pp pair with small excitation energy $E_{pp} = 0 - 3$ MeV, that provides dominance of the spin-singlet $^1S_0$ state, one could also expect a dip at beam energy $T \sim 0.8$ GeV predicted within a certain model of *pd*- backward scattering [12, 13]. However, this dip was not observed in the experiment [14, 15]. Below I will discuss what one can learn from this "non-observation".

Actually independent information on the short-range structure of nuclei can be obtained from hadron-nuclei interactions at large transferred momenta $Q$. Existing data on deep-inelastic reactions $h + A \to p + X$ in the cumulative kinematic region (i.e., in the region forbidden for $hN$-scattering at free nucleon $N$ being at rest) with final proton emitted into backward hemisphere are treated by some authors either in terms of interactions with density fluctuations of cold nuclear matter [16, 17], or short-range $NN$- correlations in nuclei [18], whose local properties are considered to be largely independent of the type nucleus and the used probe $h$. However, quantitative theory of this qualitative picture does not exist because of complicated structure of the involved nuclei and initial and final-state interactions in the nuclear medium. It is therefore important to study elementary processes in few-nucleon systems that could provide access to the short-range $NN$-interaction under more convenient conditions providing more transparent theoretical interpretation. The simplest hadron reactions, which can give information on short-range structure of the deuteron are proton-deuteron backward elastic scattering, $pd \to dp$, deuteron disintegration in inclusive $dp \to p(0°)X$ and exclusive $pd \to ppn$ kinematics. However, investigation of these processes in the GeV region with high transferred momentum in Refs.[19, 20] ($pd \to dp$), and [21, 22, 23] ($dp \to p(0°)X$) turned out to be inconclusive with respect to the short-range interaction. A clear manifestation of this failure is a longstanding problem connected with understanding the mechanism of proton-deuteron backward elastic scattering at energies above 0.5 GeV called as $T_{20}$-problem (see Ref. [24] and references therein).

## 2. $T_{20}$-Puzzle

This problem can be formulated as follows [24]. Except in the Δ–isobar region of $0.4 - 0.6$ GeV, the unpolarized differential cross section $d\sigma/d\Omega(pd \to dp)_{\theta_{cm}=180°}$ can be explained qualitatively within the impulse approximation (IA) up to large nucleon momenta in the deuteron $q \approx 1$ GeV/c, whereas the experimental values of the tensor analyzing power $T_{20}$ are in strong contradiction to the IA calculations already for $q > 0.3$ GeV/c [25, 19, 20]. Here IA means the one–nucleon–exchange (ONE) mechanism of Fig. 1a which, if it dominated the unpolarized cross section, would allow one to measure directly the high–momentum components in the deuteron wave functions.

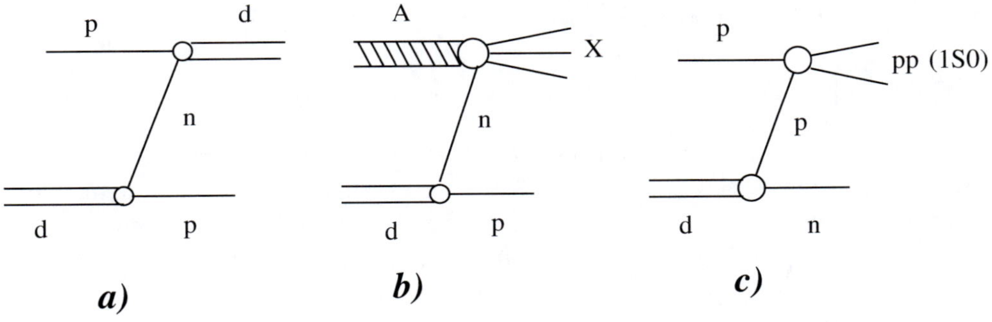


Figure 1. The one nucleon exchange (ONE) mechanisms of the reactions (a) $pd \to dp$, (b) $dA \to p(0°)X$, and (c) $pd \to \{pp\}_s n$.

A very similar problem arises in the analysis of the inclusive disintegration of the deuteron on nuclear targets, $dA \to p(0°)X$, when the ONE mechanism of Fig. 1b is used to describe the process [22, 26, 23]. In contrast, the tensor polarization $t_{20}$ of the recoil deuteron in elastic electron–deuteron scattering follows very well the IA predictions [1] up to very high transferred momenta $Q = 1.3$ GeV/c, *i.e.* at $q \sim Q/2 = 0.65$ GeV/c, if realistic phenomenological $NN$ potentials [27, 28, 9] are used to describe the deuteron using the Schrödinger equation. Corrections from meson–exchange currents are sizable, but do not change the picture qualitatively [29]. We must conclude that in exclusive and inclusive *pd*–collisions at high transferred momenta we are dealing, not only with the short-range structure of the deuteron, but also with the specific dynamics of the *pd*–interaction and that these dynamics are entirely different from those in the $ed \to ed$ process.

The above contradictions, referred to as the $T_{20}$-puzzle, can be ascribed, in part, to contributions from the excitation of nucleon isobars ($\Delta$, $N^*$) in the intermediate state, which were neglected within the IA analysis [25, 19, 22, 26, 23]. For example, the Δ–mechanism seems to dominate the large angle unpolarized $pd \to dp$ cross section in the 0.4–0.6 GeV interval [30, 31, 32]. However, the spin structure of the three–body forces related to the $\Delta$-isobar is far from being well established [33]. This therefore leads to ambiguities in any explanation of $T_{20}$ when the $\Delta$–isobar is included in the transition amplitude [30, 31, 32]. It was suggested that, in order to clarify the role of the $\Delta-$isobar, the $pd \to \{pp\}_s n$ reaction (1) should be studied [12, 34, 13]. Due to isospin invariance, the $\Delta$-mechanism is suppressed by a factor of nine in the $pd \to \{pp\}_s n$ cross section as compared to that of $pd \to dp$ (see below section 3.3.), whereas the ONE mechanism does not suffer a similar suppression [35]. Therefore, the comparison of the two reactions might allow one to get a clearer picture of the relative importance of the ONE and $\Delta$-contributions.

Here the experimental data on the reaction (1) and corresponding theoretical interpretation are reviewed.

# 3. The One+Δ+SS Model

When considering the main mechanisms of the reaction (1) (see Fig.2 ) we proceed from existing theoretical models of the $pd \to dp$ process, because the kinematics of these two processes at small $E_{pp}$ is almost the same.

## 3.1. Lessons from the $pd \to dp$ Backward Elastic Scattering

The ONE+Δ+SS model of the backward elastic pd-scattering suggested for the first time in Ref. [30] and developed later in Refs. [36], [37], seems as the most complete and consequent one. Indeed, this model is based on the Watson multiple scattering series and therefore there is no problem in this approach with relative phases of the different mechanisms. Since the ONE amplitude is very sensitive to the high momentum components of the $NN$-wave functions, the model takes into account relativistic effects for this mechanism. The parameters of the $\Delta$-amplitude of the $pd \to dp$ process were fixed by fitting the experimental data on the reaction $pp \to pn\pi^+$ just in the $\Delta$-region within the $\pi + \rho$-exchange model of the process $NN \rightleftharpoons \Delta N$ [38]. At low energies 0.1-0.3 GeV, the sum of two diagrams, ONE+SS, describes qualitatively the measured unpolarized cross section and some spin observables (see Ref.[35]). The $\Delta$- contribution is rather small in this region, nevertheless, its inclusion improves an agreement with the cross section at scattering angles $\sim 130^\circ - 150^\circ$. This fact displays a contribution of the three-nucleon forces of the N-Δ-N type [35]. At higher energies, 0.5 - 1.0 GeV, the $\Delta$ mechanism dominates in the unpolarized cross section at $\theta^{cm} = 180^\circ$ [32]. However, just in the $\Delta$-region the ONE+SS+Δ-model fails in description of the tensor analyzing power $T_{20}$. As was explained in Introduction, one can expect that the suppression of the $\Delta-$ and $N^*$-contributions in the breakup reaction (1) by the isospin conservation can improve the predictive strength of this model for the spin observables at $T_0 > 0.3$ GeV in comparison with the $pd \to dp$ process. The model ONE+Δ+SS allows one to include rescatterings in the initial and final states and new mechanisms, like $N^*$-exchanges. From this point of view this model can be used as a first step in consideration of the dynamics of the reaction (1).

The transition amplitude of the reaction (1) can be obtained from the corresponding formalism of the backward elastic pd-scattering developed in the framework of the

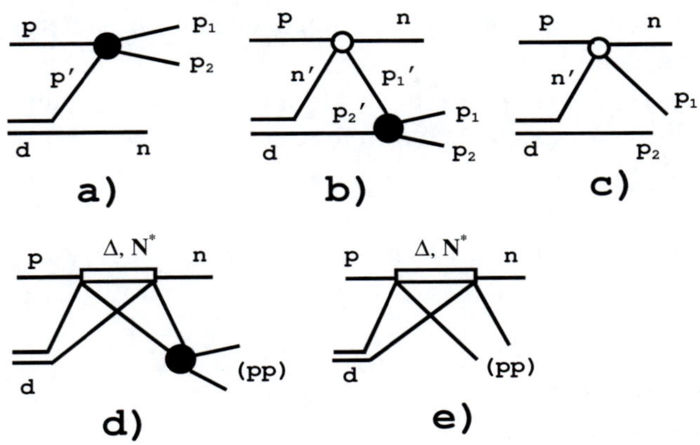


Figure 2. Mechanisms of the reaction $pd \to \{pp\}_s n$: one-nucleon exchange (ONE) (a), single scattering (SS) (b-c), Δ-isobar excitation (Δ) (d-e).

ONE+Δ+SS model in Ref.[30]. For this one should replace the final deuteron wave function $\varphi_d$ by the scattering state of the NN-system produced in the final state of the reaction $pd \to (NN)N$. We assume that the S-state dominates in the final NN-system. Therefore, neglecting the D-component of the final deuteron in the formulas for the $pd \to dp$ process from Ref.[30] one should replace the S-component according to

$$u(r) \to \sqrt{4\pi m}\,\psi_k^{(-)}(r), \tag{2}$$

where $\psi_k^{(-)}(r)$ is the S-component of the NN-scattering wave function at low relative momentum $k$ normalized as

$$\psi_k^{(-)}(r) \to \frac{\sin(kr+\delta)}{kr}, \tag{3}$$

here $\delta$ is the $^1S_0$ phase shift.

According to very common properties of the transition amplitudes of hadron-nucleus $hA$-interaction at high transferred momenta the amplitude with the deuteron formation in process $hA \to dX$ is related to the triplet $(np)_t$-pair production amplitude in the same kinematics at low energy of relative motion $E_{np} \sim 0$ in the np-pair [39, 40]. In particularly, the amplitude of the backward elastic pd-scattering is connected to the triplet pair production in the reaction $pd \to (pn)_t p$ by the following relation [40]

$$M(p\,d \to (pn)_t\,p) = -\exp\left[-i\delta_t\right] f(k) M(pd \to dp). \tag{4}$$

Here $\delta_t$ is the triplet phase shift in the pn scattering in the $^3S_1$-state. The relation (4) is based on analytical continuation of the pn-scattering S- wave function $\psi_k^{(-)}(r)$ in the triplet state over the variable $k$ to the deuteron pole. The function $f(k)$ in Eq.(4) has the following form [40] $f(k) = \sqrt{2\pi m/\alpha_t(k^2+\alpha_t^2)}$, where $k$ is the relative momentum of the final NN-pair connected to the relative energy $E$ as $E = k^2/m$, $\alpha_t = \sqrt{m\varepsilon} = 0.232 fm^{-1}$, $m$ is the average nucleon mass, $\varepsilon$ is the deuteron binding energy. As was shown in Ref.[34], the relation Eq. (4) is valid also when the D-components of the deuteron and np-scattering wave functions are taken into account.

Owing to Eq. (4) the deuteron breakup reaction with the triplet pair formation does not contain any new information as compared to the backward elastic pd-scattering. Only the channel with the singlet pair at low excitation energy $E_{NN} \sim 1-3$ MeV is an essentially new one. We will discuss below the ratio of the break-up amplitudes for the singlet $(np)_s$- and triplet $(pn)_t$-pairs for different mechanisms in the same kinematics.

### 3.2. The One Amplitude

According to general rules [41], the $p+n \to d$ vertex of the ONE diagram of the $pd \to dp$ for the S-wave of the deuteron wave function is

$$4m\sqrt{m} < \varphi_d|V_t|\mathbf{q} >= -4m\sqrt{m}(\varepsilon + \frac{q^2}{m})u(q)\frac{1}{\sqrt{4\pi}} = \\ = \sqrt{4\pi\, m}4m\int_0^\infty j_0(qr)V(r)u(r)r^2dr, \quad (5)$$

where $m$ is the nucleon mass, $\varepsilon$ is the deuteron binding energy, $V_t$ is the $NN$ interaction in the spin-triplet state $^3S_1$, $j_0(x)$ is the Bessel function of the zero order. Similarly, for the $p+d \to (pn)_s + p$ process the $p+n \to (np)_s$ vertex has the form

$$4m^2 t_s(q,k) = 4m^2 < \Psi_{\mathbf{k}}^{(-)}|V|\mathbf{q} >= -4m^2\left\{4\pi\int_0^\infty j_0(qr)V_s(r)\psi_k^{(-)*}(r)r^2dr\right\}, \quad (6)$$

where $V_s$ is the NN interaction in the spin-singlet $^1S_0$-state. One can see that Eq.(6) comes from Eq.(5) under the replacement (2). Since the $n+p \to d$ vertex has to be replaced by the vertex $p+n \to (pn)_s$ in all mechanisms under discussion, the replacement (2) is valid for the $\Delta$ and SS-mechanisms also. The t-matrix $t_t(q,k)$ of the triplet half-off-shell scattering in the $^3S_1$-state is defined as

$$t_t(q,k) = -4\pi\int_0^\infty j_0(qr)V_t(r)\psi_t^{(-)*}(r)r^2dr = \\ = \frac{1}{\sqrt{4\pi m}}f(k)\, e^{-i\delta_t}(\varepsilon + q^2/m)\, u(q); \quad (7)$$

here $\psi_t^{(-)}(r)$ is related to the bound state wave function $u(r)$ by the factor $f(k)$ according to Ref.[40]; $\delta_t$ is the triplet phase shift and $V_t$ is the triplet interaction potential.

The upper vertex of the ONE diagram of reaction (1) (Fig.2 ($a$)) with the final diproton pair $\{pp\}_s$ contains the half-off-shell amplitude of $pp$-scattering in the $^1S_0$ state, $t(q,k)$, which includes the Coulomb interaction. This T-matrix has the form of

$$t(q,k) = -4\pi\int_0^\infty \frac{F_l(qr)}{qr}V_s(r)\psi_s^{(-)*}(r)r^2dr, \quad (8)$$

where $F_l(z)$ is the Coulomb function regular at the origin, $\psi_s^{(-)}$ is the pp($^1S_0$)- scattering wave function obtained from the solution of the Schrödinger equation with the $pp(^1S_0)$ interaction potential $V_s(r)$ plus Coulomb term. (For the pn-system this function is normalized according to Eq.(3)).

The behaviour of the half-off-shell matrix $t(q,k)$ versus $q$ is shown in Fig.3 for the pp-scattering at different energies $E_{NN} = k^2/m$. The first node of $t(q,k)$ at $q = 0$ is caused by the Coulomb repulsion, whereas the second node at $q \sim 0.4$ GeV/c arises due to the short-range repulsion in the $NN$ potential of strong interaction. A similar node, available in the deuteron S-wave function in the momentum space $u(q)$ at $q \sim 0.4$ GeV/c, can be connected indirectly to the null of the measured deuteron charge form factor $G_C(Q)$ at the transferred momentum $Q \sim 4.5$ fm$^{-1}$ [42, 43]. One can see that the absolute value of $t(q,k)$ increases with decreasing of the relative energy $E_{pp}$. This is caused by the position of singlet pole near the threshold $\sim 0.07$ MeV. On the other hand, the node point caused by the strong short-range repulsion does not depend on $E_{pp}$ and is located at $q \approx 0.4$ GeV/c. The node of $u(q)$ was not yet observed directly in any reactions with the deuteron due to large contribution of the deuteron D-state. In this connection one should note that an important feature of the reaction (1) is a possibility to display the node of the amplitude $t(q,k)$ straightforwardly in the cross section. One can see it from the following formula for the cross section of the reaction (1) derived for the ONE mechanism (Fig. 2 (*a*))

$$\frac{d^5\sigma}{dp_1 d\Omega_1 d\Omega_2} = K\left[u^2(q) + w^2(q)\right] |t(q',k)|^2, \tag{9}$$

where $w(q)$ is the D-wave of the deuteron state, $K$ is the kinematic factor; the indices 1 and 2 refer to the final protons which are detected in the forward direction. Using relativistic kinematics for the ONE amplitude [34], one can find for the neutron scattering angle $\theta_{cm} = 180°$ that the node of $t(q',k)$ at $q' \sim 0.4$ GeV/c corresponds to the energy $T_p \sim 0.7 - 0.8$ GeV depending on the type of the NN interaction potential.

The triplet $t_t(q,k)$ and singlet $t_s(q,k)$ amplitudes as functions of the kinetic energy of the initial proton in the reaction (1) are shown in Fig.4, *a*. One can see from this picture that these functions are very similar both in the shape and absolute value. A small enhancement factor for the singlet pair production is expected in the ratio $R_X(ONE) = t(np)_s/t_t(np) \sim 1.2 - 1.5$ at $T_p > 1$GeV.

### 3.2.1. Higher partial waves of the $pp$-pair

The ONE mechanism of the reaction (1) was investigated in detail in Ref. [34]. The contribution of the tensor forces in the upper vertex of the diagram in Fig.2(a) corresponding to the half-off-shell NN-scattering was taken into account in Ref. [34] at the energy of the relative motion in the final NN-pair $E_{NN} = 0 - 20$ MeV. Relativistic effects were taken into account in Refs. [30, 36, 34] on the basis of the relativistic quantum mechanics for the system with fixed number of particles [44].

According to [34], where a general formalism for this mechanism was developed in the plane wave approximation, the spin-averaged squared matrix element of this reaction can be written as

$$\overline{|M_{fi}|^2} = \frac{E_d(E_p + E_n)\,\varepsilon_p(q)}{16\pi E_p^2}\left[u^2(q) + w^2(q)\right] F(\mathbf{q}', \mathbf{k}). \tag{10}$$

Here $u(q)$ and $w(q)$ are $S$- and $D$-wave components of the deuteron wave function, $E_d$, $E_p$ and $E_n$ are the total energies of the deuteron, intermediate proton and final neutron in the

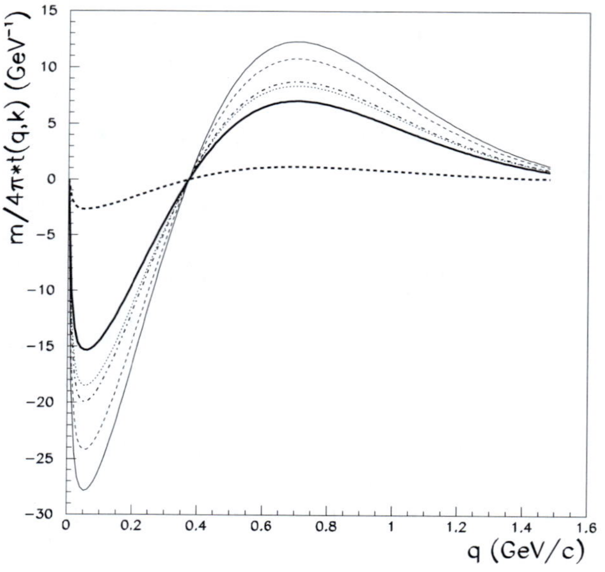


Figure 3. The half-off-shell $pp(^1S_0)$-scattering amplitude $(m/4\pi)t(q,k)$ from Ref. [13] for the RSC NN potential with Coulomb interaction included at different energies $E_{pp}$: 0.01MeV (dashed thick), 0.1 MeV (dash-dotted), 0.5 MeV (full thin), 1.0 MeV (dashed), 2.0 MeV (dotted), 3 MeV (full thick). The on-shell amplitude $t(k,k)$ is related to the Coulomb-nuclear phase shift $\delta$ in the $^1S_0$ state as $m/4\pi t(k,k) = -\frac{1}{k}\exp[i\delta]\sin\delta$. The complex factor $\exp[i\delta]$ is omitted from $t(q,k)$ shown in this picture.

cms of the reaction; $\varepsilon(q) = \sqrt{m^2+q^2}$; $q$, $q'$, $k$ are the internal momentum in the deuteron, off-shell relative momentum in the *pp*-pair, and its on-shell momentum, respectively. The internal momenta $q$ and $q'$ are related to the momenta of the initial and final particles according to relativistic kinematics. The function $F(\mathbf{q}',\mathbf{k})$ can be written via $T$-matrix of the elastic *pp*-scattering $t_{pp}(\mathbf{q}',\mathbf{k})$ in the following form

$$F(\mathbf{q}',\mathbf{k}) = \sum_{\substack{\sigma_1'\sigma_2'\\ \sigma_1\sigma_2}} \left|T_{NN\,\sigma_1\,\sigma_2}^{\,\sigma_1'\,\sigma_2'}(\mathbf{q}',\mathbf{k})\right|^2 = \sum_{JLL'\widetilde{J}\widetilde{L}\widetilde{L}'Sl} \mathcal{N}_{pp}^2(L,S)\, t_{LL'}^{JS}(q',k)\left(t_{\widetilde{L}\widetilde{L}'}^{\widetilde{J}S}(q',k)\right)^* \times$$

$$\times (L0\,l0|\widetilde{L}0\,(L'0\,l0|\widetilde{L}'0(2J+1)(2\widetilde{J}+1)(2l+1)\times$$

$$\times\sqrt{(2L+1)(2L'+1)}\left\{\begin{array}{ccc}\widetilde{L} & L & l\\ J & \widetilde{J} & S\end{array}\right\}\left\{\begin{array}{ccc}\widetilde{L}' & L' & l\\ J & \widetilde{J} & S\end{array}\right\} P_l(\mathbf{q}'\mathbf{k}/q'k), \qquad (11)$$

where $\mathcal{N}_{pp}(L,S) = 1 + (-1)^{L+S}$ is the combinatorial factor for two protons; the braces stand for usual notation of $6j$-symbols; $(LMlm|JM)$ is the Clebsh-Gordan coefficient and $P_l$ is the Legendre polynomial of the $l$-th order; $t_{LL'}^{JS}(q',k)$ is the partial $t$-matrix of the *pp*-scattering for the transition $L \to L'$ in the state with the spin $S$ and total angular momentum $J$. Note that the function (11) at $q' = k$ determines the differential cross section of *pp*-scattering.

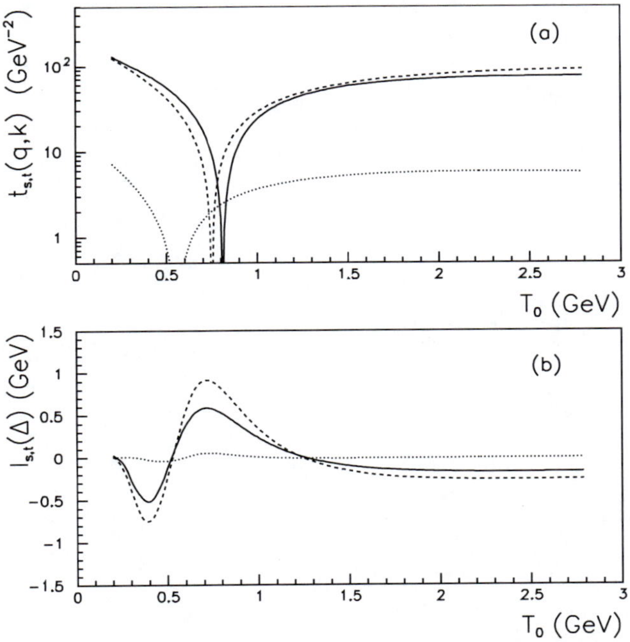


Figure 4. (*a*) The half-off-shell t-matrixes defined by Eqs. (7) and (8) for the pn-scattering in the $^1S_0$- (dashed), $^3P_0$ (dotted) and $^3S_1$- states (full) at $E_{pp} = 3$ MeV. (*b*) The integrals for the Δ mechanism defined by Eqs. (14) and (15) for the transitons to the singlet $^1S_0$ (dashed), $^3P_0$ (dotted) and triplet $^3S_1$ (full) final pn-states.

The results of numerical calculations of the differential cross section of the reaction (1) for different excitation energy $E_{pp} = 0 - 20$ MeV with Paris NN-interaction potential are shown in Fig. 5. One can see that at $E_{pp} > 10$ MeV the expected $^1S_0$ dip of the ONE mechanism almost disappears due to contribution of higher partial waves of the internal motion in the pp-pair. On the other hand, at $E_{pp} < 5$ MeV the ONE dip is well pronounced in the unpolarized cross section for the ONE mechanism.

### 3.2.2. The ONE mechanism with rescatterings

For the ONE-mechanism rescatterings in the initial and final states are taken into account in the line of works [45, 37]. We use an eikonal version of the distorted wave Born approximation (DWBA) method successfully applied in Refs. [45] to the backward elastic $p^3He$-scattering at $T_0 \geq 1$ GeV in the framework of np-pair exchange. The same method was applied to the pd- backward elastic scattering in Ref. [37] within the ONE-mechanism. The results obtained in [37] are similar to those found in Ref. [46] in the eikonal approximation written in r-space. The SS- and Δ-mechanisms are considered in the plane wave approximation. Indeed, the Δ mechanism corresponds to two-loop diagram which is complicated itself, and the SS-mechanism plays a minor role within the ONE+Δ+SS mechanism. In addition, the Δ and SS-mechanisms do not reveal a such peculiarity as the ONE-dip.

To take into account rescatterings in the initial and final states one can repeat the arguments given in Ref. [45, 47], in which the two-nucleon transfer mechanism was investigated

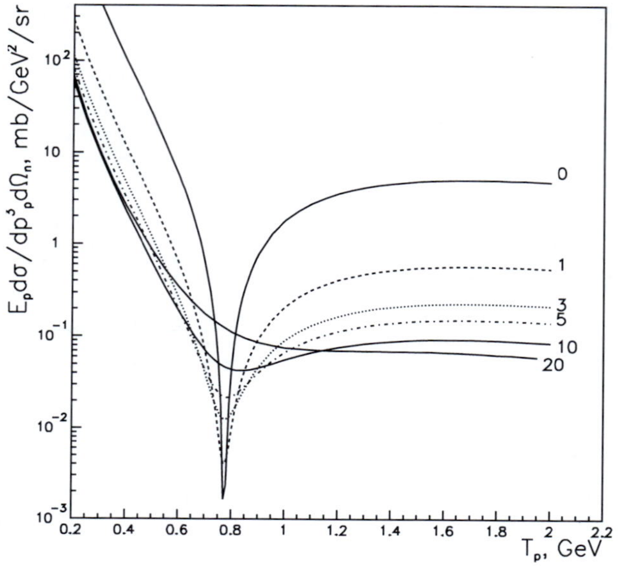


Figure 5. The cms differential cross section of the reaction $p + d \to p(0°) + p + n(180°)$ as a function of the initial proton kinetic energy in the laboratory frame, calculated in the framework of the ONE mechanism with $S$-, $P$-, $D$-, $F$- and $G$-waves in the state of the forward going *pp*-pair taken into account. The numbers at the curves denote the energy of relative motion of nucleons in the pair, $E_{pp}$, in MeV. The Paris NN-interaction potential is used in this calculations [34].

in the process $p+{}^3\mathrm{He}\to{}^3\mathrm{He}+p$. The amplitude of the off-shell np-scattering on the one proton of the diproton, $f_{np}(\mathbf{q})$, appears in the final state of the reaction (1). Here we consider the np-scattering off the diproton by analogy with pp-forward rescattering off the deuteron in the elastic process $pd \to dp$ [37]. The forward pd-scattering amplitude $F_{pd}(\mathbf{q})$ appears in the initial state of the reaction (1). Since at energies $\sim 1$ GeV the cross section of the diffraction scattering is peaked to the forward direction, the contribution of the low transferred momenta $\mathbf{q}$ is the most important. Taking into account that in the limit $\mathbf{q} \to 0$ the Glauber amplitude vanishes for inelastic transitions due to orthogonality of the different internal states of the target we take into account only elastic rescatterings. Therefore we assume that the internal energy of the diproton $E_{pp} = k^2/m$ is not changed during this rescattering. As a result one obtains the following expression for the ONE-amplitude of the reaction (1)

$$\begin{aligned} M_{DWBA} &= M_B(\mathbf{d}_{pp}, k, \mathbf{p}_f; \mathbf{d}_i, \mathbf{p}_i) + \\ &+ \frac{i}{4\pi p_i} \int d^2q\, F_{pd}(\mathbf{q}) M_B(\mathbf{d}_{pp}, k, \mathbf{p}_f; \mathbf{d}_i + \mathbf{q}, \mathbf{p}_i - \mathbf{q}) + \\ &+ \frac{i}{4\pi p_f} \int d^2q'\, f_{np}(\mathbf{q}') M_B(\mathbf{d}_{pp} - \mathbf{q}', k, \mathbf{p}_f + \mathbf{q}'; \mathbf{d}_i, \mathbf{p}_i) - \\ &- \frac{1}{(4\pi)^2 p_f p_i} \int\int d^2q d^2q'\, F_{pd}(\mathbf{q}) f_{np}(\mathbf{q}') M_B(\mathbf{d}_{pp} - \mathbf{q}', k, \mathbf{p}_f + \mathbf{q}; \mathbf{d}_i + \mathbf{q}, \mathbf{p}_i - \mathbf{q}). \quad (12) \end{aligned}$$

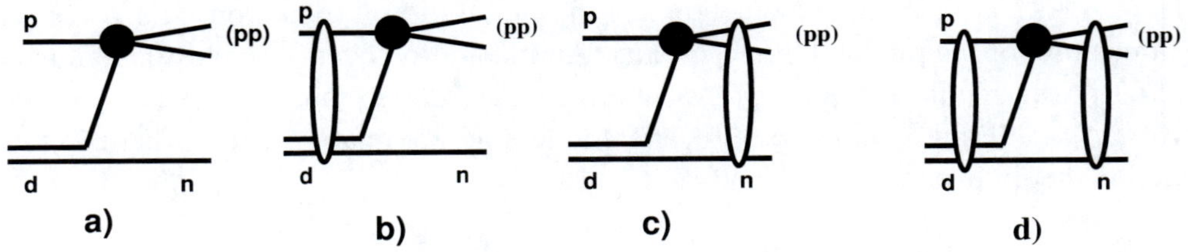


Figure 6. Rescatterings for the ONE mechanism (a): initial state interaction (b), final state interaction (c), initial and final state interactions (d).

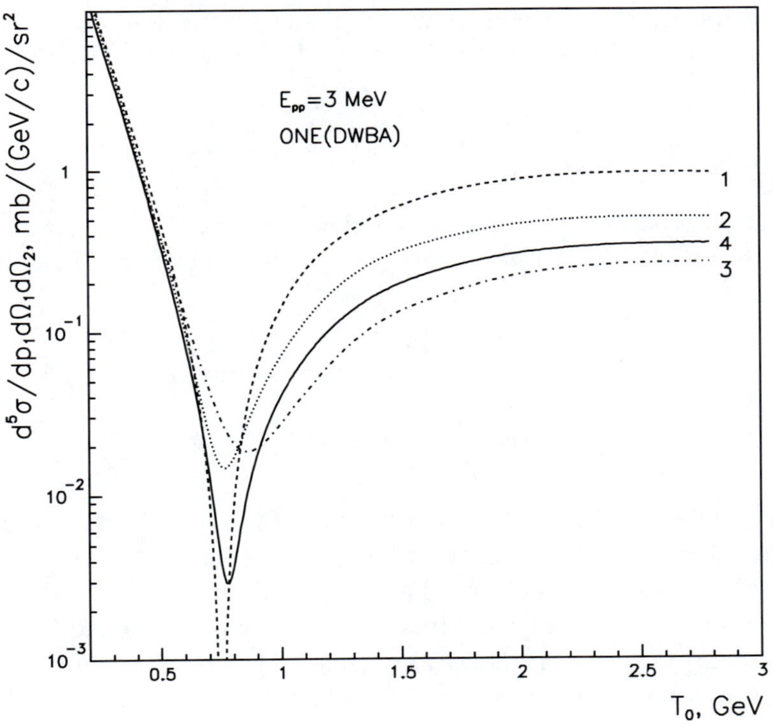


Figure 7. The laboratory cross section for the reaction $p+d \rightarrow \{pp\}_s + n$ at $E_{pp} = 3$ MeV within the ONE(DWBA) mechanism calculated with accounting for different terms from Eq.(12): 1 - the first, 2 – the first plus the second, 3 – the sum of the first, second and third, 4 – the sum of all four terms.

Here $\mathbf{p}_i(\mathbf{d}_i)$ is the c.m.s. momentum of the initial proton (deuteron) and $\mathbf{p}_f$ $(\mathbf{d}_{pp})$ is the momentum of the final neutron (pp-pair). An integration over 2-dimensional momenta $\mathbf{q}$ and $\mathbf{q}'$ transferred in $pd-$ and $pp-$rescatterings, respectively, are performed in Eq. (12). For the $pN$-scattering amplitude, $f_{pN}$, the spinless approximation is used here from the diffraction theory in a standard form [48]

$$f_{pN}(q) = \frac{k\,\sigma_{pN}}{4\pi}(i+\alpha_{pN})\exp\left(-\frac{1}{2}\beta_{pN}\,q^2\right), \tag{13}$$

where $k$ is the nucleon momentum in c.m.s. system $p+N$, $\sigma_{pN}$ is the total cross section of the $pN$-interaction; $\alpha_{pN}, \beta_{pN}$ are empirical parameters of the $pN$-scattering; $F_{pd}(q)$ is the forward $pd$-scattering amplitude calculated on the basis of the multi-step scattering theory of Glauber-Sitenko [48]. In Eq. (12) $M_B$ is the plane wave Born amplitude corresponding to the ONE mechanism.

The results of numerical calculations for experimental conditions of the COSY experiment [14, 15] are shown in Fig. 7. One can see that ISI- and FSI- rescatterings decrease the absolute value of the ONE-cross section and could shift the position of the ONE-node as compared to the Born approximation. However, in total when all four terms are taken into account the ONE-dip is still well pronounced and is located almost at the same position as for the Born term.

### 3.2.3. The $\Delta$-isobar mechanism

The following spatial matrix elements can be obtained in the S-wave approximation for the $\Delta$-mechanism of the reaction $pd \to \{pn\}_{s,t}p$ for the singlet $\{pn\}_s$ and triplet $\{pn\}_t$ NN-pairs formation [12]

$$I_s(\Delta) = \sqrt{4\pi m}\int_0^\infty dr\exp(i\mathcal{Z}r)u(r)\psi_s(r), \tag{14}$$

$$I_t(\Delta) = f(k)\int_0^\infty dr\exp(i\mathcal{Z}r)u(r)u(r). \tag{15}$$

Here $u(r)$ (or $w(r)$) is the deuteron S (D)-wave function in configuration space. The relation (4) was used here to get transition matrix element for the reaction $pd \to \{pn\}_t p$ from that for the $pd \to dp$ (the common phase $\exp(i\delta_t)$ is dropped in Eq. (15). The similar equation can be written for the D-wave of the initial deuteron, $w$, by the following substitution $u(r) \to w(r)$ in Eqs. (14), (15). The variable $\mathcal{Z}$ in Eq.(14) and (15) is expressed in Refs. [30, 12] through kinematical variables of the process (1), masses of the nucleon $m$ and the deuteron $m_d$, the mass $M_\Delta$ and width $\Gamma_\Delta$ of the $\Delta$- isobar, respectively. The detailed formula for $\mathcal{Z} \equiv \mathcal{Z}(M_\Delta, \Gamma_\Delta)$ is the following

$$\mathcal{Z}(M_\Delta,\Gamma_\Delta) = \frac{2m}{\eta(p,Q)}\left(m^2 + \frac{Q^2}{4} - M_\Delta^2 + iM_\Delta\Gamma_\Delta\right), \tag{16}$$

where

$$\eta(p,Q) = \sqrt{4Xm_d^2 + YQ^2},\ Q^2 = 4p^2\sin(\theta/2), \tag{17}$$

$$X = \frac{s\,p^2\cos^2(\theta/2)}{m_d^2 + p^2\sin^2(\theta/2)},\; Y = m^2 + X + \frac{Q^2}{4}. \tag{18}$$

Here $s$ is the invariant mass squared of the $p+d$ system, $\theta$ is the c.m.s. scattering angle of the neutron in the reaction (1) and $p = p_i \approx p_f$ is the c.m.s. momentum of the initial proton or the final neutron, **Q** is the 3-momentum transferred from the incident proton to the final *pp*-system.

For the $N^*$-mechanism the corresponding formula for the spatial integral coincides with Eqs. (14), (15) except the value $\mathcal{Z}$ includes the mass $M_{N^*}$ and width $\Gamma_{N^*}$ of the $N^*$-isobar.

#### 3.2.4. The SS-mechanism

The transition form factors of the single scattering mechanism for the reaction $pd \to \{pn\}_{s,t}p$ can be written as

$$I_s(SS) = \sqrt{4\pi m}\int_0^\infty dr\, j_0(Qr/2)u(r)\psi_s(r)\,r^2,$$
$$I_t(SS) = f(k)\int_0^\infty dr\, j_0(Qr/2)u(r)\,u(r)r^2. \tag{19}$$

Here $Q$ is the transferred momentum. For the D-wave component of the initial deuteron one should make the following replacements in Eqs.(19): $j_0 \to j_2$ and $u(r) \to w(r)$.

We can show that the singlet-to-triplet ratio of spatial factors for the $\Delta$ and SS-mechanisms are similar to that for the ONE mechanism discussed in section 3.2.: $R_X(\Delta) \sim R_X(SS) \sim R_X(ONE)$. For the $\Delta$-mechanism it can be seen from behaviour of the integrals $I_s(\Delta)$ and $I_s(\Delta)$ defined by Eqs. (14) and (15) (Fig.4, *b*). Almost the same ratio $R_X(N^*) \sim 1.2-1.4$ occurs for the $N^*$-excitation mechanism. The last result is practically independent on the type of the nucleon isobar and determined mainly by its mass $M_{N^*}$ and total width $\Gamma_{N^*}$ in Eqs. (14)-(18).

### 3.3. Isotopic Spin Relations

Let us discuss the isotopic spin factors for the mechanism with intermediate meson-nucleon rescattering [49] depicted in Fig.8. This diagram includes that part of the $\Delta-$ (and $N^*-$) mechanism depicted in Fig.2(*d,e*)), which can be described within one-meson exchange model [30, 31, 36]. Furthermore, this diagram takes into account a meson-nucleon continuum depicted in Fig. 9, *c*.

#### 3.3.1. Isovector meson exchange

For definition, when discussing here the case of isovector meson exchange we will denote the isovector meson as the $\pi$-meson. For the isovector mesons the meson-nucleon off-shell scattering amplitude can be written as the sum of two terms $M_{\frac{1}{2}}$ and $M_{\frac{3}{2}}$, corresponding to the total isospin of the meson-nucleon system $T = 1/2$ and $3/2$, respectively,

$$M(\pi_1 N_1 \to \pi_2 N_2) = \sum_T (\frac{1}{2}\tau_1\, 1\, \pi_1|TM_T)(\frac{1}{2}\tau_2\, 1\, \pi_2|TM_T)\, a_T =$$
$$= M_{\frac{1}{2}} + M_{\frac{3}{2}}, \tag{20}$$

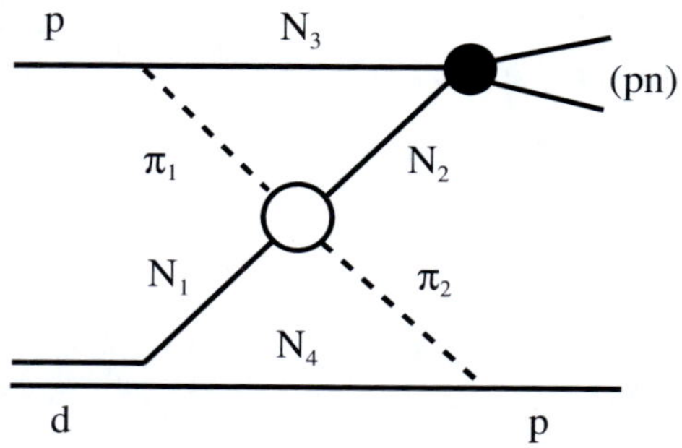


Figure 8. The two-loop mechanism of the reaction $pd \to \{pn\}p$ with intermediate meson-nucleon rescattering.

where $T$ is the total isotopic spin of the $\pi + N$ system, $\tau_i$ and $\pi_i$ are the third projection of the isotopic spin of the i-th nucleon and $\pi$-meson, respectively; $(t_1\tau_1 t_2\tau_2|TM)$ is the Clebsch-Gordan coefficient; $a_T$ is the scalar part of the off-shell $\pi N$-scattering amplitude. Dynamics of the $\pi N$ -interaction corresponding to the $\Delta$- or $N^*$-pole and continuum in the meson-nucleon system is depicted in Fig. 9. The isotopic structure of the amplitude of the process $pd \to p + (pn)_{s,t}$, corresponding to the diagram in the Fig.8 has the following form

$$M(pd \to p + (pn)_{s,t}) = \sum_{\substack{\tau_1,\,\tau_2,\,\tau_3,\,\tau_4,\\ \pi_1,\,\pi_1}} (\frac{1}{2}\tau_1\,\frac{1}{2}\tau_4|00)(1\,\pi_2\,\frac{1}{2}\tau_4|\frac{1}{2}\frac{1}{2})$$

$$\times(1\,\pi_1\,\frac{1}{2}\tau_3|\frac{1}{2}\frac{1}{2})(\frac{1}{2}\tau_3\,\frac{1}{2}\tau_2|T_{pn}0)\sum_{TM_T}(\frac{1}{2}\tau_1\,1\,\pi_1|TM_T)(\frac{1}{2}\tau_2\,1\,\pi_2|TM_T)B_T^{s,t}(T_{pn}), \quad (21)$$

where $T_{np}$ is the total isospin of the final $pn$-pair, the values $T$, $\tau_i$ and $\pi_i$ were defined after the formula (20) (the numbers of particles are shown in Fig. 8). The dynamical factor $B_T^{st}$ in Eq. (21) does not depend on the z-projections of the isospins. As it follows from the formula (21), the $pd \to p(np)_{s,t}$ amplitude for the $\Delta-$ or $N^*$-mechanism contains two terms corresponding to different values of the total isotopic spin of the intermediate meson-nucleon system, $T = \frac{1}{2}$ and $\frac{3}{2}$:

$$\begin{aligned} M(p\,d \to (pn)_s\,p) &= M^s_{T=1/2} + M^s_{T=3/2}, \\ M(p\,d \to (pn)_t\,p) &= M^t_{T=1/2} + M^t_{T=3/2}. \end{aligned} \quad (22)$$

Performing summation over the z-projections in Eq.(21) for definite values of the isotopic spins $T$ and $T_{pn}$ one has got the following isotopic spin structure of the breakup amplitude [35]

$$M_T^{s,t} = (2T+1)\sqrt{2(2T_{np}+1}(T_{np}0\frac{1}{2}\frac{1}{2}|\frac{1}{2}\frac{1}{2})\times$$
$$\times\left\{\begin{array}{ccc} 1 & \frac{1}{2} & \frac{1}{2} \\ 1 & \frac{1}{2} & T \end{array}\right\}\left\{\begin{array}{ccc} \frac{1}{2} & \frac{1}{2} & T_{np} \\ \frac{1}{2} & \frac{1}{2} & 1 \end{array}\right\} B_T^{s,t}(T_{np}). \quad (23)$$

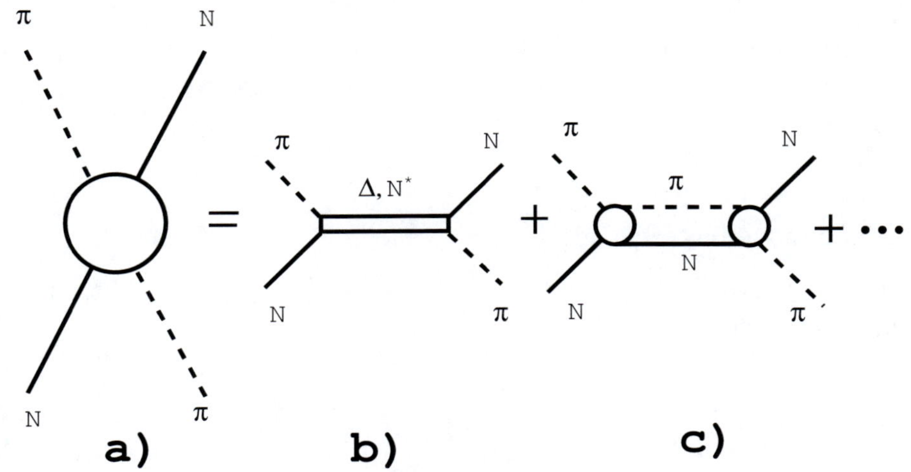


Figure 9. Contribution of the nucleon-resonance (b) and meson-nucleon continuum (c) into meson-nucleon scattering (a).

Assuming the spatial parts of the wave functions of the singlet and triplet $pn$ pairs to be the same, i.e. $B^s_T(T_{np}=1)/B^t_T(T_{np}=0)=1$ in Eq. (23), one can find from (23) the following ratios

$$\frac{M^s_{T=1/2}}{M^t_{T=1/2}}=\frac{M^s_{T=3/2}}{M^t_{T=3/2}}=-\frac{1}{3}. \tag{24}$$

Taking into account the obvious properties $\frac{a}{b}=\frac{c}{d}=\frac{a+c}{b+d}$, were $a, b$, and $c$ are complex numbers, and using Eqs. (22) and (24) one obtains the following ratio [35]

$$R^{iv}=\frac{M(p\,d\to(pn)_s\,p)}{M(p\,d\to(pn)_t\,p)}=-\frac{1}{3}. \tag{25}$$

The relative suppression of the time inverted process $p\{NN\}_s\to dN$ on the singlet $\{NN\}_s$-pair inside of the nuclei $^{6,7}Li$ was observed in the reactions of quasielastic knockout of fast deuterons $^{6,7}Li(p,Nd)$ [50] at initial energy $T_0=670$ MeV. According to the experimental data [50], the (p,nd) cross section for the transitions to the states with destroyed $\alpha$-core in the $^{6,7}Li$ nuclei is one order of magnitude smaller in comparison with the (p,pd) cross section. This fact was explained in Ref. [51] in the framework of the $\Delta$-resonance mechanism as a result of the suppression of the cross section of the process $p+\{nn\}_s\to d+n$ by the isotopic spin factor 1/9 in comparison with the process $p+\{pn\}_t\to d+p$. In particularly, this factor follows from Eq.(25).

We stress that owing to the relations (24) the ratio (25) does not depend on the unknown relative phase of the amplitudes $A_{1/2}$ and $A_{3/2}$ of the virtual process $\pi N\to\pi N$ in Eq.(22). The ratio (25) is more general in comparison with the one obtained previously in Ref. [51] for the $\Delta$- excitation and one-pion exchange mechanisms [59]. In fact, (i) all intermediate states of the meson-nucleon system both for the isotopic spin T=3/2 and T=1/2 are taken into account in Eq. (25) including the $N^*$-pole terms and continuum (Fig.9). (ii) Moreover, the relation (25) is valid not only for one definite isovector meson in the intermediate state but for the sum of the two-loop diagrams like in Fig.8 with the different isovector mesons ($\pi$, $\rho$, ...) in any combinations initial-final meson ($M_i\, M_f \equiv \pi\pi, \pi\rho, \rho\rho, \ldots$) in the process $M_iN \to M_fN$.

### 3.3.2. Isoscalar mesons exchange

For the isoscalar mesons ($\omega$, $\eta$, $\eta'$, ...) in the intermediate state in Fig.8 we come to the following isotopic relation

$$R^{is} = \frac{M(pd \to p + (pn)_s)}{M(pd \to p + (pn)_t)} = 1. \quad (26)$$

For example, excitation of the $N^*(1535)-$isobar via production and absorption of the virtual $\eta-$meson is not suppressed in the reaction (1) by the isotopic factor as compared to the $\{pn\}_t$ pair formation.

For the ONE mechanism we find the isotopic spin ratio $R_I^{ONE} = 1$.

## 3.4. The Cross Section and Spin Observables

The unpolarized cross section $d\sigma$ and spin observables $A_y^p$, $A_y^d$, $T_{20} = \frac{1}{\sqrt{2}}A_{zz}$, $A_{ij}$, $C_{i,j}$ and $C_{ij,k}$ were calculated in the framework of the ONE+Δ+SS model for the reaction (1). The unpolarized differential cross section is related to the matrix element $M_{fi}$ by the following expression

$$\frac{d^5\sigma}{dp_1 d\Omega_1 d\Omega_2} = \frac{1}{(4\pi)^5}\frac{p_1^2}{E_1}\frac{p_2^3\overline{|M_{fi}|^2}}{p_{pd}\sqrt{s}|E_n\mathbf{p}_2^2 - (\mathbf{p}_n\cdot\mathbf{p}_2)E_2|}, \quad (27)$$

where $p_{pd}$ is the c.m.s proton momentum in the initial state, $\sqrt{s}$ is the c.m.s. total energy, $p_i$ ($E_i$) is the momentum (energy) of the i-th proton (i=1,2) or the backward neutron ($i = n$).

# 4. Experiment and Theory

Experimental information on the reaction $p\,d \to (pn)_s\,p$ with formation of the singlet $(pn)_s$ pair was obtained first in semiinclusive experiments [52, 53]. The energy resolution in Refs. [52, 53] was not high enough to separate the spin-triplet final pn-state using the relation Eq.(4). The only exclusive measurement [54] of the reaction $pd \to pnp$ at 585 MeV, reanalyzed in a model independent way [55], showed a very small fraction of the singlet contribution, about a few percent. A direct measurement of the singlet pair formation with high energy resolution for the $E_{pp}$ was performed at COSY [14, 15].

## 4.1. Comparison with the ONE+Δ+SS Model

The unpolarized $pd \to \{pp\}_s n$ differential cross section was measured at large neutron c.m. angles $\approx 180°$ with respect to the proton beam and laboratory kinetic energies in the range $0.5 - 1.9\,\text{GeV}$ [14, 15]. The dominance of the $^1S_0$ state was provided by selecting events with small excitation energy of the diproton $E_{pp} < 3\,\text{MeV}$. So, the dependence on the $E_{pp}$-energy was measured within the interval $0 - 3$ MeV [15] and found in excellent agreement with the Migdal-Watson final-state-interaction prediction given by the form of $\sim \sin^2 \delta/k^2$, where $\delta$ is the ${}^1S_0$ phase shift of the *pp*-scattering at internal momentum $k$. Furthermore, angular distribution over the angle between the internal momentum **k** and beam momentum **p** was measured and found to be isotropic as it should be for the S-wave-type internal motion. The expected beam energy dependence of the differential cross section at this kinematic conditions is different for different type of NN-interaction. The reason is that for different NN potentials the relative contribution of the main mechanisms, i.e. ONE and Δ is also different. For hard NN-repulsion at short distances, like the RSC potential gives, the ONE dip is well pronounced in the unpolarized cross section and the high energy tail of the cross section is rather intensive (see Fig. 10). For softer (at short distances) NN-interaction potential, like the Paris one the dip is less visible and the high-energy tail of the spectrum becomes weaker as compared to the RSC. At last, for the soft NN-interaction potential like the CD Bonn one the ONE dip almost disappears due to increasing contribution of the Δ mechanism on the one hand, and decreasing contribution of the ONE mechanism, on the other hand, and simultaneously the high energy tail becomes much less intensive (see Fig. 10). Just the latter property demonstrates the COSY data [14, 15], as it seen from Fig. 10. The contributions of the different mechanisms in energy dependence of the cross section is shown in Fig. 11 for the case of CD Bonn NN-interaction potential in comparison with the data. One can see that the ONE dip expected at $\sim$ 0.8 GeV is filled in mainly by the Δ-mechanism contribution, which maximum is located at beam energy $\approx$ 0.6 GeV. Angular dependence at 0.5 GeV is shown in Fig. 12. At this energy the ONE contribution is comparable with the Δ-contribution, however, neither the Δ- mechanism nor the ONE mechanism alone suffices to reproduce the data, underestimating the measured cross section by factors of 3 to 5, respectively. Only their coherent sum within the ONE(DWBA)+Δ model provides good agreement at this energy.

The contribution of rescatterings in the initial and final states within the ONE mechanism is denoted here as ONE(DWBA). The cross section within the sum ONE(DWBA)+Δ +SS is shown in Fig.11 by the dotted line and for the ONE(DWBA)+Δ by full thick line. The rescatterings play an important role especially at energies above 1 GeV diminishing the high energy tail of the calculated within the ONE+Δ+SS model differential cross section at scattering angle of the neutron $\theta_n \approx 180°$.

At energies 0.4 – 0.9 GeV the reaction (1) can be used as a good instrument to study three-body forces of the N-Δ-N type in view of dominance of the Δ-mechanism and presence of the ONE node in this region. At higher beam energies $T > 1$ GeV, according to calculations within the ONE(DWBA)+Δ+SS model [56], a dominance of the ONE mechanism is expected in the unpolarized cross section. We can show [57] such a dominance also in spin observables $A_y^p$ , $A_y^d$, $T_{20}$, $C_{y,y}$ and $C_{z,z}$ (see Fig.13). The ONE+Δ+SS-model predicts at $T > 1$ GeV and $\theta_n^{cm} = 180°$ a specific for the ONE-mechanism behaviour of the spin observables under discussion. The T-odd polarization coefficient $C_{xz,y}$ depicted in Fig.13 is an exclusion from this tendency. In this observable the node of the $t(q,k)_s$-matrix is pronounced very weakly a $T \sim 0.75$ GeV and, moreover, at $T = 1 - 2$ GeV the total sum ONE+Δ+SS for $C_{xz,y}$ does not follow the ONE prediction. We can show that the value $C_{xz,y}$ is determined at $T > 1$ GeV mainly by the interference between the

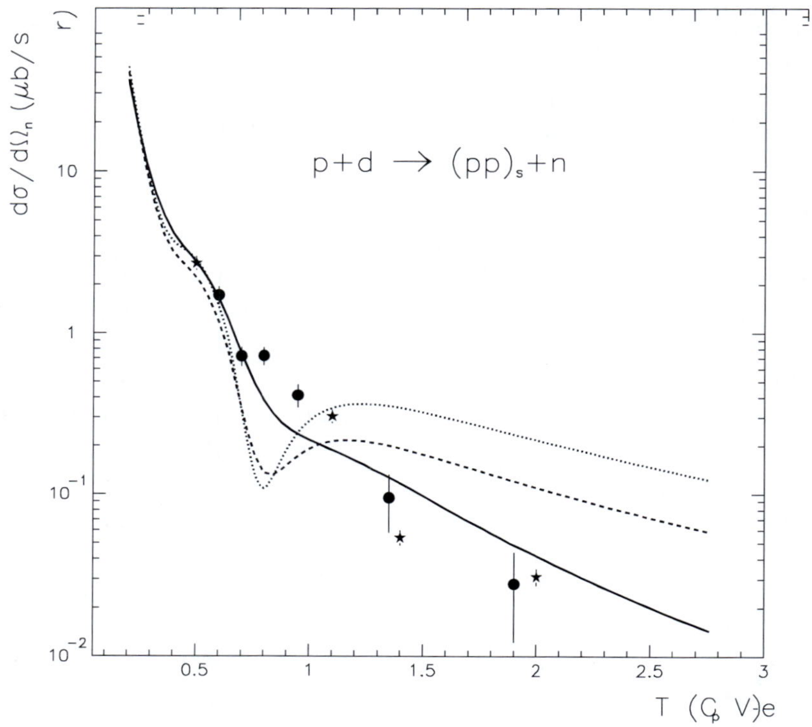


Figure 10. Cms differential cross section of the reaction (1) at neutron scattering angle $\theta_n^{cm} = 172° - 180°$ and relative energies $E_{pp} = 0 - 3$ MeV of the two forward protons. The results of calculations within the ONE(DWBA)+Δ +SS model are shown for the RSC (dotted line), Paris (dashed line) and CD Bonn (full line) NN-potentials. Data are taken from [14] (circles) and [15] (stars).

ONE and SS-mechanisms but not by the ONE rescatterings and Δ-contribution. Note, the SS-contribution at such energies is the smallest one in this model. These peculiarities display an extremely high sensitivity of the T-odd observables to the details of the reaction mechanism [58].

At last, let us consider some arguments for validity of the NN-picture of the deuteron at large enough relative momenta q between nucleons. We suppose that the non-nucleonic component of the deuteron wave function $|N_1^*N_2^* >$ arises in the transition $pn \to N_1^*N_2^*$ at sufficiently high kinetic energy of relative motion in the ordinary NN-component. Under this assumption one can estimate the minimal relative momentum $q$ in the deuteron pn-component required for such transition as

$$q_{min} = \frac{1}{2}\sqrt{s - 4m^2}, \tag{28}$$

where $\sqrt{s} = M_1 + M_2$ is the sum of the masses of the isobars $N_1^*$ and $N_2^*$, m is the nucleon mass. From Eq.(28) one can find $q_{min}$ =0.73 GeV/c, 0.796 GeV/c and 0.80 GeV/c for the $NN^*(1440)$, $\Delta(1232)\Delta(1232)$ and $NN^*(1535)$ components, respectively. Therefore, the NN-components of the deuteron dominates up to the momenta $q \sim 0.7$ GeV/c. Recent

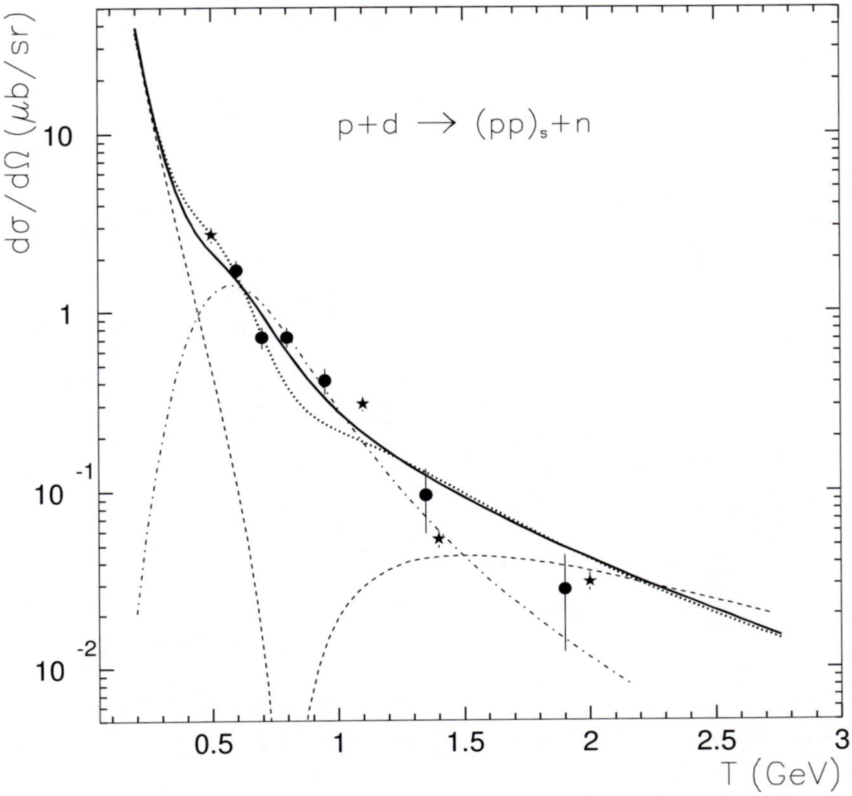


Figure 11. Contributions of the different reaction mechanisms to the cms differential cross section of the reaction (1) at neutron scattering angle $\theta_n^{cm} = 172° - 180°$ and relative energies $E_{pp} = 0 - 3$ MeV of the two forward protons. The result are shown for CD Bonn NN-potential(ONE - dashed line, Δ – dashed-dotted line, ONE(DWBA)+Δ – solid line, ONE(DWBA)+Δ +SS – dotted line). Data are taken from [14] (circles) and [15] (stars).

experimental data from JLab on $T_{20}$ in electron deuteron elastic scattering [43] confirm in general this picture up to transferred momenta $Q = 1.36$ GeV/c. It means, that the ONE mechanism of the reaction (1) based on the pn-component of the deuteron is of a physical meaning up to $q \sim \frac{1}{2}Q \sim 0.7$ GeV/c.

## 4.2. Other Approaches

*Triangle diagrams with the $pN \to d\pi$ and $pN \to \{pp\}_s\pi$ subprocess.* The mechanism of the backward elastic scattering with the subprocess $pN \to d\pi$ was studied in the Δ-isobar region in Refs. [59]. To apply the analogous mechanism to the reaction (1) one needs to know the amplitudes for both $pp \to \{pp\}_s\pi^0$ and $pn \to \{pp\}_s\pi^-$ reactions. At present, however, only unpolarized cross section for $pp \to \{pp\}_s\pi^0$ was measured at 0.8 GeV [60, 61]. In the absence of data on $\pi^-$ production, one has to make assumptions about the $pN \to \{pp\}_s\pi$ mechanism in order to add coherently the contributions from $pp \to \{pp\}_s\pi^0$ and $pn \to \{pp\}_s\pi^-$. Such ambiguity does not, however, appear for mechanisms with

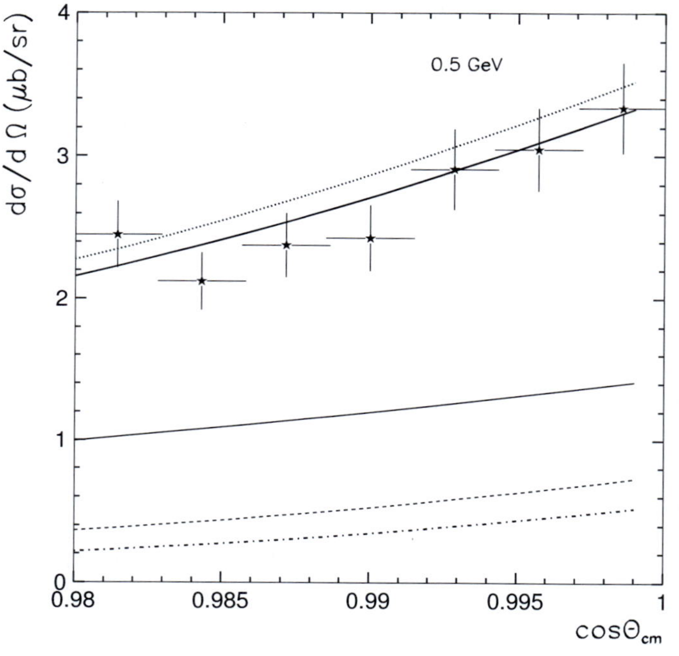


Figure 12. Angular dependence of the cms differential cross section of the reaction (1) at 0.5 GeV for different reaction mechanisms and relative energies $E_{pp} = 0 - 3$ MeV. The results are shown for CD Bonn NN-potential (ONE - dashed line, Δ- full thin line, ONE(DWBA) – dashed-dotted line, ONE(DWBA)+Δ – solid line, ONE(DWBA)+Δ +SS – dotted line. Data are taken from [14] (circles) and [15] (stars).

$\pi^0 d \to pn$ subprocess. For the reaction (1) this mechanism was applied in Ref. [24] to describe the unpolarized cross section for backward scattered neutron. The obtained results confirm the dominance of the Δ-mechanism found in Refs. [56, 15] at beam energies 0.5-1.0 GeV.

*Bethe-Salpeter approach.* With increasing beam energy, relativistic effects become more important in the reaction $pd \to \{pp\}_s n$. In the ONE+SS+Δ approach these effects were taken into account within the relativistic Hamiltonian dynamics developed for systems with fixed number of particles [44]. Another, fully covariant approach based on the Bethe-Salpeter equation was developed in [62] for the deuteron breakup reactions. This approach was applied in [63] for analysis of the reaction $pd \to \{pp\}_s n$ and semi-quantitative agreement with COSY data [14] was achieved. However, only the nucleon exchange was considered without taking into account rescatterings in the initial and final states. Very large contribution of the relativistic $P$-wave component of the $pp$-system was found resulting in filling in the dip of the ONE cross section caused by the usual $^1S_0$ $pp$-component. Inclusion of the $P$-wave component, in other language, corresponds to involvement of intermediate states $\overline{N}NNNN$. These five-baryon states are far off-shell as compared to the three-baryon intermediate $N\Delta N$ states. The NΔN configurations taken into account within the ONE+SS+Δ approach in [56, 24] and in [15], were found to dominate in the reaction $pd \to \{pp\}_s n$ at about 0.6-1.2 GeV, but were not considered in [63]. Therefore, a very

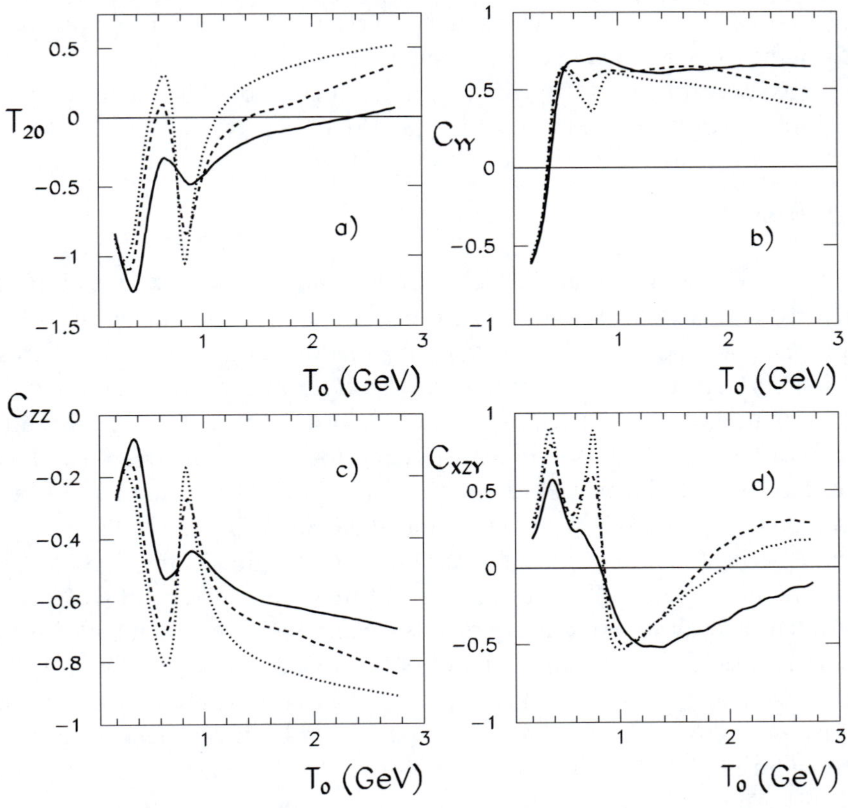


Figure 13. Tensor analysing power $T_{20}$ (a), and spin-spin correlation parameters $C_{y,y}$ (b), $C_{z,z}$ (c) and $C_{xy,z}$ (d) predicted [57] for different NN potentials within the ONE+Δ+SS model. The curves correspond to RSC (dotted), Paris (dashed), CD-Bonn (full).

important task is to include this mechanism into relativistic covarinat formalism of Ref. [63].

*Constituent counting rules.* Dimensional analysis [65, 66] and perturbative QCD [67] lead at asymptotically high energies $\sqrt{s}$ and transferred momenta $t$ to the constituent quark counting rules (CCR) for invariant cross sections of exclusive binary reactions, $d\sigma/dt \sim \frac{1}{s^n} f(t/s)$, where $n+2$ is the total number of the point-like objects in the initial and final states. For reactions with atomic nuclei the CCR behaviour of the cross sections would be considered as an indication of transition region from hadron to quark degrees of freedom in nuclear structure. Such a behaviour was observed at moderate energies for the deuteron photo-disintegration $\gamma d \to pn$ (see [5] and references therein), and also for the pure hadronic reactions $dd \to {}^3Hp$ and $dp \to dp$ in the GeV region at large scattering angles $\theta_{cm} \sim 60 - 90°$ [68] and recently for the $\gamma^3He \to dp$ reaction [69].

Fit to the data on the reaction (1) [15] gives $n = 13.3 \pm 0.5$ which is in agreement with [24]. For the *pd* $\to$ *dp* reaction fit [15] in the 0.7–1.5 GeV energy range leads to smaller value, $n = 11.4 \pm 0.3$. These numbers are not in agreement with the value $n = 16$ expected for 18 quarks available in these *pd* reactions. The obtained difference in the value $n$ for

$pd \to dp$ and $pd \to \{pp\}_s n$ suggests that the observed $s^{-n}$ dependence most likely do not originate from the CCR scaling. Indeed, the internal momenta $q > 0.5$ GeV/c, at which the CCR is expected according to [68], can be probed in collinear kinematics only above 1.5 GeV, i.e. mainly out of the region studied in [14, 15].

## 5. Conclusion

The study of hard *pd* interactions in the GeV region induced in the past several fruitful ideas. When studying the quasi-elastic knock-out of fast deuterons from nuclei by protons at 675 MeV, for instance, a hypothesis about density fluctuations of nuclear matter was supposed in Ref. [70]. This idea stimulated an intensive search for fluctons (or, in modern language, multiquark configurations) in nuclei and lead to the observation of cumulative effects [16] (for review, see Ref.[17]). The possibility to search for $NN^*$ components in the deuteron had been related also to *pd* backward elastic scattering [71]. Furthermore, the possible formation of other exotic objects in the intermediate state, like three-baryon resonances [30] and color strings [72] was also discussed in conjunction with $pd$ backward elastic scattering in the GeV region. The contribution of these ingredients, however, is strongly model dependent and, despite the many efforts undertaken, well-established results were not obtained. However, the excitation of well-known resonances, like the $\Delta$ isobar, plays an important role in these collisions. The expected contribution of three-baryon resonances, introduced as an attempt to solve the $T_{20}$ puzzle in *pd* backward elastic scattering, was significantly reduced after a more careful analysis of the $\Delta(1232)$ isobar contribution was carried out [36]. The $\Delta$-isobar excitation, caused by medium-range interaction, complicates the study of short-range properties of few-nucleon systems. Studies of the breakup reaction with the diproton in the final state (1), in which the $\Delta$ and $N^*$ contributions are suppressed by isospin invariance, could reveal the aforementioned aspects of the short-range dynamics in $pd$-collisions. Very similar arguments were applied to motivate investigations of other reactions with formation of a diproton in the final state (see Ref. [15] for references). These are $pN \to \{pp\}_s\pi$ and $pp \to \{pp\}_s\gamma$.

The results of the analysis performed within the conventional meson-baryon picture in Refs. [56, 15] show that exotic degrees of freedom, if any do exist, yield rather small contributions in the GeV region. The ONE+$\Delta$+SS model showed that only for a soft $NN$ potential, such as that of CD Bonn [9], and with absorptions taken into account in the initial and final states, can a qualitative agreement with data be achieved [56]. In the other extreme, harder $NN$-potentials, *e.g.* the Paris [8] or especially the Reid soft core [7], generate intense high–momentum components in the $NN$ wave functions and therefore lead to very large ONE contributions that are in strong disagreement with the $pd \to \{pp\}_s n$ data [14, 15]. This is the most interesting observation resulting from the $pd \to \{pp\}_s n$ analysis of Ref. [56, 15]. At moderate energies, but large enough to test the region of the NN core via the ONE mechanism, the $\Delta(1232)$ still dominates. This mechanism realizes contribution of a special type of three-body force, whose relative contribution at lower energies is much smaller and rather uncertain [73, 74]. One may conclude that the reaction (1) at 0.5-1.0 GeV offers a new tool for study of this type of three-body force, as discussed in Ref. [75]. At higher energies the agreement is worse; nevertheless, one observes that better phenomenological NN potentials lead to a better agreement with the experimental data on the reaction (1). One should note that new data on momentum distribution of nucleons in the deuteron obtained at JLab in the reaction $^2H(e,e'p)n$ at specific conditions providing supperssion of the final state interaction also shows that the CD Bonn NN interaction model is preferrable in describing this data [76].

Under the assumption that the ONE mechanism dominates at energies above $\Delta(1232)$-isobar region ($T > 1.5$ GeV) the observed small ratio $d\sigma(pd \to \{pp\}_s n)/d\sigma(pd \to dp) \approx 0.01$ implies that triplet $np$ pairs (deuterons) with high internal momenta $q = 0.5 - 0.6$ GeV/c occur much more often than the singlet $\{pp\}_s$ pairs. A similar finding was reported with respect to correlated $NN$ pairs in nuclei in the analysis of nuclear reactions $(p, ppn)$ in Refs.[77, 78] and in the $(e,e'pp)$ and $(e,e'pn)$ reactions at high transferred momenta [79].

Like in the study of the reaction (1), the Δ-isobar contribution was expected to be significantly suppressed in $pp \to \{pp\}_s\pi^0$ and $pp \to \{pp\}_s\gamma$ as compared the $pp \to d\pi^+$ and $\gamma d \to pn$ (see Ref.[80] and references therein). But measurements show pronounced enhancement of the cross sections in the Δ-region. Presumably, these results also indicate that high momentum components in the NN- wave functions are rather weak [80].

On the experimental side, the next step towards unravelling the dynamics of the $pd \to \{pp\}_s n$ reaction will be the measurement of the deuteron tensor analyzing power $T_{20}$ at 0.5-2 GeV [81]. The $T_{20}$-problem is not yet resolved. Other spin observables, like $C_{y,y}$, $C_{z,z}$, $C_{xy,z}$ demonstrate very different behavior for different NN-interaction potentials. Thus, measurement of those observables would get more insight into underlying dynamics of the reaction (1). On the theoretical front, an important task is to study the spin structure of the $NN \to N\Delta$ subprocess beyond the $\pi + \rho$ exchange model. Recent data on the charge-exchange reaction $dp \to \{pp\}_s X$ with formation of the Δ isobar in the final state at 1.6 – 2.2 GeV and rather small transferred momenta shows clearly that $\pi + \rho$ exchange fails to describe $T_{20}$ in this reaction, although the unpolarized cross section and $T_{22}$ are explained well [82].

**Acknowledgments**. Author would like to thank V. Komarov, F. Rathmann, H. Seyfarth, and C. Wilkin for fruitful discussions.

# References

[1] Gilman, R.; Gross, F., *J. Phys. G: Nucl. Part. Phys.* 2002, vol. 28, R37 –R116.

[2] Garson, M.; Van Orden, J.W. *Adv. Nucl.Phys.*, 2001, vol. 26, 293 – 378.

[3] Alexa, L.C. et al.*Phys. Rev. Lett.*, 1999, vol. 82, 1374 – 1378.

[4] Blomqvist, K.I. et al.*Phys.Lett.*, 1998, vol. B424 33 – 38.

[5] Rossi, P. et al., *Phys. Rev. Lett.*,2005, vol. 94, 012301-1 –012301-5.

[6] Nikolenko, D.M. *et al.*, *Phys. Rev. Lett.*, 2003, vol. 90, 072501-1 – 072501-4.

[7] Reid, Jr. R.V. , *Ann. Phys. (N.Y.)*, 1968, vol. 50, 411 – 448.

[8] Lacombe, M. *et al.*, *Phys. Rev. C* , 1980, vol. 21, 861 – 873.

[9] Machleidt, R., *Phys. Rev. C*, 2000, vol. 63, 024001-1 –024001-32.

[10] Mathiot, J.-F., *Phys. Rep.*, 1989, vol. 173 63 – 175.

[11] Auffert, S. *et al.*, *Phys. Rev. Lett.*, 1985, vol. 55 1362 – 1365.

[12] Imambekov, O. and Uzikov,Yu.N., *Yad. Fiz.* , 1990 vol.52 1361 –1371 .

[13] Uzikov, Yu.N., *J. Phys. G*, 2002, vol. **28**, B13 – B18.

[14] Komarov, V.I. et al., *Phys. Lett. B* 2003, vol. **553**, 179-185.

[15] Dymov, S. et al. *Phys. Rev. C* 2010, vol. 81, 044001-1 –044001-19.

[16] Baldin, A.M. *Nucl.Phys.*, 1985, vol. A434, 695C – 704C.

[17] Leksin, G.A., *Yad. Fiz.*, 2002, vol. 65, 2042 – 2051 [*Phys.Atom. Nucl.*, 2002, vol, 65, 1985 – 1994.]

[18] Strikman, M.I.; Frankfurt, L.L., *Phys. Rep.*, 1981, vol. 76, 215 – 347.

[19] Punjabi, V. *et al.*, *Phys. Lett. B*, 1995, vol. **350**, 178 – 183.

[20] Ahzgirey, L.S. et al., *Phys. Lett.*, 1997, vol. **B**391, 22 –28; *Yad. Fiz.* 1998, vol. **61**, 494 – 510 [*Phys. Atom. Nucl.* ., 1998, vol. 61, 432-447].

[21] Ableev, V.G. et al., *Nucl. Phys.*, 1983, vol. **A**393, 491 – 501.

[22] Perdrisat, C.F. *et al.*, *Phys. Rev. Lett.*, 1987, vol. 59, 2840 – 2843.

[23] Azhgirey, L.S. *et al.*, *Phys. Lett. B*, 1996, vol. **3**87, 37 – 42.

[24] Uzikov, Yu. N.; Haidenbauer, J.; Wilkin, C. *Phys. Rev. C*, 2007, vol. **7**5, 014008-1 –014008-10.

[25] Arvieux, J. et al., *Phys. Rev. Lett.*, 1983, vol. **50** 19 –22; *Nucl. Phys. A*, 1984, vol. **431** 613 – 636.

[26] Ableev, V.G. et al., *Pis'ma Zh. Eksp. Teor. Fiz.*, 1988, vol. **47**, 558 – 561.

[27] Stoks, V.G.J.; Klompt, R.A.M.; Terheggen, C.P.F.; and de Swart, J.J., *Phys. Rev. C*, 1994, vol. **49** 2950 – 2962.

[28] Wiringa, R.B.; Stoks, V.G.J.; and Schiavilla, R.,*Phys. Rev. C*, 1995, vol. **51**, 38 – 51.

[29] Arenhövel, H.; Ritz, F.; and Wilbois, T.,*Phys. Rev. C*, 2000, vol.**61**, 034002-1 – 034002-10.

[30] Kondratyuk, L.A.; Lev, F.M.; Schevchenko, L.V., *Yad. Fiz.*, 1981, vol.**3**3, 1208 – 1223.

[31] Boudard, A.; Dillig, M. *Phys. Rev. C*, 1985, vol. 31, 302-305.

[32] Uzikov, Yu.N., *Fiz. Elem. Chast. At. Yadra*, 1998, vol. 29, 1405 –1455. *Phys. Atom., Nucl.*, 1998, vol. 29, 583 – 605.

[33] Sakai, H. et al., *Phys. Rev. Lett.*, 2000, vol. **84**, 5288 – 5291.

[34] Smirnov, A.V.; and Uzikov, Yu.N., *Yad. Fiz.*, 1998, vol. **6**1, 421 – 434 [*Phys. Atom. Nucl.*, 1998, vol. 61, 361 – 374].

[35] Uzikov, Yu.N., *Pisma Zh. Eksp. Theor. Fiz.*, 2002, vol. **75**, 7 – 19 [*JETP Lett*, 2002, vol. 75, 5-8].

[36] Imambekov,O; Uzikov, Yu.N.; Schevchenko, L.V., *Z.Phys. A*, 1989, vol. **332**, 349 – 357.

[37] Uzikov, Yu.N.,*Yad. Fiz.*, 1997, vol. **6**0, 1603 – 1614 [*Phys. Atom. Nucl.*, 1997, vol. **60** 1458 – 1469].

[38] Imambekov, O; Uzikov, Yu.N., *Yad. Fiz.*, 1988, vol. **47**, 1089 – 1093 [ *Sov.J. Nucl. Phys.*, 1988, vol. **47**, 695 – 697].

[39] Anisovich, V.V.; Dakhno, L.G.; Makarov, M.M., *Yad. Fiz.*, 1980, vol. **32**, 1521 – 1526.

[40] Boudard, A; Fäldt, G; Wilkin, C. *Phys. Lett. B* , 1996, vol. 389, 440 – 444.

[41] Blokhintsev, L.D.; Truhlik, E., *Izv. AN SSSR, ser. fiz.*, 1969, vol. 33, iss.1, 88-91.

[42] Bouwhuis, M.; Alarcon, R; Botto, T. et al., *Phys. Rev. Lett.*, 1999, vol. **82**, 3755 – 3758.

[43] Abbott, D. et al., *Phys. Rev. Lett.*, 2000, vol. 84, 5053 – 5057.

[44] Bakker, B.L.G.; Kondratyuk, L.A.; Terentjev, M.V., *Nucl. Phys.*, 1979, vol. **B158**, 497 – 519.

[45] Blokhintsev, L.D.; Lado, A.V.; Uzikov, Yu.N., *Nucl. Phys. A*, 1996, vol. **597**, 487 – 514; Uzikov, Yu.N., *Nucl. Phys. A*, 1998, vol. **644**, 321 – 344 ; Lado, A.V.; Uzikov, Yu.N., *Phys. Lett. B*, 1992, vol. **279**, 16 – 19.

[46] Levitas, M.; Noble, J.V., *Nucl. Phys.*, 1975, vol. **A251**, 385 – 400.

[47] Uzikov, Yu.N., *Fiz. Elem. Chast. At. Yadr.*, 1988, vol. 29, iss. 4, 1010 – 1052.

[48] Franco, V.; Glauber, R., *Phys. Rev.* 1966, vol. 142, 1195 –1214; Sitenko, A.G., *Fiz. Elem. Chast. At. Yadr.*, 1973, vol.**4**, 546 – 584 .

[49] Anjos, J.C.; Santoro, A.; Simao, F.R.A.; Levy, D., *Nucl. Phys. A*, 1981, vol. **356**, 383 – 400.

[50] Albrecht, D; Csatlos, M.; J.Erë, J. et al., *Nucl. Phys. A*, 1980, vol. **338**, 477 – 494.

[51] Imambekov, O.; Uzikov, Yu.N., *Izv. AN SSSR, ser. fiz.*, 1987, vol. 51, N 5, 947 – 951 .

[52] Azhgirey, L.S.et al., *Yad. Fiz.*, 1998, vol. **61**, 494 – 510.

[53] Boudard, A. These. CEA-N-2386, Saclay, 1984.

[54] Witten, T. et al. *Nucl. Phys.*, 1975, vol. **A254**, 269 – 280.

[55] Uzikov, Yu.N. et al. *Phys. Lett.*, 2002, vol. **B524**, 303 – 307.

[56] Haidenbauer, J.; Uzikov, Yu.N., *Phys. Lett. B*, 2003, vol. 562, 227 – 233.

[57] Uzikov, Yu.N., Seyfarth, H. in Proc. X-th Adv., Res/ Workshop on High Enegy Spin Physics, Dubna, 2003 (Eds. Efremov A.V. and Teryaev O.V.) p.214.

[58] Rekalo, M.P.; Sitnik., I.M., *Phys. Lett.*, 1995, vol. **B356**, 434 – 438.

[59] Craigie, N.S.; Wilkin, C., *Nucl.Phys. B*, 1969, vol.**14**, 477 – 488; Kolybasov, V.M.; Smorodinskaya, N.Ya., *Yad. Fiz.*, 1973, vol. **17**, 1211 – 1221; Nakamura, A; L. Satta, L. *Nucl.Phys. A* , 1985, vol.**445**, 706 – 716.

[60] Dymov, S. et al. *Phys. Lett. B* 2006, vol. 635, 270-274;

[61] Kurbatov, V. et al. *Phys. Lett.* 2008, vol. B 661, 22-27.

[62] Kaptari, L. et al., *Phys. Lett.*, 1997, vol. **B**404, 8 – 14.

[63] Kaptari, L. et al., *Eur. Phys. J.*, 2004, vol. **A**19, 301 – 306.

[64] Kaptari, L. et al., *Phys. Rev. C*, 1998, vol.**5**7, 1097 – 1111.

[65] Matveev, A.V.; Muradyan, R.M.; and Tavkhelidze, A.N., *Lett. Nuovo Cimento*, 1973, vol. **7**, 719 – 723.

[66] Brodsky, S.J. and Farrar, G.R.,*Phys. Rev. Lett.*, 1973, vol.**3**1, 1153 – 1156.

[67] Lepage, G.P. and Brodsky, S.J., *Phys. Rev. D*, 1980, vol.**2**2, 2157 – 2198.

[68] Uzikov, Yu.N., *Pisma Zh. Exp. Theor. Phys.*, 2005, vol. 81, 387 – 390 [*JETP Lett.*, 2005, vol. 81, 303-306].

[69] Ilieva, Y., *Few-Body Syst.*, 2013, vol. 54, 989-995; Pomerantz I. et al. *Phys. Rev. Lett.*, 2013, vol. 110, 242301-1 – 242301-7.

[70] Blokhintsev, D.I., *J. Exp. Theor. Phys.*, 1957, vol. **33**, 1295 – 1299.

[71] Kerman, A. and Kisslinger,L., *Phys. Rev.*, 1969, vol. 180, 1483 – 1489.

[72] Kopeliovich, B.Z. and Niedermayer, F., *Phys. Lett.*, 1982, vol. B117, 101 – 104; [*Sov. Phys. JETP*, 1984, vol. **6**0, 640 – 650.

[73] Witala, H.; Golak, J.; Skibinski, R. et al., *Phys. Rev. C*, 2006, vol. **7**3, 044004-1 – 044004-16.

[74] Kalantar-Nayestanaki, N.; Epelbaum, E.; Messchendorp, J.G.; Nogga, A. *Rep. Prog. Phys.* 2012, vol. 75, 016301 (41pp).

[75] Uzikov, Yu.N., *Few-Body Syst.*, 2008, vol. 44, 211 - 213.

[76] Boeglin, W.U. et al.*Phys. Rev. Lett.*, 2011, vol. **107**, 262501-1 – 2625011-6.

[77] Tang, A. et al. *Phys. Rev. Lett.*, 2007, vol. **90** 042301-1 – 042301-4.

[78] Piasetzky, E. et al., *Phys. Rev. Lett.*, 2006, vol. **97**, 162504-1 – 162504-4.

[79] Shneor, R. et al., *Phys. Rev. Lett.*, 2007, vol. **99**, 072501-1 – 072501-5.

[80] Uzikov, Yu.N., In: *Proc. of the XIX Int. Baldin Seminar on High Energy Physics Problem (September 2008, Dubna), Conference: C08-09-29.12, arXiv:0811.3288*, Eds. A.N. Sissakian et al. JINR, Dubna, Russia, 2009), Vo. 2, 307 - 312.

[81] Kacharava, A.; Rathmann, F.; Wilkin, C. (For the ANKE Collaboration) *Spin Physics from COSY to FAIR*, COSY Proposal, Jülich, 2005; arXiv:nucl-ex/0511028.

[82] Mchedlishvili, D. et al. *Phys. Lett. B*, 2013, vol. **726**, 145 – 150.

In: NN and 3N Interactions
Editors: Leonid Blokhintsev and Igor Strakovsky
ISBN: 978-1-63321-053-0


*Chapter 8*

# BINDINGS AND SPECTRA OF LIGHT NUCLEI WITH JISP16

***Andrey M. Shirokov*[1], *Vasily A. Kulikov*[1], *Pieter Maris*[2] *and James P. Vary*[2]**
[1]Skobeltsyn Institute of Nuclear Physics,
Moscow State University, Moscow, Russia
[2]Department of Physics and Astronomy,
Iowa State University, Ames, IA, US

**Abstract**

We present here a comprehensive overview of *ab initio* results for binding energies and spectra of light nuclei obtained with the JISP16 $NN$ interaction. The calculations were performed by means of the No-Core Full Configuration approach (NCFC) which extends the No-Core Shell Model (NCSM) to infinite basis space through extrapolation.

## 1. Introduction

An *ab initio* description of nuclear structure and reactions, i. e. high-precision simulations of many-nucleon systems based on realistic inter-nucleon interactions, is one of the mainstreams of modern nuclear theory [1]. It is based on a rapid development of supercomputer facilities and recent advances in the utilization of high-performance computing systems [2]. The modern *ab initio* nuclear theory has opened a wide range of nuclear phenomena that can be evaluated to high precision using realistic nucleon-nucleon ($NN$) and three-nucleon ($NNN$) interactions. In particular, *ab initio* approaches like the No-Core Shell Model (NCSM) [3], the Green's Function Monte Carlo (GFMC) [4] and the Coupled-Cluster Theory [5] are able to reproduce properties of a large number of atomic nuclei with mass up to $A = 16$ and a few heavier nuclear systems around closed shells. Very important progress has been achieved in the *ab initio* description of reactions with light nuclei, in particular, by combining the NCSM with the Resonating Group Method [6].

The *ab initio* theory requires a high-quality realistic strong interaction providing an accurate description of $NN$ scattering data and predictions for binding energies, spectra and other observables in light nuclei. A number of meson-exchange potentials sometimes supplemented with phenomenological terms to achieve high accuracy in fitting $NN$ data (CD-Bonn [7], Nijmegen [8], Argonne [9]) have been developed that should be used together with modern $NNN$ forces (Urbana[10, 11], Illinois [12], Tucson–Melbourne [13, 14, 15]) to reproduce properties of many-body nuclear systems. A very important step in the theory of inter-nucleon interactions in nuclei is the emergence of realistic $NN$ and $NNN$ interactions tied to QCD via Chiral Perturbation Theory [16, 17, 18, 19].

Three-nucleon forces require a significant increase of computational resources needed to diagonalize a many-body Hamiltonian matrix since the $NNN$ interaction increases the number of non-zero matrix elements approximately by a factor of 30 in the case of $p$-shell nuclei [20, 21]. As a result, one needs to restrict the basis space in many-body calculations when $NNN$ forces are involved that makes the predictions less precise. *Ab initio* many-body studies benefit from the use of recently developed purely two-nucleon interactions of INOY (Inside Nonlocal Outside Yukawa) [22, 23] and JISP ($J$-matrix Inverse Scattering Potential) [24, 25, 26, 27] types fitted not only to the $NN$ data but also to binding energies of $A = 3$ and heavier nuclei. At the fundamental level, these $NN$ interactions are supported by the work of Polyzou and Glöckle [28] who demonstrated that a realistic $NN$ interaction is equivalent at the $A = 3$ level to some $NN + NNN$ interaction where the new $NN$ force is related to the initial one through a phase-equivalent transformation (PET). It seems reasonable then to exploit this freedom and work to minimize the need for the explicit introduction of three- and higher-body forces. Endeavors along these lines have resulted in the design of INOY and JISP strong interaction models.

Conventional realistic meson-exchange $NN$ interactions [7, 8, 9] and $NN$ interactions obtained in Chiral Perturbation Theory [18] present convergence challenges in many-body calculations. To improve the convergence, these $NN$ interactions are usually softened either using the Okubo–Lee–Suzuki (OLS) [29, 30] method reviewed in detail in Ref. [3] or the Similarity Renormalization Group (SRG) technique [31, 32]. Unfortunately, the convergence of many-body calculations with OLS-softened $NN$ interactions is not monotonic as the basis space is enlarged, and hence the respective results cannot be extrapolated to the infinite basis space. The SRG softening of $NN$ interaction induces $NNN$ forces.

We present here the results for binding energies and energies of narrow (with width less than 300 keV) excited states of $s$- and $p$-shell nuclei based on NCSM calculations with the JISP16 [27, 33] $NN$ interaction. The JISP interactions provide a fast convergence of NCSM calculations and are tuned with PETs to reduce the need for $NNN$ forces. We extrapolate the NCSM results for energies to the infinite basis space following the No-Core Full Configuration (NCFC) approach [34]. The reliability of this approach was demonstrated by the prediction of the binding energy and spectra of the exotic proton-excess nucleus $^{14}$F based on the JISP16 $NN$ interaction [35]. The accuracy of the predictions was later confirmed by the first experimental observation of this nucleus [36]. Improved extrapolation techniques have been suggested recently [37, 38, 39, 40], and we use one of these techniques to obtain some of the results presented here.

After a brief description of the JISP16 $NN$ interaction, we discuss the *ab initio* NCSM approach and various methods of extrapolation of the NCSM results for energies of ground and excited states to the infinite basis space (NCFC approach). Next we present the results for binding energies and spectra of narrow levels of $s$- and $p$-shell nuclei obtained with

JISP16 obtained up to now. We do not discuss results for both nuclei that are mirror pairs but present calculations only for the one with $Z < N$. The binding energies and spectra of some nuclei have been already published [21, 34, 35, 41, 42, 43, 44, 45, 46, 47, 48, 49]. We present here all these results which are improved in some cases due to the use of larger basis spaces in NCSM calculations together with previously unpublished results. We compare our description of binding energies and spectra of light nuclei based on JISP16 with the GFMC description based on Argonne AV18 $NN$ interaction supplemented by meson-exchange IL7 $NNN$ force.

## 2. JISP16 $NN$ Interaction

The $J$-matrix inverse scattering approach was introduced in Ref. [50]. It was further developed and used to design a high-quality JISP $NN$ interaction in Ref. [24]. A nonlocal interaction obtained in this approach is in the form of a matrix in the oscillator basis in each of the $NN$ partial waves. To reproduce scattering data in a wider energy range, one needs to increase the size of the interaction matrix and/or the $\hbar\Omega$ parameter of the oscillator basis. From the point of view of shell model applications, it is desirable however to reduce the size of interaction matrices and to use $\hbar\Omega$ values in the range of few tens of MeV. A compromise solution is to use $\hbar\Omega = 40$ MeV with $N_{\max} = 9$ truncation of interaction matrices, i. e., the JISP $NN$ interaction matrices include all relative $NN$ motion oscillator states with excitation quanta up to 8 or 9 depending on parity. In other words, we use interaction matrices of the rank $r = 5$ in $s$ and $p$ partial waves, $r = 4$ matrices in $d$ and $f$ partial waves, etc. In case of coupled waves, the rank of the interaction matrix is a sum of the respective ranks, e. g., the rank of the coupled $sd$-wave matrix is $r = 5 + 4 = 9$. The $N_{\max} = 9$ truncated JISP interaction with $\hbar\Omega = 40$ MeV provides an excellent description of $np$ scattering data with $\chi^2/\text{datum} = 1.03$ for the 1992 $np$ data base (2514 data), and 1.05 for the 1999 $np$ data base (3058 data) in the laboratory energy range up through 350 MeV [51]. The definition of the 1992 and 1999 $np$ data bases is discussed in the review paper [52]. The $NN$ data bases and analyses can be obtained also by the online SAID facility [53]. The JISP16 $np$ phase shifts description is discussed in detail in Ref. [24]. The interaction is assumed to be charge-independent.

PETs originating from unitary transformations of the oscillator basis proposed in Refs. [54, 55], give rise to ambiguity of the interaction obtained in the $J$-matrix inverse scattering approach. This ambiguity is eliminated at the first stage by postulating the simplest tridiagonal form of the $NN$ interaction in uncoupled and quasi-tridiagonal form in coupled $NN$ partial waves [24]. At the next stage, PETs are used to fit the JISP interaction to various nuclear properties. First of all, the $sd$ component of the $NN$ interaction is modified with the help of PETs to reproduce the deuteron quadrupole moment $Q$ and rms radius without violating the excellent description of scattering data. It is worth noting here that the deuteron binding energy $E_d$ and asymptotic normalization constants are used as an input in the inverse scattering approach and are not affected by PETs.

Next, PETs were employed in other $NN$ partial waves attempting to improve the description of binding energies and spectra of light nuclei in NCSM calculations. Following this *ab exitu* route, the JISP6 $NN$ interaction fitted to properties of nuclei with masses $A \leq 6$, was proposed in Refs. [25, 26]. It was found out later that JISP6 strongly overbinds nuclei with $A \geq 10$. Therefore a new fit of PET parameters was performed that resulted in the JISP16 $NN$ interaction [27, 33] fitted to binding energies of nuclei with masses up through $A \leq 16$.

Generally, the JISP16 $NN$ interaction provides a reasonable description of binding energies and spectra of nuclei with mass $A \leq 16$. However the fit of JISP16 to properties of light nuclei was performed in 2005–2006 [27] utilizing the OLS technique to improve the convergence of NCSM calculations. Since that time, new generations of supercomputers have become available which made it possible to perform NCSM calculations in larger basis spaces and the NCFC technique [34] was developed to include extrapolations with quantified uncertainties. It became clear [35, 41] that the OLS transformation does not produce a monotonic sequence in the binding energy at fixed oscillator energy which had been inferred from smaller space calculations. As we shall see, the new results obtained in larger basis spaces (without the OLS technique so the variational principle applies) and extrapolated to the infinite basis space clearly demonstrate that the JISP16 $NN$ interaction is reasonably accurate in the description of spectra and binding energies of nuclei with $A \leq 12$ but provides overbinding of nuclei at the end of $p$ shell.

## 3. No-Core Shell Model

We adopt here the NCSM with the NCFC extension discussed in the the next section for the *ab initio* modeling of light nuclei. The energies $E_i$ and respective wave functions $\Psi_i(\vec{r}_1, \dots, \vec{r}_A)$ of the ground and excited states of a nucleus with $Z$ protons and $N$ neutrons of mass $A = Z + N$ are obtained by solving an eigenproblem for the $A$-body Schrödinger equation

$$H\,\Psi_i(\vec{r}_1, \dots, \vec{r}_A) = E_i\,\Psi_i(\vec{r}_1, \dots, \vec{r}_A). \tag{1}$$

The wavefunction $\Psi$ is expanded in an $A$-body basis of Slater determinants $\Phi_k$ of single-particle wavefunctions $\phi_{nljm}(\vec{r})$,

$$\Psi(\vec{r}_1, \dots, \vec{r}_A) = \sum c_k\,\Phi_k(\vec{r}_1, \dots, \vec{r}_A), \tag{2}$$

where

$$\Phi_k(\vec{r}_1, \dots, \vec{r}_A) = \mathscr{A}\,[\phi_{n_1 l_1 j_1 m_1}(\vec{r}_1)\,\phi_{n_2 l_2 j_2 m_2}(\vec{r}_2) \dots \phi_{n_A l_A j_A m_A}(\vec{r}_A)], \tag{3}$$

and $\mathscr{A}$ is the antisymmetrization operator. Conventionally, harmonic oscillator wave functions are used as the single-particle basis functions $\phi_{nljm}(\vec{r})$, and we present below the results obtained with such a basis. However other single-particle basis functions can be also utilized in the NCSM calculations [56, 57]. All single-particle functions $\phi_{nljm}(\vec{r})$ are obtained with the same oscillator frequency $\hbar\Omega$ and are labelled by the quantum numbers $n$, $l$, $j$, and $m$, where $n$ and $l$ are the radial and orbital oscillator quantum numbers (with $N_i = 2n_i + l_i$ being the number of single-particle oscillator quanta), $j$ is the total single-particle angular momentum, and $m$ is its projection. The many-body basis states $\Phi_k$ have a well-defined parity and total angular momentum projection, $M = \sum m_i$, but they do not have a well-defined total angular momentum $J$. Thus, in two runs, one for each parity, we can obtain the complete low-lying spectrum for a given value of the basis parameter $\hbar\Omega$, including the ground state, even if the spin of the ground state is not known a priori.

No model assumptions are used while constructing the many-body basis (3), it includes all possible many-body configurations. The only restriction that limits the accuracy of *ab initio* NCSM calculations is that the infinite-dimensional many-body basis (3) should be truncated in practical applications. Any finite truncation of the complete basis (3) produces NCSM results that due to the variational principle, provide a strict upperbound for the lowest state of a given spin and parity for a given nuclear potential. We are interested in the convergence with increasing basis space dimensions with the goal of obtaining, to within quantifiable uncertainties, by means of the NCFC extrapolations discussed below, the results corresponding to the complete basis. Following this route, we have to address eigenvalue problems for increasingly large sparse matrices with dimensions in some cases exceeding a billion. Improved algorithms to construct these matrices and to determine their lowest eigenstates, as well as efficient use of increasing computational resources, are critical for the success of this approach [58, 59, 60].

Here we use the so-called $N_{\max}$ truncation, which is a truncation on the total number of oscillator quanta of the many-body basis: the basis is limited to all many-body basis states with $\sum_i^A N_i \leq N_0 + N_{\max}$, where $N_i$ is the number of quanta of each single-particle state in a many-body basis state, $N_0$ is the minimal number of quanta for the nucleus under consideration, and $N_{\max}$ is the truncation parameter. Note that for the many-body harmonic oscillator basis, this truncation makes it possible to factorize exactly the center-of-mass wavefunction and the relative wavefunction [61].

The many-body Hamiltonian $H$ in Eq. (1) can be expressed as

$$H = H_A + \lambda\Big(H_{\text{CM}} - \frac{3}{2}\hbar\Omega\Big), \tag{4}$$

where

$$H_A = T_{\text{rel}} + \mathcal{V} = \frac{1}{A}\sum_{i<j}^{A}\frac{(\vec{p}_i - \vec{p}_j)^2}{2m} + \sum_{i<j}^{A} V_{ij} + \sum_{i<j<k}^{A} V_{ijk} + \ldots \tag{5}$$

is the Hamiltonian describing internal degrees of freedom of nucleus and the last term in Eq. (4) with a large enough parameter $\lambda$ is responsible for shifting up in energy the $A$-body center-of-mass excited states. The net result is a set of low-lying eigenstates free from the spurious center-of-mass excitations (the so-called Lipkin–Lawson projection method [62]). In Eq. (5), $m$ is the nucleon mass, $p_i$ is the single-nucleon momentum operator, $V_{ij}$ is the $NN$ interaction including the Coulomb interaction between protons, $V_{ijk}$ is the three-nucleon interaction, and we can allow for higher-body interactions as well if desired. However we primarily focus here on results obtained with the realistic two-body $NN$ interaction JISP16 which does not require the use of $NNN$ and higher-body interactions.

## 4. NCFC Extrapolations

The NCFC approach suggested in Ref. [34] adopted an empirical dependence

$$E(\hbar\Omega, N_{\max}) = E_\infty + A\, e^{-bN_{\max}} \tag{6}$$

of eigenenergies $E(\hbar\Omega, N_{\max})$ of some state obtained in NCSM calculations with the basis truncation $N_{\max}$ and a given value of $\hbar\Omega$. The fitting parameters $A$ and $b$ in Eq. (6) are functions of $\hbar\Omega$ and may be different for different ranges of $N_{\max}$ used to estimate $E_\infty$, the energy of the state under consideration in the limit of infinite basis. Two extrapolation methods were proposed in Ref. [34]: a global extrapolation based on the results obtained in four successive basis spaces with five $\hbar\Omega$ values from a 10 MeV interval (Extrapolation A); and Extrapolation B based on the results obtained at various fixed $\hbar\Omega$ values in three successive basis spaces and defining the most reliable value for the extrapolation. These two extrapolations provided results consistent with each other and were carefully tested in a number of light nuclei where high-precision, essentially converged, result was achieved. A detailed description of these extrapolations and methods to evaluate their uncertainties illustrated by numerous examples can be found in Ref. [34]. We adopt here also an improvement in Extrapolation B proposed in Ref. [46] which leads in some cases to a larger and more realistic uncertainty estimate.

A detailed study of ultraviolet and infrared divergences of harmonic oscillator expansions [37] and further studies of Refs. [38, 39] resulted in the extrapolation formula

$$E(\hbar\Omega, N_{\max}) = E_\infty + A\, e^{-4\sqrt{\frac{E'}{\hbar\Omega}}\sqrt{N_0^{sp}+N_{\max}+i+\frac{3}{2}}}, \qquad (7)$$

where $N_0^{sp} = 2n + l$ is the number of quanta of the highest single-particle oscillator state in the minimal possible many-body oscillator configuration of a given nucleus. According to the theory of Refs. [37, 38], $E'$ should be equal to the energy in the infinite basis space limit $E_\infty$ in a one-body or two-body problem. However setting $E' = E_\infty$ results in a poor quality of extrapolation and $E'$ is conventionally used as an independent fitting parameter. The theory of Refs. [37, 38] suggests $i = 0$. Later it was found in Ref. [39] that a phenomenological value $i = 2$ better fits the data. The value $i = 2$ is consistent with the $J$-matrix formalism in scattering theory (see Ref. [63]) and we use it in our extrapolations. The extrapolation based on Eq. (7) is referred to below as Extrapolation C.

The extrapolation formula (7) is applicable when the ultraviolet cutoff parameter

$$\Lambda_{\mathrm{UV}} = \sqrt{\left(N_0^{sp} + N_{\max} + \frac{3}{2}\right) m\hbar\Omega} \qquad (8)$$

takes large enough values [37, 38] depending on the $NN$ interaction. According to recommendations of Ref. [37], in the case of JISP16 $NN$ interaction, we can use the extrapolation (7) when

$$\Lambda_{\mathrm{UV}} \gtrsim 500\ \mathrm{MeV}/c. \qquad (9)$$

We however use here a stronger condition

$$\Lambda_{\mathrm{UV}} \gtrsim 600\ \mathrm{MeV}/c. \qquad (10)$$

We use the NCSM calculations of the $^7$Li ground state with JISP16 $NN$ interaction to illustrate in Fig. 1 how the extrapolation formula (7) works and how the Extrapolation C is designed using a least squares fitting strategy on data satisfying specified criteria. The results obtained by NCSM with various $\hbar\Omega$ values are shown by symbols (closed and open circles and diamonds), solid lines connect the results obtained with the same $N_{\max}$ which values are ranging from 2 through 14. Diamonds depict the results violating the inequality (10)

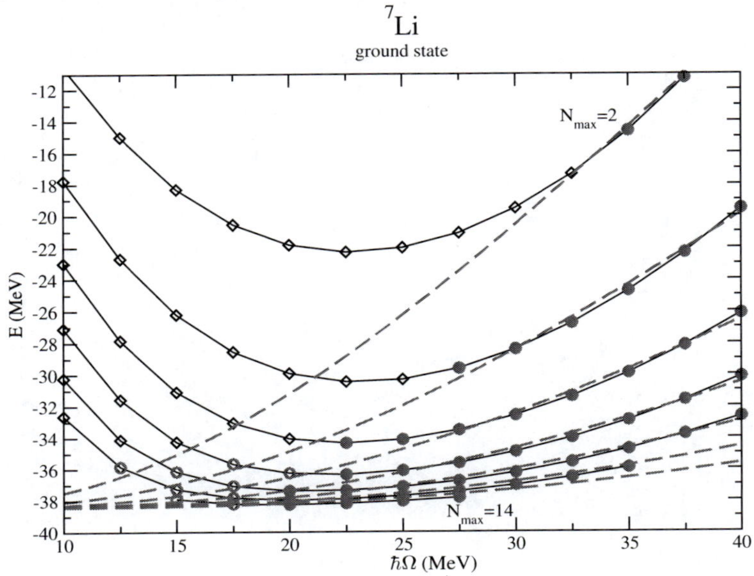


Figure 1. Ground state energy of $^7$Li obtained in NCSM calculations with JISP16 $NN$ interaction with various $\hbar\Omega$ values and $N_{\max}$ ranging from 2 to 14 in steps of 2: energies used in Extrapolation C are shown by closed circles; energies corresponding to $\Lambda_{\rm UV} < 600$ MeV/$c$ and hence not used in Extrapolation C are shown by diamonds; open circles depict energies corresponding to $\Lambda_{\rm UV} \geq 600$ MeV/$c$ but not used in the Extrapolation C since they are obtained with $\hbar\Omega$ values below the value associated with minimum of $E(\hbar\Omega, N_{\max})$ for the given $N_{\max}$. Dashed lines are calculations by Eq. (7) with parameters obtained by the fit.

which cannot be approximated by Eq. (7). The NCSM energies $E(\hbar\Omega, N_{\max})$ obtained with the same $N_{\max}$ have a minimum at some $\hbar\Omega$ value. On the other hand, Eq. (7) describes the energies $E(\hbar\Omega, N_{\max})$ monotonically increasing with $\hbar\Omega$ for a given $N_{\max}$. The NCSM results obtained with small enough values of $\hbar\Omega$ before the minimum of the function $E(\hbar\Omega, N_{\max})$ should be excluded from the fit of Extrapolation C. Those of these excluded results which still fit the inequality (10) are depicted by open circles in Fig. 1. Closed circles show the NCSM energies fitting the inequality (10) and increasing with $\hbar\Omega$ — only these results should be used in the fit by Eq. (7) to obtain values of parameters $A$, $E'$ and $E_\infty$ of the Extrapolation C. The dashed lines in Fig. 1 demonstrate calculations by formula (7) for various $N_{\max}$ using the least squares fitted values of the parameters.

Most of the energies $E_\infty$ presented below obtained by Extrapolation C were calculated with the same prescription: we use for the fit all NCSM energies at the minimum of their $\hbar\Omega$ dependance and larger $\hbar\Omega$ values if they fit inequality (10). In addition, we exclude from the fit NCSM results obtained with small $N_{\max}$ values, i. e., results obtained with $N_{\max} < 4$ for positive and $N_{\max} < 3$ for negative parities.

According to the theory of Refs. [37], the ultraviolet cutoff (9) is the only restriction for the use of approximation (7) for the NCSM eigenenergies $E(\hbar\Omega, N_{\max})$. We use a stronger condition (10), however, the open circles in Fig. 1 clearly demonstrate that this condition is

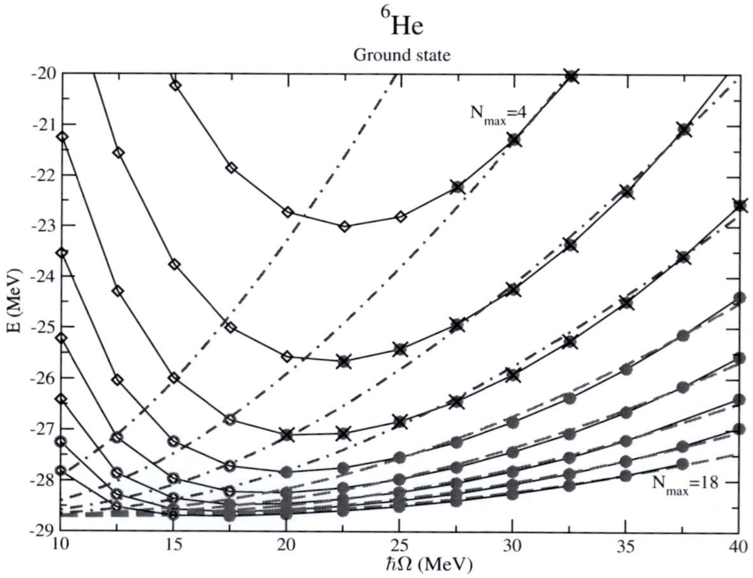


Figure 2. Ground state energy of $^6$He obtained in NCSM calculations with JISP16 $NN$ interaction with various $\hbar\Omega$ values and $N_{\max}$ ranging from 2 to 16 in steps of 2. Eigenenergies marked by crosses were additionally excluded from the fit to obtain a reasonable prediction for $E_\infty$, dash-dotted lines show approximation of these energies by Eq. (7) with parameters obtained by the fit. See Fig. 1 for other details.

insufficient and should be supplemented by additional restrictions. We additionally exclude the eigenenergies $E(\hbar\Omega, N_{\max})$ before the minimum of their $\hbar\Omega$ dependence, however the need for even stronger restrictions become more obvious when the NCSM results obtained with larger $N_{\max}$ are available. The Extrapolation C is seen to reproduce well the NCSM results at sufficiently large $\hbar\Omega$ values. However, it fails around minima of $\hbar\Omega$ dependences for larger $N_{\max}$ where the NCSM eigenenergies lie above the Extrapolation C predictions as seen from Fig. 1. As a result, the Extrapolation C tends to underbind a nucleus predicting sometimes the $E_\infty$ values above the strict variational upperbounds obtained in NCSM calculations. To obtain the Extrapolation C predictions below the variational upperbound, we used additional restrictions for a few levels in some light nuclei where the NCSM results obtained with large $N_{\max}$ values where available. For example, in the case of the $^6$He ground state, we additionally excluded from the fit all states with $N_{\max} < 10$ shown by crosses in Fig. 2. It is seen from Fig. 2 that Extrapolation C reproduces the NCSM eigenenergies obtained with $N_{\max} < 10$ not included in the fit which lie outside the vicinity of the minima. We note however, that inclusion of the NCSM results obtained with small $N_{\max}$ values generally stabilizes the parameter fit of the Extrapolation C especially in cases when the results with large $N_{\max}$ values are not available: note, the vicinities of the minima are excluded for small $N_{\max}$ by the inequality (10). Clearly, the theory of Refs. [37] needs further development to formulate more adequate restrictions for the selection of NCSM eigenstates.

It is very important to estimate the uncertainties of Extrapolation C. It seems natural

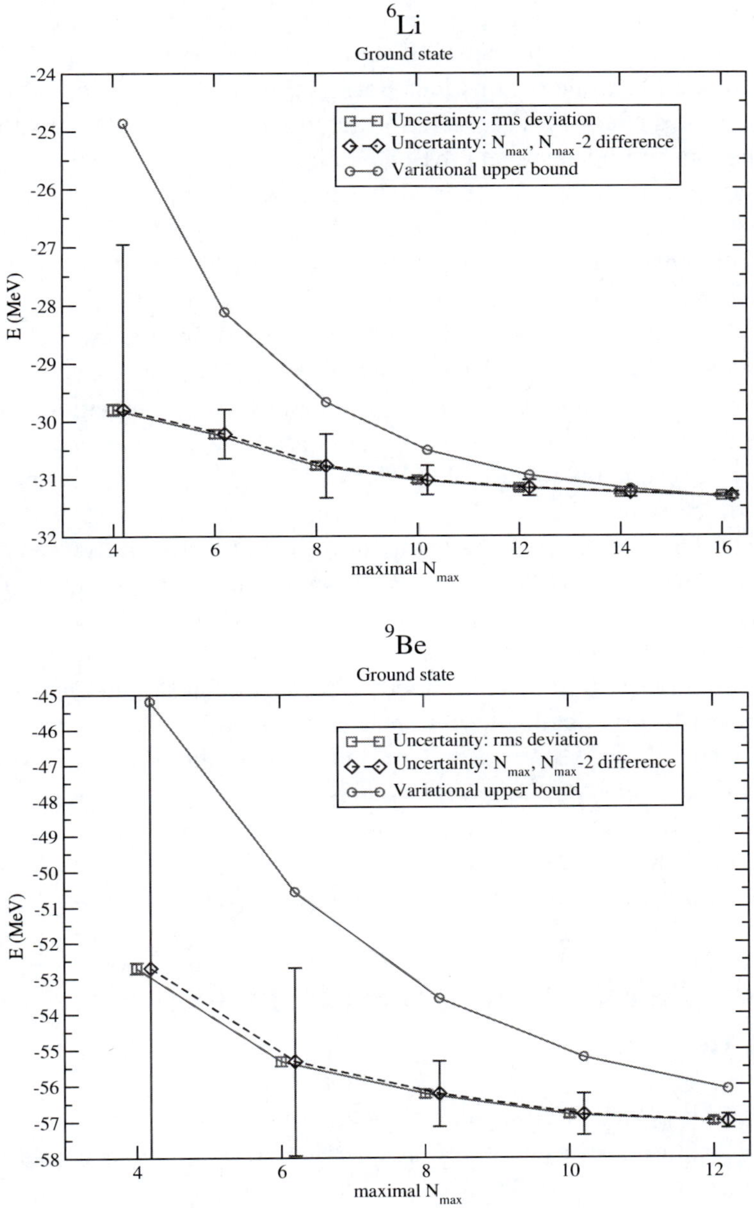


Figure 3. Variational upper bound (circles) and Extrapolation C for the ground state energies of $^6$Li (upper panel) and $^9$Be (lower panel): squares — uncertainties evaluated as a rms deviation of the NCSM results from calculations by Eq. (7) with parameters obtained by the fit; diamonds — uncertainties evaluated as a difference of $E_\infty$ values from the fits to NCSM results obtained with all $N_{\max} \leq$ maximal $N_{\max}$ and with all $N_{\max} \leq$ maximal $N_{\max} - 2$.

to evaluate this uncertainty as a rms difference between the NCSM results and the values of the energy calculated with the same $\hbar\Omega$ and $N_{\max}$ by means of Eq. (7) with parameters obtained by the fit. However this idea does not work. The extrapolation formula (7) accurately reproduces energies acceptable for the fit and the suggested method is found in our test applications to underestimate the uncertainty of the extrapolation. Specifically, if we first calculate $E_\infty$ including in the fit all acceptable NCSM results for a given nucleus obtained with all $N_{\max}$ less or equal to some maximal $N_{\max}$ value $N_{\max}^M$ and next add the NCSM results obtained with $N_{\max} = N_{\max}^M + 2$ to obtain $E_\infty$ by a new fit, we see that the difference between these two $E_\infty$ values is essentially larger than the above estimate of uncertainty (see Fig. 3). In other words, this method of evaluation of extrapolation uncertainty is inconsistent with results obtained by increasing the NCSM basis space.

Therefore we suggest another approach to evaluation of extrapolation uncertainty that is close to the uncertainty evaluation in Extrapolation B. We obtain $E_\infty$ in a fit using all acceptable NCSM results obtained with all $N_{\max} \leq N_{\max}^M$. Next we exclude from the fit the results obtained with $N_{\max} = N_{\max}^M$ to obtain a new value of $E_\infty$. The difference between these two $E_\infty$ values is our estimate for the extrapolation uncertainty. In this case, we get a realistic uncertainty estimate consistent with increasing the NCSM basis space. This is illustrated by extrapolations of the ground state energies of $^6$Li and $^9$Be in Fig. 3.

A more elaborate extrapolation formula was suggested in Ref. [38]. In addition to the exponent involved in Eq. (7), it contains another exponent that should improve the description of the NCSM results obtained with smaller $\hbar\Omega$ values and is expected to describe the $\hbar\Omega$ dependance of $E(\hbar\Omega, N_{\max})$ before and after the minimum. Our numerous attempts to use this formula for extrapolating energies of a large enough set of light nuclei were unsuccessful. It appears that the two-exponent extrapolation formula of Ref. [38] is able to reproduce either the $\hbar\Omega$ dependence of $E(\hbar\Omega, N_{\max})$ at a given $N_{\max}$ or its $N_{\max}$ dependence at fixed $\hbar\Omega$ but fails to reproduce both these dependencies simultaneously.

## 5. Binding Energies and Spectra of Light Nuclei with JISP16 $NN$ Interaction

We present in Tables I and II all available results for energies of states of $s$- and $p$-shell nuclei with the width not exceeding 300 keV — we do not attempt to describe here the broader resonances using an oscillator basis without coupling to the continuum. We however present the results only for one of each pair of mirror nuclei, the nucleus with $Z < N$ since mirror nuclei have similar properties. The tables list nearly all known $s$- and $p$-shell nuclei with $Z \leq N$ which have states with $\Gamma \leq 300$ keV. All experimentally known states with $\Gamma \leq 300$ keV are listed for each nucleus in order to also include the levels that we cannot describe. The only exception is $^{16}$O which has many narrow excited states but we describe only its ground state. In some cases, our failure in description of a particular level is due to the limited supercomputer time — we need to perform more Lanczos iterations to obtain its converged energy in NCSM calculations at large $N_{\max}$ values. However there are levels whose description is not possible due to more fundamental reasons. A well-known example is the so-called Hoyle state, the first excited $0^+$ state in $^{12}$C. The Hoyle state was shown [70, 71] to have a strong contribution of many-body oscillator components with very large $N_{\max}$ values which are impossible to include in calculations with modern computer facilities in the NCSM approach employing the $N_{\max}$ basis truncation. The impacts of the Hoyle state can be found in other nuclei. For example, we find states in $^{13}$C which are formed by coupling of a single-neutron state to the Hoyle state in the $^{12}$C cluster. Such states are, of course, also irreproducible in our calculations but may be eventually treated in a symmetry adapted basis such as demonstrated in Ref. [72].

The $N_{\max}^{M}$ column in the tables shows the maximal $N_{\max}$ values used in NCSM calculation. However, we sometimes present two $N_{\max}^{M}$ values, one of them in square brackets; the respective extrapolated results are also given in square brackets. In some cases, e. g., for $^{8}$Be, we have a very limited set of results with different $\hbar\Omega$ with the largest $N_{\max}$ to perform Extrapolations A and C which is still adequate to can get a reasonable estimate of $E_{\infty}$ by means of Extrapolation B, and we would like to present the Extrapolation B results using the largest NCSM calculations. In other cases, e. g., in $^{11}$B we performed calculations with the largest $N_{\max}$ with the number of Lanczos iterations insufficient to get all excited states converged. We have the results for those states only with smaller $N_{\max}$ values. It seems reasonable to calculate excitation energies of such states using the results for the excited and ground states obtained in the same model space and hence to show these smaller basis space results for the ground state in the table.

## 5.1. Binding Energies

Binding energies are presented in the first row for each nucleus in Table I. The results obtained by Extrapolations A, B and C are consistent.

The uncertainties of Extrapolations A and B are known [34] to be approximately the same for a wide range of nuclei. We see that the uncertainty estimates of Extrapolation C for ground states in $s$ and $p$ shells are often similar to those of A and B. Extrapolation C has a theoretical foundation in contrast with the phenomenological Extrapolations A and B. However the Extrapolation C can be used only for large enough values of $\hbar\Omega$ where ground state energies obtained in NCSM increase with $\hbar\Omega$ and fails to reproduce the $\hbar\Omega$ dependance of eigenenergies $E(\hbar\Omega, N_{\max})$ in the vicinity of their minimum at fixed $N_{\max}$. The vicinity of the minimum of $E(\hbar\Omega, N_{\max})$ is important since the NCSM eigenenergies around this minimum are better approximations to $E_{\infty}$. Extrapolations A and B describe the minimum of $E(\hbar\Omega, N_{\max})$, and this feature seems to compensate the lack of their theoretical grounding. As a result, the Extrapolation C does not seem to be especially advantageous as compared to Extrapolations A and B.

The JISP16 interaction correctly reproduces in most cases the experimental spin-parities of ground states. An exception is $^{11}$Be. This nucleus demonstrates the so-called parity inversion: its ground state $\frac{1}{2}^{+}$ has an unnatural parity. In our calculations the natural parity state $\frac{1}{2}^{-}$ extrapolates below the $\frac{1}{2}^{+}$ state and appears to be the ground state. However the difference in our predictions for energies of $\frac{1}{2}^{+}$ and $\frac{1}{2}^{-}$ is smaller than the uncertainty estimates. Therefore we cannot definitely state that JISP16 does not describe the parity inversion in $^{11}$Be (see also Refs. [43, 46]). Experimentally these states are also found to have very close energies.

**Table 1. Binding energies and spectra of natural parity states. All energies and widths are in MeV if other units are not indicated. The first line for each nucleus gives binding energies, all the remaining lines for this nucleus show excitation energies of excited states in the spectrum. The only exception is $^{11}$Be whose ground state given in the first line is of unnatural parity, the next line gives the binding energy of the lowest state of natural parity and the remaining lines describe excitation energies relative to the lowest state of natural parity. The question mark in the second column indicates states whose isospin is not known (or doubtful) experimentally. The third column shows the maximal $N_{\max}$ value used in our calculations. Columns A, B and C show the results of respective extrapolations. The column Av. $\hbar\Omega$ shows excitation energies obtained by averaging excitation energies from the NCSM calculations with the largest available $N_{\max}$ value (indicated in the third column) over a set of reasonable $\hbar\Omega$ values. The column Av. $N_{\max}$ shows excitation energies obtained by averaging excitation energies from the NCSM calculations with $N_{\max} = N^M_{\max}$, $N_{\max} = N^M_{\max} - 2$, $N_{\max} = N^M_{\max} - 4$, $N_{\max} = N^M_{\max} - 6$. The AV18/IL7 column presents the results of GFMC calculations with Argonne AV18 $NN$ and Illinois IL7 $NNN$ interactions [64, 65, 66, 67, 68]. The last two columns present experimental data from [69]. Uncertainties of the last significant digit are given in round brackets after the value; see text for the method of evaluating uncertainties**

| | $(J^\pi, T)$ | $N^M_{\max}$ | A | B | C | Av. $\hbar\Omega$ Av. $N_{\max}$ | AV18/IL7 | Exp. | Width |
|---|---|---|---|---|---|---|---|---|---|
| $^3$H | $(\frac{1}{2}^+, \frac{1}{2})$ | 20 | 8.368(1) | 8.369(2) | $8.368(^{+0.022}_{-0.001})$ | – | 8.47(0) | 8.482 | 12.32 years |
| $^4$He | $(0^+, 0)$ | 18 | 28.299(4) | 28.299(0) | 28.300(0) | – | 28.43(0) | 28.296 | stable |
| $^6$He | $(0^+, 1)$ | 18 | 28.80(5) | 28.803(6) | 28.706(3) | – | 29.20(3) | 29.269 | 807 ms |
| $^6$He | $(2^+, 1)$ | 18 | 2.2(1) | 2.26(12)(6) | $2.39(^{+0.01}_{-0.03})$ | 2.55(4) 2.63(2) | – | 1.80(3) | 0.113 |
| $^6$Li | $(1^+, 0)$ | 16 | 31.47(9) | 31.49(6) | 31.42(5) | – | 31.93(3) | 31.995 | stable |
| $^6$Li | $(3^+, 0)$ | 16 | 2.56(4) | 2.552(65)(5) | $2.60(^{+0.04}_{-0.05})$ | 2.550(5) 2.50(1) | 2.34(3) | 2.186(2) | 0.024 |
| $^6$Li | $(0^+, 1)$ | 16 | 3.8(1) | 3.68(8)(3) | 3.72(6) | 3.75(2) 3.75(2) | 3.50(3) | 3.563 | $8.2 \cdot 10^{-6}$ |
| $^7$Li | $(\frac{3}{2}^-, \frac{1}{2})$ | 14 | 38.6(1) | 38.6(1) | 38.42(8) | – | 39.0(1) | 39.245 | stable |
| $^7$Li | $(\frac{1}{2}^-, \frac{1}{2})$ | 14 | 0.5(2) | 0.52(6)(2) | 0.6(1) | 0.60(1) 0.73(2) | 0.1(1) | 0.478 | 73 fs |
| $^7$Li | $(\frac{7}{2}^-, \frac{1}{2})$ | 14 | 5.2(2) | 5.25(10)(2) | 5.4(1) | 5.288(7) 5.26(1) | 4.9(1) | 4.630(9) | 0.093 |
| $^7$Li | $(\frac{5}{2}^-, \frac{1}{2})$ | 14 | 7.1(4) | 7.1(2)(1) | 7.4(1) | 7.45(6) 7.88(7) | 6.5(1) | 6.68(5) | 0.88 |
| $^7$Li | $(\frac{5}{2}^-, \frac{1}{2})$ | 14 | 8.0(5) | 8.10(20)(5) | 8.4(1) | 8.36(5) 8.53(3) | 7.7(2) | 7.46(1) | 0.089 |
| $^7$Li | $(\frac{3}{2}^-, \frac{3}{2})$ | 14 | 12.2(2) | 12.1(2)(1) | 12.3(2) | 11.60(4) 11.1(2) | – | 11.24(3) | 0.26 |
| $^8$He | $(0^+, 2)$ | 14 | 29.9(2) | 29.9(2) | 29.4(2) | – | 31.06(15) | 31.409 | 119 ms |
| $^8$Li | $(2^+, 1)$ | 12 | 40.3(3) | 40.3(3) | 40.1(4) | – | 41.5(2) | 41.278 | 840 ms |
| $^8$Li | $(1^+, 1)$ | 12 | 1.5(4) | 1.50(20)(5) | 1.6(4) | 1.71(5) 1.96(6) | 1.4(2) | 0.981 | 8.2 fs |
| $^8$Li | $(3^+, 1)$ | 12 | 2.8(3) | 2.77(20)(1) | 2.8(4) | 2.752(7) 2.65(2) | 3.0(3) | 2.255(3) | 0.032 |
| $^8$Li | $(4^+, 1)$ | 12 | 7.1(5) | 7.1(2)(1) | 7.3(5) | 7.59(8) 8.1(1) | 6.7(3) | 6.53(2) | 0.035 |
| $^8$Li | $(0^+, 2)$ | – | – | – | – | – | – | 10.822 | <0.012 |
| $^8$Be | $(0^+, 0)$ | 10 [12] | 55.2(5) | 55.35(3) [55.5(3)] | 54.8(9) | – | 56.3(1) | 56.500 | $6 \cdot 10^{-6}$ |
| $^8$Be | $(2^+, 1?)$ | 10 [12] | 16.8(5) | 16.86(30)(4) [16.7(1)] | 16.7(9) | 16.66(1) 16.49(4) | 16.8(2) | 16.626(3) | 0.108 |
| $^8$Be | $(2^+, 0?)$ | 10 [12] | 17.2(7) | 17.1(5)(2) [17.3(5)] | 17.4(9) | 17.40(3) 17.47(2) | 16.8(2) | 16.922(3) | 0.074 |
| $^8$Be | $(1^+, 1?)$ | 10 [12] | 18.3(7) | 18.23(50)(7) [18.2(2)] | 18.3(9) | 18.60(7) 18.69(5) | 17.5(2) | 17.640(1) | 0.0107 |
| $^8$Be | $(1^+, 0)$ | 10 [12] | 18.9(9) | 18.7(7)(4) [18.8(3)] | 19(2) | 20.1(2) 20.9(2) | 18.0(2) | 18.150(4) | 0.138 |
| $^8$Be | $(3^+, 1?)$ | 10 [12] | 19.5(5) | 19.6(3)(1) [19.5(1)] | 19.5(8) | 19.33(3) 19.05(7) | 19.4(2) | 19.07(1) | 0.271 |
| $^8$Be | $(3^+, 0?)$ | 10 [12] | 21(1) | 20.0(9)(3) [19.8(2)] | 21.1(8) | 21.7(2) 22.0(3) | 19.9(2) | 19.24(1) | 0.227 |
| $^8$Be | $(0^+, 2)$ | – | – | – | – | – | – | 27.494 | 0.0055 |
| $^9$Li | $(\frac{3}{2}^-, \frac{3}{2})$ | 12 | 43.6(3) | 43.7(3) | 43.4(4) | – | 45.2(3) | 45.342 | 178 ms |
| $^9$Be | $(\frac{3}{2}^-, \frac{1}{2})$ | 10 [12] | 57.0(6) | 57.09(5) [57.2(3)] | 56.8(6) [57.0(2)] | – | 58.1(2) | 58.167 | stable |
| $^9$Be | $(\frac{5}{2}^-, \frac{1}{2})$ | 10 [12] | 2.82 | 2.9(4)(1) [2.8(1)] | 2.9(6) | 2.960(9) 3.12(3) | 2.4(2) | 2.429(1) | 0.001 |
| $^9$Be | $(\frac{3}{2}^-, \frac{3}{2})$ | 10 | 15.1(6) | 15.26(33)(5) | 15.4(7) | 14.96(5) 14.5(1) | – | 14.392(2) | $365 \cdot 10^{-6}$ |
| $^{10}$Li | $(1^+, 2?)$ | 10 | 42(1) | 42.0(2) | 41(1) | – | | 45.11(4) | 0.1 |
| $^{10}$Be | $(0^+, 1)$ | 10 | 64.1(5) | 64.0(4) | 64.0(4) | – | 64.1(3) | 64.979 | $1.5 \cdot 10^6$ years |
| $^{10}$Be | $(2^+, 1)$ | 10 | 3.5(6) | 3.53(30)(4) | 3.6(5) | 3.69(2) 3.89(4) | 3.4(3) | 3.368 | 125 fs |
| $^{10}$Be | $(2^+, 1)$ | 10 | 6.2(9) | 6.1(6)(2) | 7(1) | 7.2(1) 8.3(2) | 5.3(3) | 5.958 | < 55 fs |
| $^{10}$Be | $(0^+, 1)$ | – | – | – | – | – | – | 6.179 | 1 ps |
| $^{10}$Be | $(2^+, 1)$ | 10 | 9.5(7) | 9.6(5)(1) | 9.1(3) | 10.13(5) 11.2(2) | – | 7.542 | 0.006 |
| $^{10}$Be | $(2^+, 1?)$ | 10 | 11(1) | 10.8(9)(7) | 12(2) | 11.2(5) 10.5(4) | – | 9.560 | 0.141 |

| | | | | | | | | | |
|---|---|---|---|---|---|---|---|---|---|
| $^{10}$B | $(3^+,0)$ | 10 [8] | 63.8(5) [63(1)] | 63.7(3) [64.0(2)] | 63.6(4) [63(2)] | – | 64.8(9) | 64.753 | stable |
| $^{10}$B | $(1^+,0?)$ | 10 | 0.4(8) | 0.9(2.4)(0.9) | 1(1) | 2.4(8) 5.3(7) | 0.7(8) | 0.718 | 0.707 ns |
| $^{10}$B | $(0^+,1?)$ | 8 | 2(1) | 1.78(1.5)(0.03) | 2(1) | 2.3(2) 2.4(1) | 2.2(2) | 1.74 | 4.9 fs |
| $^{10}$B | $(1^+,0?)$ | 10 | 2(1) | 4.1(9)(3) | 2.9(9) | 3.4(4) 4.5(4) | 3.0(6) | 2.154 | 1.48 ps |
| $^{10}$B | $(2^+,0?)$ | 10 | 3.4(8) | 3.8(8)(4) | 3.7(5) | 4.30(6) 5.9(2) | 3.9(5) | 3.587 | 102 fs |
| $^{10}$B | $(3^+,0)$ | 10 | 6(1) | 5.7(9)(5) | 6.2(8) | 7.3(2) 9.0(2) | – | 4.774 | $8 \cdot 10^{-6}$ |
| $^{10}$B | $(2^+,1)$ | 8 | 6(1) | 5.3(2.0)(0.3) | 5(2) | 6.1(1) 6.30(9) | 5.8(3) | 5.164(1) | $2 \cdot 10^{-6}$ |
| $^{10}$B | $(1^+,0?)$ | – | – | – | – | – | – | 5.182(8) | 0.11 |
| $^{10}$B | $(2^+,0)$ | 10 | 5(1) | 5.2(9)(5) | 5(1) | 7.0(6) 10.1(7) | – | 5.920 | 0.006 |
| $^{10}$B | $(4^+,0?)$ | 10 | 5.3(7) | 5.7(4)(1) | 5.5(6) | 5.72(8) 6.5(1) | 5.6(5) | 6.025 | 0.00005 |
| $^{10}$B | $(3^+,0)$ | 10 | 9(1) | 9.5(9)(6) | 9.1(7) | 7.0(6) 10.1(7) | – | 7.004(5) | 0.098 |
| $^{10}$B | $(2^+,?)$ | – | – | – | – | – | – | 7.470(4) | 0.065 |
| $^{10}$B | $(0^+,1)$ | – | – | – | – | – | – | 7.5599(4) | 0.003 |
| $^{10}$B | $(2^+,1?)$ | 8 | 8(3) | 8.2(2.4)(0.5) | 9(2) | 10.3(3) 11.2(2) | – | 8.895 | 0.039 |
| $^{11}$Li | $(\frac{3}{2}^-,\frac{5}{2})$ | 8 | 42(3) | 42.4(1) | 40(2) | – | | 45.65(2) | 8.75 ms |
| $^{11}$Be | $(\frac{1}{2}^+,\frac{3}{2})$ | 11 [9] | [62(3)] | 63.3(8) | 60(3) | – | | 65.483 | 13.8 s |
| $^{11}$Be | $(\frac{1}{2}^-,\frac{3}{2}?)$ | 10 | 64.0(9) | 64.0(6) | 63(1) | – | | 65.163 | 115 fs |
| $^{11}$Be | $(\frac{3}{2}^-,\frac{3}{2}?)$ | 10 | 3(1) | 2.7(7)(2) | 4(2) | 3.36(7) 3.61(8) | | 2.33(1) | 0.206 |
| $^{11}$Be | $(\frac{3}{2}^-,\frac{3}{2}?)$ | 10 | – | – | – | – | | 3.080(6) | 0.122 |
| $^{11}$Be | $(\frac{3}{2}^-,\frac{3}{2}?)$ | 10 | – | – | – | – | | 3.635(1) | 0.01 |
| $^{11}$Be | $(\frac{5}{2}^-,\frac{3}{2}?)$ | 10 | 3.4(9) | 3.53(40)(5) | 3(1) | 3.2(1) 3.1(1) | | 4.935(3) | 0.45 |
| $^{11}$Be | $(\frac{3}{2}^-,\frac{3}{2}?)$ | 10 | – | – | – | – | | 7.70(2) | 0.23 |
| $^{11}$Be | $(\frac{3}{2}^-,\frac{3}{2}?)$ | 10 | 6(1) | 10(3)(2) | 9(2) | 8.9(4) 10.4(4) | | 8.49(3) | 0.2 |
| $^{11}$Be | $(\frac{5}{2}^-,\frac{3}{2}?)$ | 10 | 9(1) | 9.2(5)(2) | 9(1) | 9.69(2) 10.2(1) | | 10.27(5) | 0.21 |
| $^{11}$B | $(\frac{3}{2}^-,\frac{1}{2})$ | 10 [8] | 76.0(4) [75(1)] | 76.1(4) [76(1)] | 76.1(8)[75(2)] | – | | 76.206 | stable |
| $^{11}$B | $(\frac{1}{2}^-,\frac{1}{2}?)$ | 10 | 0.8(6) | 0.7(4)(1) | 1(1) | 1.41(7) 2.5(2) | | 2.125 | 3.8 fs |
| $^{11}$B | $(\frac{5}{2}^-,\frac{1}{2}?)$ | 10 | 3.2(6) | 3.2(4)(1) | 2.9(7) | 3.73(4) 4.6(2) | | 4.445 | 0.82 fs |
| $^{11}$B | $(\frac{3}{2}^-,\frac{1}{2}?)$ | 10 | 4.3(7) | 4.3(4)(2) | 4(1) | 5.7(1) 7.4(3) | | 5.020 | 0.24 fs |
| $^{11}$B | $(\frac{7}{2}^-,\frac{1}{2}?)$ | 10 | 5.9(6) | 5.9(4)(1) | 6(1) | 6.57(6) 7.6(2) | | 6.743(2) | 15 fs |
| $^{11}$B | $(\frac{5}{2}^-,\frac{1}{2}?)$ | 8 | 10(2) | 10.0(1.0)(0.4) | 10(3) | 11.8(2) 13.1(3) | | 8.920(2) | 0.82 fs |
| $^{11}$B | $(\frac{3}{2}^-,\frac{1}{2}?)$ | 8 | 12(2) | 12(4)(1) | 13(3) | 14.1(1) 15.7(4) | | 10.262(8) | 0.163 |
| $^{11}$B | $(\frac{5}{2}^-,\frac{1}{2}?)$ | – | – | – | – | – | | 10.33(1) | 0.112 |
| $^{11}$B | $(\frac{5}{2}^-,\frac{1}{2}?)$ | – | – | – | – | – | | 11.89(1) | 0.194 |
| $^{11}$B | $(\frac{1}{2}^-,\frac{1}{2}?)$ | 8 | 12(2) | 11.3(2.8)(0.7) | 12(3) | 13(1) 14(1) | | 12.92(1) | 0.23 |
| $^{12}$Be | $(0^+,2)$ | 10 [8] | 66.9(8) [66(4)] | 67(1) [67(2)] | 65.2(4) [64(1)] | – | | 68.651 | 21.3 ms |
| $^{12}$Be | $(2^+,2)$ | 8 | 3(3) | 3.03(3.20)(0.02) | 2(3) | 2.78(9) 2.64(7) | | 2.102(12) | ? |
| $^{12}$B | $(1^+,1)$ | 10 [8] | [79(2)] | 80.0(5) [80(2)] | 78.80(7) [79(3)] | – | | 79.577(2) | 20.2 ms |
| $^{12}$B | $(2^+,1?)$ | 10 [8] | [-0.2(1.6)] | 0.2(7)(3) | -0.3(4) | -1.0(2) -1.5(2) | | 0.953 | 180 fs |
| $^{12}$B | $(2^+,1?)$ | 10 [8] | [3(2)] | 3.1(1.2)(0.4) | 3.2(3) | 3.9(1) 5.4(2) | | 3.759(6) | 0.040 |
| $^{12}$B | $(1^+,1?)$ | 8 | 4(2) | 3.3(1.8)(0.5) | 4(4) | 5.32(8) 6.7(3) | | 5.00(2) | 0.050 |
| $^{12}$B | $(3^+,1?)$ | 8 | 6(2) | 5.1(3.1)(0.7) | 6(4) | 7.5(1) 8.9(3) | | 5.612(8) | 0.110 |
| $^{12}$B | $(1^+,1?)$ | 8 | 7(3) | 7.4(1.3)(0.7) | 7(4) | 11.3(4) 13.5(6) | | 6.600 | 0.140 |
| $^{12}$B | $(0^+,2)$ | 6 | – | 17(3)(1) | 18(5) | 20.9(8) 21.9(7) | | 12.75(5) | 0.085 |
| $^{12}$C | $(0^+,0)$ | 10 [8] | 94.9(4) [94.1] | 94.8(4) [95(3)] | 95.3(3) [95(2)] | – | 93.3(4) | 92.163 | stable |
| $^{12}$C | $(2^+,0)$ | 10 | 3.8(5) | 3.86(40)(6) | 3.7(3) | 4.19(3) 5.1(2) | 3.9(9) | 4.439 | $10^{-8}$ |
| $^{12}$C | $(0^+,0)$ | – | – | – | – | – | 10.4(5) | 7.654 | $8.5 \cdot 10^{-6}$ |
| $^{12}$C | $(1^+,0)$ | 10 | 12.7(5) | 12.76(60)(6) | 12.7(4) | 12.91(6) 14.0(3) | | 12.710(6) | $18 \cdot 10^{-6}$ |
| $^{12}$C | $(4^+,0)$ | 10 | 13.5(8) | 13.6(9)(2) | 13.3(5) | 14.95(9) 17.4(4) | | 14.083 | 0.258 |
| $^{12}$C | $(1^+,1)$ | 10 | 17.8(7) | 17.5(6)(3) | 18(1) | 19.2(2) 20.6(3) | | 15.110(3) | $44 \cdot 10^{-6}$ |
| $^{12}$C | $(2^+,1)$ | 10 | 17(3) | 18.6(1.0)(0.9) | 17.8(8) | 18.0(2) 19.2(2) | | 16.106(1) | 0.005 |
| $^{12}$C | $(0^+,1)$ | 10 [8] | [19(2)] | [18(5)(1)] | 20(1) | 20.9(6) 24.0(6) | | 17.76(2) | 0.08 |
| $^{12}$C | $(2^+,1)$ | 8 | 21(2) | 20.3(2.9)(0.7) | $24(^{+3}_{-6})$ | 24.4(4) 26.2(5) | | 18.80(4) | 0.1 |
| $^{12}$C | $(1^+,?)$ | – | – | – | – | | – | – 19.69 | 0.23 |
| $^{12}$C | $(0^+,2)$ | – | – | – | – | – | – | 27.595(24) | 0.08 |
| $^{13}$B | $(\frac{3}{2}^-,\frac{3}{2})$ | 8 | 85(2) | 87(4) | 86(3) | – | | 84.455(1) | 17.33 ms |
| $^{13}$C | $(\frac{1}{2}^-,\frac{1}{2})$ | 8 [6] | 101(1) | 102(3) [98(5)] | 102(3) [99(6)] | – | | 97.110 | stable |
| $^{13}$C | $(\frac{3}{2}^-,\frac{1}{2}?)$ | 8 | 4(1) | 3.3(5.2)(0.5) | 4(4) | 6.0(2) 7.5(3) | | 3.685 | 1.10 fs |
| $^{13}$C | $(\frac{5}{2}^-,\frac{1}{2}?)$ | 8 | 6(1) | 5.5(5.2)(0.9) | 9.5(8) | 7.89(8) 9.1(2) | | 7.547(3) | 0.0012 |
| $^{13}$C | $(\frac{1}{2}^-,\frac{1}{2}?)$ | 8 | 11(1) | 10.7(2.9)(0.6) | 10(4) | 13.3(3) 15.1(5) | | 8.86(2) | 0.15 |
| $^{13}$C | $(\frac{3}{2}^-,\frac{1}{2})$ | – | – | – | – | – | | 9.897(5) | 0.026 |
| $^{13}$C | $(\frac{7}{2}^-,\frac{1}{2})$ | – | – | – | – | – | | 10.753(4) | 0.055 |
| $^{13}$C | $(\frac{1}{2}^-,\frac{1}{2}?)$ | 6 | – | 14(13)(1) | 15(13) | 14.2(3) 15.3(5) | | 11.080(5) | < 0.004 |
| $^{13}$C | $(\frac{3}{2}^-,\frac{1}{2}?)$ | 6 | – | 12.0(8.0)(0.6) | 7(2) | 15.12(8) 16.0(3) | | 11.75(1) | 0.11 |
| $^{13}$C | $(\frac{5}{2}^-,\frac{1}{2}?)$ | 6 | – | 18.3(10.0)(0.5) | 19(11) | 28.0(7) 30(1) | | 12.13(5) | 0.08 |
| $^{13}$C | $(\frac{3}{2}^-,\frac{1}{2}?)$ | 6 | – | 14(15)(1) | 19(10) | 23.3(6) 25.4(9) | | 12.19(1) | 0.15 |
| $^{13}$C | $(\frac{7}{2}^-,\frac{1}{2})$ | 6 | – | 14.8(12.4)(0.5) | 16(12) | 21.7(7) 23.5(8) | | 12.44(1) | 0.14 |
| $^{14}$B | $(2^-,2)$ | 8 | 84(3) | 87(4) | 83(5) | – | | 85.43(2) | 12.5 ms |
| $^{14}$C | $(0^+,1)$ | 8 | 112(1) | 114(6) | 112(3) | – | | 105.284 | 5700 years |
| $^{14}$C | $(0^+,1)$ | 8 | – | – | – | – | | 6.589 | 66 ps |
| $^{14}$C | $(2^+,1)$ | 8 | 9(1) | 9(7)(1) | 8(6) | 12.2(4) 14.0(5) | | 7.012 | 9.0 fs |
| $^{14}$C | $(2^+,1)$ | 8 | – | – | – | – | | 8.318 | 0.003 |
| $^{14}$C | $(4^+,1)$ | 8 | – | – | – | – | | 10.736 | 0.02 |
| $^{14}$C | $(1^+,1)$ | 8 | 13(1) | 12(7)(1) | 12(6) | 16.7(5) 18.8(5) | | 11.306 | 0.046 |

**Table 1. (Continued)**

| | | | | | | | | | |
|---|---|---|---|---|---|---|---|---|---|
| $^{14}$N | $(1^+,0)$ | 8 | 114(2) | 116(5) | 114.8(7) | – | | 104.659 | stable |
| $^{14}$N | $(1^+,0)$ | 8 | 7(2) | 6.7(5.3)(0.9) | 7(5) | 10.4(5) 12.3(6) | | 3.948 | 4.8 fs |
| $^{14}$N | $(1^+,0)$ | 8 | – | – | – | – | | 6.2035 | 111 fs |
| $^{14}$N | $(3^+,0)$ | 8 | – | – | – | – | | 6.4462 | 430 fs |
| $^{14}$N | $(2^+,0)$ | 8 | 8(2) | 8(8)(1) | 9(4) | 10.39(8) 12.0(3) | | 7.029 | 3.7 fs |
| $^{14}$N | $(5^+,0)$ | 8 | – | – | – | – | | 8.964 | 73 fs |
| $^{14}$N | $(2^+,0)$ | 8 | – | – | – | – | | 8.980 | 0.008 |
| $^{15}$N | $(\frac{1}{2}^-,\frac{1}{2})$ | 8 | 128(1) | 131(8) | 129(2) | – | | 115.494 | stable |
| $^{16}$O | $(0^+,0)$ | 8 | 144(1) | 145(8) | 145(3) | – | | 127.619 | stable |

A similar situation is observed in $^{12}$B. Its ground state is known from experiment to be a $1^+$ state, the first excited state is a $2^+$ state with an excitation energy of about 1 MeV. In our calculations these states extrapolate to nearly the same energy with uncertainties exceeding the energy difference between the levels.

It is conventionally supposed [73, 74, 19] that one cannot describe the spin of the $^{10}$B ground state without the use of $NNN$ forces. The $^{10}$B ground state is known to be a $3^+$ state and there are two low-lying $1^+$ states with excitation energies about 700 keV and 2.15 MeV in the spectrum. These $1^+$ states are mixed in the NCSM calculations and they cross each other as $\hbar\Omega$ increases. One should be very careful in separating these states and assigning them to two different sets of states that should be extrapolated independently. A small change in this state assignment can result in large changes in their extrapolated energies. In our calculations with a purely two-nucleon interaction JISP16, we obtain the correct spin of the ground state and one of the $1^+$ states lies just above it. However, due to the extrapolation uncertainties, we cannot make an unambiguous statement that JISP16 generates the $1^+$ state above the $3^+$; moreover, the change in state assignment can change the ordering of these states. It is interesting to note that in the GFMC calculations with AV18 $NN$ and IL7 $NNN$ forces, the uncertainties of theoretical results also prevent unambiguous confirmation of the correct ordering of these states (see Table I).

Generally, the JISP16 $NN$ interaction provides a good description of binding energies of nuclei with $N \approx Z$ up to $A = 12$ and overbinds nuclei at the end of $p$ shell. This is illustrated by Fig. 4 where we present some of the results from Table I. It is seen that $s$-shell nuclei and those at the beginning of the $p$ shell are slightly underbound, the difference with the experimental binding does not exceed a few percent. Starting from $A = 12$, nuclei with $Z \approx N$ are overbound and the difference with experiment increases with $A$. In particular, $^{16}$O is overbound by more than 13 %. Another drawback of JISP16 interaction is that it predicts a faster decrease of binding energies in a set of isobars at a fixed value of $A$ with increase of the isospin $I$ of their ground state.

Rms deviations of the JISP16 binding energies predictions from the experiment are presented in Table 3. The rms deviation of absolute energies $E_i^{th}$ from experimental values $E_i^{exp}$ for nuclei with $A \leq 12$ is about 1.7 MeV for Extrapolation B and about 2.3 MeV for Extrapolation C. These deviations essentially increase by adding all other examined nuclei with $12 < A \leq 16$ to approximately 5.5 MeV. However we study nuclei in a wide range of masses. A 1 MeV discrepancy in description of $^3$H which has a binding energy of about 8 MeV will be interpreted as a significant failure of the theory while the same 1 MeV discrepancy in description of $^{16}$O bound by approximately 130 MeV will be treated

**Table 2. Binding energies and spectra of unnatural parity states. The first line for each nucleus gives binding energies for the ground state which is a state of natural parity; the next line gives binding energies for the lowest state of unnatural parity; all the remaining lines for this nucleus show excitation energies of excited states of unnatural parity relative to the lowest state of the same parity. The only exception is $^{11}$Be whose ground state given in the first line is of unnatural parity. See Table 1 for other details**

| | $(J^\pi,T)$ | $N_{\max}^M$ | A | B | C | Av. $\hbar\Omega$ Av. $N_{\max}$ | AV18/IL7 | Exp. | Width |
|---|---|---|---|---|---|---|---|---|---|
| $^{8}$Be | $(0^+,0)$ | 10 [12] | 55.2(5) | 55.35(3) [55.5(3)] | 54.8(9) | – | 56.3(1) | 56.500 | $6\cdot10^{-6}$ |
| $^{8}$Be | $(2^-,0?)$ | 11 | 36(1) | 35.9(3) | 34(2) | – | | 37.59 | 0.122 |
| $^{9}$Be | $(\frac{3}{2}^-,\frac{1}{2})$ | 10 [12] | 57.0(6) | 57.09(5) [57.2(3)] | 56.8(6) [57.0(2)] | – | 58.1(2) | 58.167 | stable |
| $^{9}$Be | $(\frac{1}{2}^+,\frac{1}{2}?)$ | 11 | 54(1) | 53.8(6) | 52(1) | – | | 56.49(2) | 0.214 |
| $^{9}$Be | $(\frac{5}{2}^+,\frac{1}{2}?)$ | 11 | 2(1) | 1.62(80)(1) | 2(1) | 1.47(2) 1.32(4) | | 1.372(9) | 0.282 |
| $^{10}$Be | $(0^+,1)$ | 10 | 64.1(5) | 64.0(4) | 64.0(4) | – | 64.1(3) | 64.979 | $1.5\cdot10^{6}$ years |
| $^{10}$Be | $(3^-,1)$ | – | – | – | – | – | | 57.608(1) | 0.016 |
| $^{10}$Be | $(3^-,1?)$ | 9 | 54(3) | 54(2) | 52(3) | – | | 54.83(2) | 0.296 |
| $^{10}$B | $(3^+,0)$ | 10 | 63.8(5) | 63.7(3) | 63.6(4) | – | 64.8(9) | 64.753 | stable |
| $^{10}$B | $(2^-,0)$ | 9 | 56(3) | 56(2) | 55(3) | – | | 59.643 | 0.001 |
| $^{10}$B | $(3^-,0?)$ | 9 | 1(3) | 1.34(2.14)(0.04) | 2(3) | 1.40(2) 1.35(2) | | 1.019(3) | 0.0015 |
| $^{10}$B | $(1^-,0?)$ | 9 | 2(3) | 1.5(1.5)(0.5) | 2(3) | 2.82(7) 3.07(6) | | 1.765(5) | 0.120 |
| $^{10}$B | $(1^-,1?)$ | 9 | 3(3) | 3.4(1.7)(1.3) | 2(3) | 2.8(2) 2.8(1) | | 2.318(5) | 0.094 |
| $^{10}$B | $(2^-,0?)$ | 9 | 3(3) | 3.2(2.0)(0.2) | 3(3) | 3.87(5) 4.10(5) | | 2.369(2) | 0.074 |
| $^{10}$B | $(2^-,1?)$ | 9 | 3(3) | 3.35(1.40)(0.04) | 3(3) | 3.2(1) 3.1(1) | | 2.64(3) | 0.21 |
| $^{10}$B | $(1^-,0?)$ | – | – | – | – | – | | 2.70(2) | 0.26 |
| $^{10}$B | $(3^-,1?)$ | 9 | 5(3) | 4.59(1.58)(0.02) | 4(3) | 4.32(9) 4.15(7) | | 3.777(3) | 0.096 |
| $^{10}$B | $(2^-,1)$ | 9 [7] | – | 4(2)(1) | [8.5(5)] | 6.0(2) 6.9(2) | | 14.18 | 0.19 |
| $^{11}$Be | $(\frac{1}{2}^+,\frac{3}{2})$ | 11 [9] | [62(3)] | 63.3(8) | 60(3) | – | | 65.483 | 13.8 s |
| $^{11}$Be | $(\frac{5}{2}^+,\frac{3}{2}?)$ | 11 [9] | [2(3)] | 1.7(1.9)(0.2) | 1(2) | 1.58(8) 1.43(7) | | 1.783(4) | 0.1 |
| $^{11}$Be | $(\frac{1}{2}^+,\frac{3}{2}?)$ | 11 [9] | [8(3)] | 8(2)(1) | 8(4) | 9.3(3) 9.8(2) | | 5.85(1) | 0.139 |
| $^{11}$B | $(\frac{3}{2}^-,\frac{1}{2})$ | 10 | 76.0(4) | 76.1(4) | 76.1(8) | – | | 76.206 | stable |
| $^{11}$B | $(\frac{1}{2}^+,\frac{1}{2}?)$ | 9 | 66(3) | 66(2) | 64(3) | – | | 69.414 | $39\cdot10^{-8}$ |
| $^{11}$B | $(\frac{5}{2}^+,\frac{1}{2}?)$ | 9 | 0.4(3.0) | 0.7(1.5)(0.2) | 1(2) | 0.114(9) -0.09(5) | – | 0.494 | $1\cdot10^{-6}$ |
| $^{11}$B | $(\frac{3}{2}^+,\frac{1}{2}?)$ | 9 | 2(3) | 1.36(3.09)(0.03) | 2(3) | 1.93(2) 2.03 (3) | – | 1.186 | $1\cdot10^{-6}$ |
| $^{11}$B | $(\frac{7}{2}^+,\frac{1}{2}?)$ | 9 | 2(3) | 1.9(1.3)(0.3) | 1(2) | 0.82(8) 0.4(1) | | 2.392(1) | $1\cdot10^{-6}$ |
| $^{11}$B | $(\frac{5}{2}^+,\frac{1}{2}?)$ | 9 | 2(3) | 1.6(2.0)(0.2) | 2(2) | 1.38(2) 1.25(3) | – | 2.480(1) | 0.004 |
| $^{11}$B | $(\frac{3}{2}^+,\frac{1}{2}?)$ | – | – | – | – | – | – | 3.081(4) | 0.109 |
| $^{11}$B | $(\frac{7}{2}^+,\frac{1}{2}?)$ | 9 | 4(3) | 4.0(2.4)(0.1) | 4(2) | 4.15(4) 4.21(2) | | 3.810(4) | 0.091 |
| $^{11}$B | $(\frac{9}{2}^+,\frac{1}{2}?)$ | 9 | 4(3) | 3.9(1.7)(0.3) | 4(2) | 2.89(9) 2.4(1) | | 4.47(1) | 0.11 |
| $^{11}$B | $(\frac{5}{2}^+,\frac{1}{2}?)$ | – | – | – | – | – | – | 4.80(2) | 0.18 |
| $^{11}$B | $(\frac{1}{2}^+,\frac{3}{2}?)$ | 9 | 5(2) | 5.7(1.6)(0.2) | 6(4) | 4.9(1) 4.5(2) | | 5.75(1) | 0.205 |
| $^{11}$B | $(\frac{5}{2}^+,\frac{3}{2})$ | – | – | – | – | – | | 7.54(2) | 0.253 |
| $^{12}$B | $(1^+,1)$ | 8 [10] | 79(2) | 80(2) [80.0(5)] | 78.80(7) [79(3)] | – | | 79.577(2) | 20.2 ms |
| $^{12}$B | $(2^-,1?)$ | 9 | – | 76(2) | 73(6) | – | | 77.903(1) | <35 fs |
| $^{12}$B | $(1^-,1?)$ | 9 | – | 1.0(2.1)(0.2) | 1(3) | 1.54(4) 1.78(7) | | 0.94 | <49 fs |
| $^{12}$B | $(3^-,1?)$ | 9 | – | 1.4(2.1)(0.1) | 1(2) | 1.50(5) 1.43(5) | | 1.715(2) | $3\cdot10^{-6}$ |
| $^{12}$B | $(1^-,1?)$ | 9 | – | 1.5(1.7)(0.3) | 1(4) | 3.27(4) 4.4(2) | | 2.627(7) | 0.009 |
| $^{12}$B | $(4^-,1?)$ | 9 | – | 3.1(2.4)(0.2) | 3(2) | 3.19(4) 3.08(5) | | 2.844(8) | 0.11 |
| $^{12}$B | $(3^-,1?)$ | 9 | – | 3.53(2.51)(0.05) | 3(2) | 4.40(7) 5.1(2) | | 4.052(8) | 0.05 |
| $^{12}$B | $(2^-,1?)$ | 9 | – | 5.5(2.3)(0.4) | 4(6) | 7.1(1) 8.0(2) | | 5.99 | 0.045 |
| $^{12}$B | $(1^-,1?)$ | 9 | – | 5.0(2.6)(0.5) | 6(4) | 7.64(5) 9.2(4) | | 6.16(2) | 0.06 |
| $^{12}$B | $(3^-,1?)$ | 9 | – | 6.6(2.4)(0.2) | 7(4) | 8.35(7) 9.3(2) | | 6.566(3) | 0.065 |
| $^{12}$B | $(1^-,1?)$ | 9 | – | 5.9(4.4)(0.4) | 8(2) | 7.51(3) 7.84(6) | | 7.36(2) | 0.095 |
| $^{12}$B | $(3^-,1?)$ | 9 | – | 7.0(2.3)(0.4) | 7(4) | 8.1(9) 8.6(6) | | 7.911(5) | 0.034 |
| $^{12}$C | $(0^+,0)$ | 10 | 94.9(4) | 94.8(4) | 95.3(3) | – | 93.3(4) | 92.163 | stable |
| $^{12}$C | $(3^-,0)$ | 9 | 79(3) | 80(2) | 79(4) | – | | 82.522(5) | 0.034 |
| $^{12}$C | $(2^-,0)$ | 9 | 3(2) | 2.5(2.4)(0.1) | 2(2) | 2.8(1) 2.8(1) | | 2.19(2) | 0.26 |
| $^{12}$C | $(2^-,1)$ | 9 | 7(2) | 6.7(1.8)(0.1) | 6(3) | 5.88(4) 5.48(7) | | 6.93 | 0.3 |
| $^{12}$C | $(3^-,1)$ | 9 | 8(2) | 8.1(2.2)(0.1) | 8(3) | 7.35(6) 6.86(9) | | 8.71(5) | 0.22 |
| $^{12}$C | $(1^-,1)$ | 9 | 8(2) | 7.7(2.2)(0.2) | 8(4) | 7.44(8) 7.28(8) | | 12.76(4) | 0.275 |
| $^{12}$C | $(1^-,1)$ | – | – | – | – | – | | 13.88(3) | 0.23 |

**Table 3. Rms deviations of theoretical binding energies from experimental data. For comparison with GFMC AV18/IL7 results, we select only nuclei for which the AV18/IL7 results are available (see Tables 1 and 2)**

| | Number of nuclei | JISP16 | | AV18/IL7 |
|---|---|---|---|---|
| | | Extrap. B | Extrap. C | |
| Comparison of JISP16 and AV18/IL7 results | | | | |
| Absolute energies (MeV) | 13 | 1.16 | 1.44 | 0.43 |
| Relative energies | 13 | 0.023 | 0.029 | 0.007 |
| Energies per nucleon (MeV) | 13 | 0.12 | 0.16 | 0.04 |
| $A \leq 12$ ground states | | | | |
| Absolute energies (MeV) | 19 | 1.68 | 2.33 | – |
| Relative energies | 19 | 0.033 | 0.047 | – |
| Energies per nucleon (MeV) | 19 | 0.16 | 0.22 | – |
| All ground states | | | | |
| Absolute energies (MeV) | 26 | 5.63 | 5.49 | – |
| Relative energies | 26 | 0.055 | 0.059 | – |
| Energies per nucleon (MeV) | 26 | 0.39 | 0.39 | – |

as a significant success of the theory. Therefore we present in Table 3 also the rms deviations of the energy per nucleon predictions, i. e., rms deviations of $E_i^{th}/A$ from the respective experimental data. In addition, we also present in Table 3 the rms for relative energies $(E_i^{th} - E_i^{exp})/E_i^{exp}$. We believe that these rms deviations are more instructive for the statistical analysis of energies when nuclei differ widely in mass.

Table 3 also compares the JISP16 binding energy predictions with those of GFMC obtained with AV18 $NN$ and IL7 $NNN$ interactions. For this comparison, we use only 13 nuclei for which the AV18/IL7 results are available (see Table 1). The AV18 + IL7 strong interaction model appears to be more accurate in description of light nuclei bindings. However the JISP16 predictions seem reasonable: the rms deviations of relative energies for all 26 nuclei studied does not exceed 6 %.

## 5.2. Spectra of Excited States

The energies of excited states can be obtained by the same Extrapolations A, B and C as of the ground states. These results are summarized in Tables 1 and 2. Note that, for excited states, we present in the Tables excitation energies which are, of course, the difference between the energy of the given state and the ground state energy. For natural parity states in $^{11}$Be where the ground state has unnatural parity and for unnatural parity states in other nuclei, we present excitation energies relative to the lowest state of the same parity. The numbers in brackets after excitation energies shown in the Tables are evaluated as uncertainties of the respective absolute energy extrapolation of the excited state.

Our statistical analysis is presented in Table 4. We see that JISP16 provides a less accurate description than the AV18 + IL7 interaction model of 38 states in light nuclei for which AV18/Il7 results are available. However, even for those states, the rms deviation of relative

**Table 4. Rms deviations of theoretical level energies from experimental data. For comparison with GFMC AV18/IL7 results, we select only states for which the AV18/IL7 results are available (see Tables 1 and 2)**

| | Number of levels | JISP16 | | AV18/IL7 |
|---|---|---|---|---|
| | | Extrap. B | Extrap. C | |
| Comparison of JISP16 and AV18/IL7 results | | | | |
| Absolute energies (MeV) | 38 | 1.38 | 1.8 | 0.55 |
| Relative energies | 38 | 0.03 | 0.04 | 0.009 |
| Energies per nucleon (MeV) | 38 | 0.16 | 0.21 | 0.05 |
| Natural parity states | | | | |
| Absolute energies (MeV) | 96 | 3.71 | 3.97 | – |
| Relative energies | 96 | 0.043 | 0.05 | – |
| Energies per nucleon (MeV) | 96 | 0.28 | 0.31 | – |
| Natural and unnatural parity states | | | | |
| Absolute energies (MeV) | 135 | 3.54 | 4.04 | – |
| Relative energies | 135 | 0.05 | 0.056 | – |
| Energies per nucleon (MeV) | 135 | 0.28 | 0.34 | – |

energies for JISP16 predictions does not exceed 4 %. For all 133 calculated states obtained with JISP16, the rms deviation of relative energies is about 5 % for Extrapolation B and about 6% for Extrapolation C. Hence JISP16 can be regarded as a realistic $NN$ interaction providing a reasonable description of level energies in light nuclei.

The JISP16 predictions for the level ordering in the spectra is correct in most cases. However, evaluated uncertainties of level energies are sometimes close to or even larger than the splitting between levels. We have already discussed this issue while examining predictions for ground state spins of a few nuclei. One can however expect that the extrapolation uncertainties in Tables 1 and 2 overestimate the uncertainties of excitation energies. All states of the same parity are calculated in the same NCSM run. Truncation of the basis space in each run can cause an overall shift of the whole spectrum with a smaller effect in energy difference between levels, i. e., in excitation energies. In this case the extrapolation uncertainties may mostly reflect this overall shift of the spectrum rather than the uncertainties of excitation energies. Unfortunately, we cannot prove this expectation and we are unable to suggest an accurate estimate of excitation energy uncertainties. As a rough estimate of these uncertainties, we suggest in the second set of brackets after Extrapolation B excitation energy results in Tables 1 and 2 the rms deviation from the quoted value of the excitation energy obtained by the same extrapolation with the largest available $N_{\max}$ by varying $\hbar\Omega$ in a reasonable interval that is a 10 MeV interval of $\hbar\Omega$ values which includes the minimum of the NCSM results for the absolute energy of the given state. A small number in these second set of brackets and hence the stability of Extrapolation B with regard to the $\hbar\Omega$ variation confirms our expectation of the overall shift of the spectrum and possible reduction of the excitation energy uncertainty. We see from the Tables that in many cases the number in the second set of brackets is smaller, sometimes by an order of magnitude, than the uncertainty estimate based on absolute energy given in the first set of brackets.

**Table 5. Rms deviations of theoretical excitation energies from experiment. For comparison with GFMC AV18/IL7 results, we select 25 states for which the AV18/IL7 results are available (see Table 1). For unnatural parity states, we calculate excitation energy relative to the lowest state of the same parity**

| | Number of levels | JISP16 | | | | AV18/IL7 |
|---|---|---|---|---|---|---|
| | | Extrap. B | Extrap. C | Av. $\hbar\Omega$ | Av. $N_{max}$ | |
| Natural parity states | | | | | | |
| Excitation energies (MeV) | 25 | 0.61 | 0.74 | 0.96 | 1.60 | 0.42 |
| Excitation energies (MeV) | 70 | 1.62 | 2.88 | 3.49 | 4.43 | – |
| Unnatural parity states | | | | | | |
| Excitation energy (MeV) | 31 | 2.25 | 1.57 | 2.05 | 2.32 | – |

However, there are states for which these numbers do not differ essentially, and in these cases we should not expect a reduction of the uncertainty estimated from the absolute energy extrapolation.

Extrapolations A and B provide, as a rule, consistent predictions for excitation energies which sometimes differ from Extrapolation C predictions. This is illustrated by Fig. 5 where we plot spectra of a few light nuclei. Our impression is that Extrapolations A and B provide more reliable results at least in cases of sufficiently large $N_{\max}$ values. Note, Extrapolations A and B employ information from NCSM results around the minima of their $\hbar\Omega$ dependence which lie close to the extrapolated energy. Hence their predictions may be expected to be more accurate than the predictions of Extrapolation C based on NCSM results which lie higher in energy. Rms deviations of excitation energies of natural parity states from Extrapolation B are essentially smaller than those from Extrapolation C as is seen in Table 5. For excitation energies of unnatural parity states, we calculate excitation energies relative to the lowest state of the same parity. Surprisingly, Extrapolation C results in much smaller rms deviations of excitation energies of unnatural parity states than Extrapolation B (see Table 5). A possible explanation may rely on the fact that unnatural parity states were generally calculated with smaller $N_{\max}$ than natural parity states: there are no narrow unnatural parity states in nuclei with $A < 8$, most of examined unnatural parity states are in nuclei with $A \geq 10$ where we cannot use large $N_{\max}$ values. In this case, the minima of the $\hbar\Omega$ dependence of the NCSM results are further from the extrapolated energy and the Extrapolation B become less reliable. On the other hand, the Extrapolation C is founded on theory and its predictions may be less affected by the decrease of the basis space.

As compared to the AV18/IL7 predictions, the JISP16 results in larger rms deviations of excitation energies (see Table 5) but still is rather accurate in the description of the spectra of light nuclei.

As was already noted, we obtain all states of a given parity in each NCSM run and thus obtain excitation energies of all excited states of this parity directly from the NCSM results. These excitation energies may depend on $\hbar\Omega$ and $N_{\max}$. However, if these dependences are weak, one can get a reasonable estimate of excitation energies (see discussions in Refs. [35, 46]). We present in Tables 1 and 2 excitation energies obtained by averaging for a

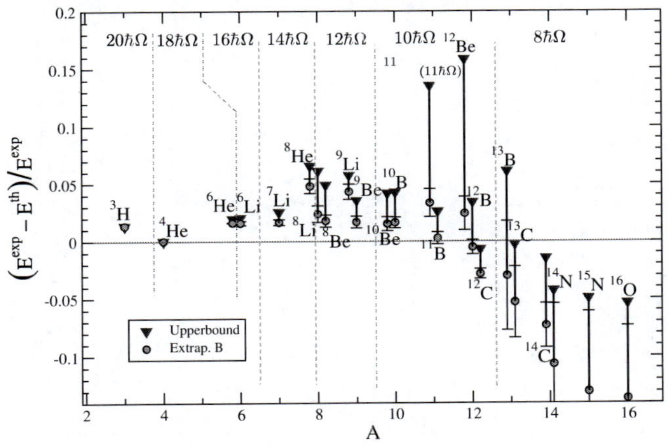

Figure 4. Relative error in description of binding energies (as compared to experiment) of light nuclei. Triangles — variational upperbounds for the ground state energies obtained directly from NCSM calculations; circles — Extrapolation B results with estimated uncertainties.

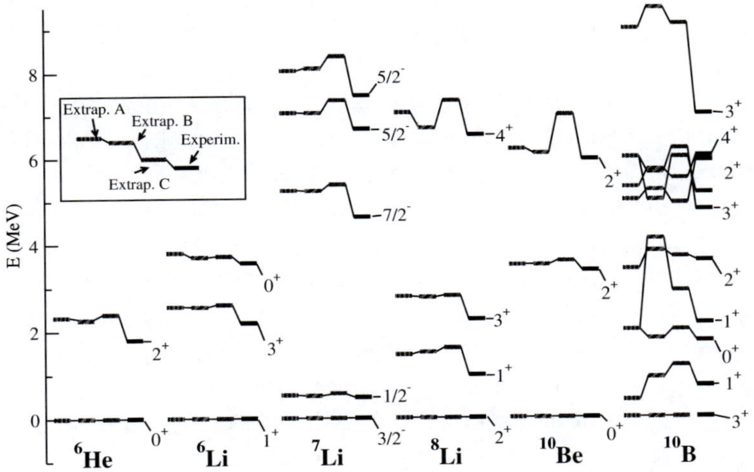

Figure 5. Spectra of a few light nuclei obtained by Extrapolations A, B and C in comparison with experiment. The ground state energies are all shifted to zero energy in order to compare excitation energies.

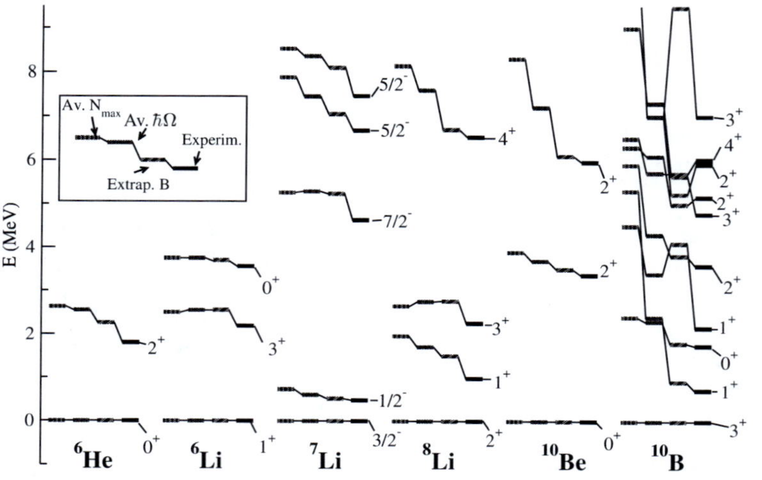


Figure 6. Spectra of a few light nuclei obtained by averaging excitation energies of a given level obtained directly in the NCSM calculations within a reasonable interval of $\hbar\Omega$ values with the largest available $N_{\max}$ value, $N_{\max} = N^M_{\max}$ (Av. $\hbar\Omega$), and with all $N_{\max} = N^M_{\max}$, $N_{\max} = N^M_{\max} - 2$, $N_{\max} = N^M_{\max} - 4$, $N_{\max} = N^M_{\max} - 6$ (Av. $N_{\max}$) and by Extrapolation B in comparison with experiment.

given level the NCSM excitations from calculations with largest available $N_{\max} = N^M_{\max}$ in a reasonable interval of $\hbar\Omega$ values (usually a 10 MeV interval around minima of $\hbar\Omega$ dependencies for the given and lowest state of the same parity) in the 'Av. $\hbar\Omega$' column with a rms deviation of these excitation energies from the average in round brackets. As it was shown in Ref. [75] (see Ref. [63] for a more detailed discussion), wide resonances should have correspondingly strong $\hbar\Omega$ dependence of excitation energies that should result in larger values of these rms deviations. Therefore the number in brackets is useful for selecting narrow resonances. In the 'Av. $N_{\max}$' column of Tables 1 and 2, we present the result of averaging excitation energies of a given level obtained directly in the NCSM calculations with $N_{\max} = N^M_{\max}$, $N_{\max} = N^M_{\max} - 2$, $N_{\max} = N^M_{\max} - 4$ and $N_{\max} = N^M_{\max} - 6$ within a reasonable interval of $\hbar\Omega$ values selected in the same manner as in Extrapolation A with a rms deviation of these excitation energies from the average in round brackets. A large number in the round brackets indicates a strong $N_{\max}$ dependence which means that the respective estimate of the excitation energy is inaccurate. Even more instructive is a difference between values from the columns 'Av. $\hbar\Omega$' and 'Av. $N_{\max}$' — this estimate of the excitation energy is reliable only if that difference is small.

We illustrate this method of calculation of excitation energies in Fig. 6 where we compare the spectra of a few light nuclei obtained by it and by Extrapolation B. It is seen that levels for which the excitation energies from 'Av. $\hbar\Omega$' and 'Av. $N_{\max}$' columns are close are described consistently with Extrapolation B. Otherwise we have a large difference with predictions based on Extrapolation B. Generally, the results of averaging excitation energies obtained with maximal attainable $N_{\max}$ only are closer to the predictions based on extrapolations than the results of averaging states obtained with a set of $N_{\max}$ values.

In the case of natural parity states, the numbers in the columns 'Av. $\hbar\Omega$' and 'Av. $N_{\max}$' are significantly different for many states. As a result, the statistical analysis of all excited states based on these numbers without special selection of states brings us to large rms deviations with experiment shown in Table 5. Remarkably, in the case of unnatural parity, the numbers in the columns 'Av. $\hbar\Omega$' and 'Av. $N_{\max}$' of Table 2 are consistent nearly for all states. Therefore averaging NCSM excitation energies is more suitable for studying spectra of unnatural parity states. Statistical analysis shows that rms deviations with experiment for spectra of unnatural states based on numbers from columns 'Av. $\hbar\Omega$' and 'Av. $N_{\max}$' are even smaller than rms deviations for Extrapolation B.

## 6. Conclusion

We present here *ab initio* results for binding energies and spectra of 26 $s$- and $p$-shell nuclei with the JISP16 $NN$ interaction. The calculations are performed using the NCSM with further extrapolation of energies to the infinite basis space, along with uncertainty quantification, in accord with NCFC. Our results are compared wherever possible with GFMC results obtained with AV18 $NN$ and IL7 $NNN$ forces.

Generally, JISP16 provides a reasonable description of $s$- and $p$-shell nuclei. Binding energies are well described with the exception of few nuclei at the end of $p$ shell where JISP16 overbinds nuclei by approximately 10 %. The spin of ground states is correctly reproduced with a few possible exceptions where the excitation energy of the lowest excited state is less than the uncertainty of theoretical predictions. The spectra of narrow states are also reasonably well reproduced. However, we see that the AV18 + IL7 strong interaction model provides a somewhat better description of energies in light nuclei. On the other hand, JISP16 does not require three-nucleon forces that makes it possible to use larger basis spaces in calculations thus reducing uncertainties of the theory and making it possible to explore heavier nuclear systems.

This work was supported in part by the Ministry of Education and Science of the Russian Federation through the contract 14.V37.21.1297 and by US DOE Grants DE-FG02-87ER40371, DESC0008485 (SciDAC-3/NUCLEI), and by US NSF grant 0904782. A portion of the computational resources were provided by the National Energy Research Scientific Computing Center (NERSC), which is supported by the US DOE Office of Science, and by an INCITE award, "Nuclear Structure and Nuclear Reactions", from the US DOE Office of Advanced Scientific Computing. This research also used resources of the Oak Ridge Leadership Computing Facility at ORNL, which is supported by the US DOE Office of Science under Contract DE-AC05-00OR22725, and resources of the Argonne Leadership Computing Facility at ANL, which is supported by the DOE Office of Science under Contract DE-AC02-06CH11357. Some calculations were performed on the supercomputer Lomonosov at Moscow State University.

## References

[1] W. Leidemann and G. Orlandini, Progr. Part. Nucl. Phys. **68** (2013) 158.

[2] S. K. Bogner *et al.*, Comp. Phys. Comm. **184** (2013) 2235.

[3] B. R. Barrett, P. Navrátil and J. P. Vary, Prog. Part. Nucl. Phys. **69** (2013) 131.

[4] S. Pieper and R. B. Wiringa, *Annu. Rev. Nucl. Part. Sci.* **51** (2001) 53.

[5] K. Kowalski, D. J. Dean, M. Hjorth-Jensen, T. Papenbrock and P. Piecuch, Phys. Rev. Lett. **92** (2004) 132501.

[6] P. Navrátil, S. Quaglioni, I. Stetcu and B. R. Barrett, J. Phys. G **36** (2009) 083101.

[7] R. Machleidt, Phys. Rev. C **63** (2001) 024001.

[8] V. G. J. Stoks, R. A. M. Klomp, C. P. F. Terheggen and J. J. de Swart, Phys. Rev. C **49** (1994) 2950.

[9] R. B. Wiringa, V. G. J. Stoks and R. Schiavilla, Phys. Rev. C **51** (1995) 38.

[10] J. Carlson, V. R. Pandharipande and R. B. Wiringa, Nucl. Phys. A **401** (1983) 59.

[11] B. S. Pudliner, V. R. Pandharipande, J. Carlson, S. C. Pieper and R. B. Wiringa, Phys. Rev. C **56** (1997) 1720.

[12] S. C. Pieper, V. R. Pandharipande, R. B. Wiringa and J. Carlson, Phys. Rev. C **64** (2001) 014001.

[13] S. A. Coon, M. D. Scadron, P. C. McNamee, B. R. Barrett, D. W. E. Blatt and B. H. J. McKellar, Nucl. Phys. A **317** (1979) 242.

[14] J. L. Friar, D. Hüber and U. van Kolck, Phys. Rev. C **59** (1999) 53.

[15] D. Hüber, J. L. Friar, A. Nogga, H. Witała and U. van Kolck, Few-Body Syst. **30** (2001) 95.

[16] P. F. Bedaque, H.-W. Hammer and U. van Kolck, Phys. Rev. Lett. **82** (1999) 463.

[17] E. Epelbaum, A. Nogga, W. Glö ckle, H. Kamada, U.-G. Meißner and H. Witała, Phys. Rev. C **66** (2002) 064001.

[18] D. R. Entem and R. Machleidt, Phys. Lett. B **524** ( 2002) 93; Phys. Rev. C **68** (2003) 041001(R).

[19] P. Navrátil, V. G. Gueorguiev, J. P. Vary, A. Nogga and W. E. Ormand, Phys. Rev. Lett. **99** (2007) 042501.

[20] J. P. Vary, P. Maris, E. Ng, C. Yang and M. Sosonkina, J. Phys. Conf. Ser. **180** (2009) 012083.

[21] P. Maris, H. M. Aktulga, M. A. Caprio, U. V. Çatalyürek, E. G. Ng, D. Oryspayev, H. Potter, E. Saule, M. Sosonkina, J. P. Vary, C. Yang and Z. Zhou, J. Phys. Conf. Ser. **403** (2012) 012019.

[22] P. Doleschall, Phys. Rev. C **69** (2004) 054001.

[23] P. Doleschall, I. Borbély, Z. Papp and W. Plessas, Phys. Rev. C **67** (2003) 064005.

[24] A. M. Shirokov, A. I. Mazur, S. A. Zaytsev, J. P. Vary and T. A. Weber, Phys. Rev. C **70** (2004) 044005.

[25] A. M. Shirokov, J. P. Vary, A. I. Mazur, S. A. Zaytsev and T. A. Weber, Phys. Lett. B **621** (2005) 96.

[26] A. M. Shirokov, J. P. Vary, A. I. Mazur, S. A. Zaytsev and T. A. Weber, J. Phys. G **31** (2005) S1283.

[27] A. M. Shirokov, J. P. Vary, A. I. Mazur and T. A. Weber, Phys. Lett. B **644** (2007) 33.

[28] W. N. Polyzou and W. Glöckle, Few-Body Syst. **9** (1990) 97.

[29] S. Okubo, Progr. Theor. Phys. **12** (1954) 603 .

[30] K. Suzuki and S. Y. Lee, Progr. Theor. Phys. **64** (1980) 2091.

[31] S. D. Glazek and K. G. Wilson, *Phys. Rev. D* **48** (1993) 5863.

[32] F. Wegner, Ann. Phys. (NY) **506** (1994) 77.

[33] A Fortran code generating the JISP16 interaction matrix elements is available at http://nuclear.physics.iastate.edu.

[34] P. Maris, J. P. Vary, A. M. Shirokov, Phys. Rev. C **79** (2009) 014308.

[35] P. Maris, A. M. Shirokov, J. P. Vary, Phys. Rev. C **81** (2010) 021301(R).

[36] V. Z. Goldberg, B. T. Roeder, G. V. Rogachev, G. G. Chubarian, E. D. Johnson, C. Fu, A. A. Alharbi, M. L. Avila, A. Banu, M. McCleskey, J. P. Mitchell, E. Simmons, G. Tabacaru, L. Trache and R. E. Tribble, Phys. Lett. B **692** (2010) 307.

[37] S. A. Coon, M. I. Avetian, M. K. G. Kruse, U. van Kolck, P. Maris and J. P. Vary, Phys. Rev. C **86** (2012) 054002.

[38] R. J. Furnstahl, G. Hagen and T. Papenbrock, Phys. Rev. C **86** (2012) 031301.

[39] S. N. More, A. Ekström, R. J. Furnstahl, G. Hagen and T. Papenbrock, Phys. Rev. C **87** (2013) 044326.

[40] R. J. Furnstahl, S. N. More and T. Papenbrock, arXiv:1312.6876 [nucl-th] (2013).

[41] A. M. Shirokov, V. A. Kulikov, P. Maris, A. I. Mazur, E. A. Mazur and J. P. Vary, EPJ Web Conf. **3** (2010) 05015.

[42] C. Cockrell, J. P. Vary and P. Maris, Phys. Rev. C **86** (2012) 034325.

[43] P. Maris, J. Phys. Conf. Ser. **402** (2012) 012031.

[44] P. Maris, J. Phys. Conf. Ser. **445** (2013) 012035.

[45] K. D. Launey, T. Dytrych, J. P. Draayer, G. K. Tobin, M. C. Ferriss, D. Langr, A. C. Dreyfuss, P. Maris, J. P. Vary and C. Bahri, in *Fission and Properties of Neutron-Rich Nuclei*, Proc. Fifth Int. Conf. ICFN5. Sanibel Island, Florida, USA, 4–10 November 2012. World Scientific, 2013, p. 29; doi: 10.1142/9789814525435_0003.

[46] P. Maris and J. P. Vary, Int. J. Mod. Phys. E **22** (2013) 1330016.

[47] M. A. Caprio, P. Maris and J. P. Vary, Phys. Lett. B **719** (2013) 179.

[48] J. P. Vary, Proc. Int. Conf. Nucl. Theor. Supercomputing Era — 2013 (NTSE-2013). Eds. A. M. Shirokov and A. I. Mazur. Pacific National University, Khabarovsk, Russia, 2014, p. 15, http://www.ntse-2013.khb.ru/Proc/JPVary.pdf.

[49] P. Maris, Proc. Int. Conf. Nucl. Theor. Supercomputing Era — 2013 (NTSE-2013). Eds. A. M. Shirokov and A. I. Mazur. Pacific National University, Khabarovsk, Russia, 2014, p. 37, http://www.ntse-2013.khb.ru/Proc/Maris.pdf.

[50] S. A. Zaytsev, Theor. Math. Phys. **115** (1998) 575.

[51] R. Machleidt, *private communication* (2006).

[52] R. Machleidt and I. Slaus, J. Phys. G **27** (2001) R69.

[53] R. A. Arndt, W. J. Briscoe, I. I. Strakovsky and R. L. Workman, Phys. Rev. C **76**, (2007) 025209, http://gwdac.phys.gwu.edu.

[54] Yu. A. Lurie and A. M. Shirokov, Bull. Rus. Acad. Sci. Phys. **61** (1997) 1665.

[55] Yu. A. Lurie and A. M. Shirokov, Ann. Phys. (NY) **312** (2004) 284.

[56] M. A. Caprio, P. Maris and J. P. Vary, Phys. Rev. C **86** (2012) 034312.

[57] G. A. Negoita, J. P. Vary, A. M. Shirokov, P. Maris and M. Sosonkina, *in preparation*.

[58] P. Sternberg, E. G. Ng, C. Yang, P. Maris, J. P. Vary, M. Sosonkina and H. V. Le, Proc. 2008 ACM/IEEE Conf. Supercomputing, SC '08. IEEE Press, Piscataway, NJ, USA, 2008, p. 15:1.

[59] P. Maris, M. Sosonkina, J. P. Vary, E. G. Ng and C. Yang, Proc. Comput. Sci. **1** (2010) 97.

[60] H. M. Aktulga, C. Yang, E. G. Ng, P. Maris and J. P. Vary, Lecture Notes in Computer Science **7484** (2012) 830.

[61] M. Moshinsky and Yu. F. Smirnov, *The Harmonic Oscillator in Modern Physics*. Harwood Academic Publishers, 1996.

[62] D. H. Gloeckner and R. D. Lawson, Phys. Lett. B **53** (1974) 313.

[63] A. I. Mazur, A. M. Shirokov, J. P. Vary, P. Maris and I. A. Mazur, Proc. Int. Workshop Nucl. Theor. Supercomputing Era NTSE-2012. Eds. A. M. Shirokov and A. I. Mazur. Pacific National University, Khabarovsk, Russia, 2013, p. 146, http://ntse-2012.khb.ru/files/proceedings/A_Mazur.pdf.

[64] I. Brida, S. C. Pieper and R. B. Wiringa, Phys. Rev. C **84** (2011) 024319.

[65] E. A. McCutchan, C. J. Lister, S. C. Pieper, R. B. Wiringa, D. Seweryniak, J. P. Greene, P. F. Bertone, M. P. Carpenter, C. J. Chiara, G. Gürdal, C. R. Hoffman, R. V. F. Janssens, T. L. Khoo, T. Lauritsen and S. Zhu, Phys. Rev. C **86** (2012) 014312.

[66] S. Pastore, S. C. Pieper, R. Schiavilla and R. B. Wiringa, Phys. Rev. C **87** (2013) 035503.

[67] R. B. Wiringa, S. Pastore, S. C. Pieper and G. A. Miller, arXiv:1308.5670 [nucl-th] (2013).

[68] S. C. Pieper and R. B. Wiringa, *private communication*.

[69] http://cdfe.sinp.msu.ru/services/unifsys/index.html.

[70] T. Neff, J. Phys. Conf. Ser. **403** (2012) 012028.

[71] A. C. Dreyfuss, K. D. Launey, T. Dytrych, J. P. Draayer and C. Bahri, Phys. Lett. B **727** (2013) 511.

[72] T. Dytrych, K. D. Launey, J. P. Draayer, P. Maris, J. P. Vary, E. Saule, U. Catalyurek, M. Sosonkina, D. Langr and M. A. Caprio, Phys. Rev. Lett. **111** (2013) 252501.

[73] S. C. Pieper, K. Varga and R. B. Wiringa, Phys. Rev. C **66** (2002) 044310.

[74] P. Navrátil and W. E. Ormand, Phys. Rev. C **68** (2003) 034305.

[75] A. M. Shirokov, A. I. Mazur, J. P. Vary and E. A. Mazur, Phys. Rev. C **79** (2009) 014610.

In: NN and 3N Interactions
Editors: Leonid Blokhintsev and Igor Strakovsky
ISBN: 978-1-63321-053-0


*Chapter 9*

# HADRON SCATTERING FROM BOUND, INTERACTING NUCLEONS

***R. J. Peterson***
Department of Physics, University of Colorado,
Boulder, CO, US

## ABSTRACT

The strong interactions of hadron beams with nucleons bound within a complex nucleus need not be the same as seen in scattering in free space, since the strong nuclear fields may induce any of a wide range of alterations of the nucleons, their interactions, and even the properties of the incoming hadron. Quasi-free scattering, which is as closely related to free scattering as is possible, is the simplest way to investigate these possible changes. The conditions for such a simple scattering mechanism are demanding, and their validity must be demonstrated. This test is most readily carried out by combining a wide range of intermediate energy hadron-nucleus data by means of ``scaling relations", which would unify these data if the conditions are met and the parameters correct. If such methods are established as reliable, hadron beam experiments could give access to all six combinations of nuclear spin and isospin responses, far beyond than the two responses available to simpler electron scattering experiments. The methods and a few examples will be given in this chapter. In addition, the special roles of spin observables, deltas in nuclei, and ``swollen nucleons" will be reviewed.

**PACS** 25.80.Ls, 25.40.Ve, 25.40.Kv
**Keywords**: Quasi-free hadron reactions, scaling relations, nuclear correlations

## 1. INTRODUCTION

Much of what we have learned about the strong interaction has come from scattering experiments, with beams of hadrons incident upon proton targets. In this work, this approach will be expanded to the scattering of hadrons from nucleons, both protons and neutrons, within complex nuclei. Indeed, early studies of the properties of neutrons relied upon electron scattering from deuterium, with some appropriate procedure to isolate that part of the scattering due to the magnetism of the neutron [1]. Such methods continue to be used on deuterium samples as avatars for neutrons in many experiments.

A first need for such an effort to scatter from individual bound nucleons is to establish that the scattering is truly one-on-only-one and incoherent. Coherent scattering from a nucleus adds amplitudes, then squares these for the cross sections. For instance, the Rutherford equation includes the charge Z as the square, in recognition of coherent scattering from all Z charges in a nucleus. Coherent scattering may also be defined by noting that the recoil from the scattering is the entire nuclear system, of mass A, as on page 810 of Ref. [2]. Incoherent scattering from a single nucleon is noted by the recoil energy of only the one struck nucleon, with the corresponding larger energy loss from the scattered beam particle.

The requirement that scattering from a single constituent of a bound system be related to free-space scattering demands that the impulse approximation be valid (page 683 of Ref. [2]) and quasi-free (page 693 of Ref. [2]) (not as commonly misused ``quasi-elastic", which is defined in an altogether different way). These criteria place limits on the range of beam energies and momentum transfers that would enable experiments clear access to single nucleons within nuclei.

The impulse approximation allows us to consider all that we have learned from truly free scattering to be applied to collisions with nucleons bound in a potential U, which is then ignored. This is the ``weak binding" condition, with the beam kinetic energy T much larger than both the binding potential U and the average kinetic energy of nucleons bound in that potential $K_{av}$ [2]. With nuclear potentials near 45 MeV deep and with $K_{av}$ near 23 MeV, this condition is met for intermediate energy hadron beams. The numerical example to illustrate this condition in Ref [2] is for 1000 MeV pions incident on carbon; similar modern pion data near this energy will be used below as examples.

Although there are no formal upper limits to the beam energy for a quasi-free analysis, the strength and very strong forward peaking in the lab frame for high energy protons means that a high energy scattering finds it more probable to achieve moderate momentum transfers by several more likely smaller angle scatterings. This has been demonstrated for 18.6 GeV protons scattered from nuclei [3]. This proton beam energy is too high for useful studies of single scatterings.

Corrections to the weak-binding limit are of order ($2\pi$ f R/$\lambda$ T), with $\lambda$ = h/p for beam momentum p, f the scattering amplitude, and R the "level shift", the difference between the beam kinetic energy and that energy including the potential sensed by the beam [2]. For protons [4] and pions [5] above 300 MeV, these real local potentials are small. Quasi-free electron scattering, with small cross sections and scattering amplitudes, has a much smaller correction effect than will be encountered with hadrons. The need to keep this correction term small also places a limit on the hadron beam momentum.

A numerical example for this correction term will use the optical theorem with the 500 MeV pion total cross section on carbon (290 mb [6]) for f, the momentum for this beam p and a level shift from the local optical potential sensed by such 500 MeV pions (3 MeV) [5]. This gives the correction factor a magnitude of 0.09, allowing use of the impulse approximation, with due caution. For lead, the larger total cross section triples this correction factor, and the impulse approximation will be less reliable for heavy nuclei.

The quasi-free condition, which allows the in-medium scattering to be incoherent, sets conditions on the three-momentum transfer q to the struck nucleon (page 693 of Ref. [2]). First, $q^2/2M$, with M the free nucleon mass, must be much greater than the potential sensed by the bound nucleon. This requires q >> 290 MeV/c. The second condition is that $q^2/2M$ is much greater than 4/3 times the average kinetic energy of the bound nucleon (23 MeV), or q >> 209 MeV/c.

To some extent the Pauli Principle will limit the recoil of any struck nucleon when q is less than twice the Fermi momentum, which imposes a limit for unblocked quasi-free scattering from carbon to be q > 456 MeV/c. An upper limit to the range of momentum transfers of interest is set by considering the length scales appropriate for the study of nucleons. The size and spacing of nucleons in a saturated complex nucleus suggest that a momentum transfer of 250 MeV/c is sufficient to sense nucleons, but above 1 GeV/c, it is the details within nucleons that become relevant.

All of these criteria are more readily met when the nuclear target is deuterium, and hadron scattering from the deuteron may serve as a calibration point for reactions with more complex systems. A recent striking example is found in the electron scattering work of Ref. [7].

We then can set demands for the data to be considered for an analysis of quasi-free, incoherent, impulse approximation scattering of hadrons from individual bound nucleons to be beam energies greater than 300 MeV, beam momentum greater than 600 MeV/c, and momentum transfers of 400 MeV/c for validity. This range includes many experimental results from intermediate energy pion, kaon, and proton scattering from complex nuclei. A large data base is available to check or to demonstrate the validity of scattering models based on the quasi-free conditions.

We are familiar with the distorting effects of complex nuclear potentials from many studies of hadron elastic coherent scattering. These do complicate the analysis of quasi-free scattering. The strongest assumption made in the path described above and followed below is the factorization of the single scattering from the nuclear response. A recent extension of the simple quasi-free mechanism includes relativistic distorted waves for the beam and ejectile hadrons [8]. As with the simple model, the projectile-nucleon interaction and the target response separate, allowing access to the spin and isospin nuclear responses, and confirming the basic idea to be followed more simply in the present chapter.

One can also extend the quasi-free scattering hypothesis to observations of the recoil single struck nucleon, in (p,2p) reactions. This doubles the hard part of the analysis, but useful results have been obtained from such experiments [9], as well as from simpler (e,ep) studies [10,11,12]. Only inclusive (single-arm) data will be considered in this chapter.

Under appropriate kinematic conditions, we may then use inclusive continuum hadron scattering spectra from complex nuclei to understand fundamental hadron-nucleon scattering, this time from bound nucleons, to gain access to medium effects that may be present.

Electron scattering has access to only two couplings to nucleons, the non-spin charge ("longitudinal") scattering from protons and the nearly isovector magnetic ("transverse") spin coupling to nucleons. Hadron scattering experiments, using polarized proton beams and making observations of the polarization of the final nucleons, offer access to the complete set of six couplings with bound nucleons, isoscalar ($\Delta T = 0$) and isovector ($\Delta T = 1$), nonspin $\Delta S = 0$, spin transverse ($\boldsymbol{\sigma}\mathbf{x}\mathbf{q}$), and spin longitudinal ($\boldsymbol{\sigma}.\mathbf{q}$). If we can learn how to use these data in appropriate studies we could learn much about interactions of hadrons with bound nucleons.

## 2. METHODS

Since the conditions for quasi-free incoherent scattering of intermediate energy hadron beams from nucleons within complex nuclei are at least in some doubt, a system of testing the data for consistency must be established. The method used will be "scaling", as familiar from the large industry of continuum scattering of intermediate and high energy electrons from complex nuclei [13]. Only the simplest nonrelativistic y-scaling system will be used in this chapter to present the consistency which would demonstrate the validity of a scaling analysis for hadrons. This, in turn, would imply the validity of the kinematic assumptions described in the previous section, allowing the desired studies of interactions with bound nucleons from data which have demonstrated good scaling.

If the scattering has been incoherent and only from a single nucleon, "billiard ball" kinematics suggests that the momentum transfer q and the energy transfer ω to that struck nucleon can be combined to a single variable

$$y = \sqrt{[\omega\,(\omega + 2M)]} - q, \tag{1}$$

with y the least momentum held by the struck nucleon that can scatter the beam particle to the corresponding momentum transfer and energy loss. Scattering from a bound nucleon at rest will be found at y = 0, and scattering with lower energy loss ω will occur for head-on collisions, at negative y.

In comparing scaling responses of different nuclei it is convenient to use a dimensionless scaling variable using the nuclear Fermi momentum $k_F$ [14]:

$$Y = y / k_F. \tag{2}$$

In practice, since the hadron beam energies will not be large and hadrons of both signs will be used, the momentum transfer is taken to include the Coulomb interaction, using the effective momentum transfer

$$q_{eff} = q\,(1 \pm 4\,Z\,e^2/3\,T\,r_0\,A^{1/3}), \tag{3}$$

with T the beam kinetic energy, $r_0$ = 1.2 fm, and Z, A the atomic number and mass of the target nucleus.

The energy loss ω to the bound nucleon is determined from the energy loss of the beam particle ε–ε', a (positive) binding energy (SE) for the nucleon, the Q-value for the hadron reaction (in the case of charge exchange), the Coulomb energy change (again for charge changing reactions), and the recoil of the (A – 1) daughter nucleus. For ω << 2M, the energy loss to the one struck nucleon is then

$$\omega = [(2q^2 + A\,X) + 2q\,\sqrt{(q^2 + AX)}]\,/\,2\,M\,A^2, \tag{4}$$

with the term

$$X = 2M\,(A - 1)\,(\varepsilon - \varepsilon' - SE) - q^2. \tag{5}$$

Binding energies (SE) for the hadron reactions are taken to be as determined from electron scattering [15].

In terms of this y-scaling variable, the inclusive continuum doubly-differential cross section data ($d^2\sigma/d\omega d\Omega$) are presented as single-nucleon responses:

$$F(y) = d^2\sigma/d\omega d\Omega\; q \,/\, d\sigma/d\Omega\; A_{eff} \sqrt{M^2 + (q + y)^2}, \tag{6}$$

or a dimensionless response

$$f(Y) = k_F\, F(y). \tag{7}$$

These responses have been normalized to the number of nucleons struck once and only once, $A_{eff}$. This is the practice for similar analyses of continuum electron scattering, with the complete number of target nucleons used for normalization. Good examples are found in Refs [16-20], where the full nuclear charge is indeed found in quasi-elastic charge ("longitudinal") scattering to 3 ± 12% [18].

For hadrons, another model must be invoked to compute the number of nucleons sensed once and only once. The simplest is the eikonal or Glauber model, as described for hadron scattering in Ref. [21]. This calculation assumes only small angle scattering, which might then be inconsistent with the demands on momentum transfer as described above. Input parameters are the distributions of target nucleons, taken to be the same for neutrons and protons [22] and the total cross sections encountered by the beam hadron passing through the nucleus. Since in-medium scattering is forbidden for low momentum transfers, this total cross section is taken to be 70% of the free-space values from Ref. [23]. A deeper analysis confirms this simple usage [24]. A more thorough discussion of this choice is found in Ref. [25]. On a carbon sample, $A_{eff}$ ranges from 5.33 for 367 MeV $K^+$, to 5.16 for 500 MeV pions and 3.53 for 795 MeV protons.

Also noted in the definition of F(y) is the hadron-nucleon singly-differential cross section $d\sigma/d\Omega$, as allowed by use of the impulse approximation. Since the continuum scattering is from moving nucleons, this entry will depend upon the energy loss $\omega$ at values of y other than zero. There is an optimum reference frame in which to evaluate free space (static target nucleons) differential cross sections [26]; this reference frame has been used for several detailed quasi-free analyses [27]. For each $\omega$, a new scattering angle and beam energy are calculated, and free-space scattering cross sections are evaluated for each $\omega$ across the continuum spectra. These do not vary rapidly with $\omega$ at intermediate energies, but can have a strong effect at lower energies in pion scattering.

An experiment that measured the yields for the quasi-free knockout reaction $^{12}C(\pi^+,p)$ at pion momenta near 490 MeV/c (T = 370 MeV) measured cross sections much larger than expected for a collision of a pion with a static nucleon [28]. An analysis including only the effects of the nucleon binding and internal momentum was able to account for this larger yield, as some of the collisions occurred at the energy of the delta at this low beam energy at large positive y. Any such effects in the present analysis would be included in the off-shell prescription [26], and would be smaller because of the higher pion energies used here. This chapter emphasizes data at negative y, where the optimum frame prescription gives free space cross sections at higher beam energies than the experimental value.

The presentation in this chapter will also include pion and proton charge exchange continuum cross sections. The standard compilation [23] computes pion charge exchange on protons, and charge symmetry implies the same for charge exchange (CX) on neutrons. For (p,n) CX reactions, the amplitudes from the standard compilation [23] may be recombined in the Bystricki, Lehar, and Winteritz formulation [29] to yield the CX part of the full proton-neutron cross section. Only this CX cross section is used in this chapter for CX responses.

The expressions above and the usages described thus give a set of tools that will allow any hadron continuum scattering spectra from complex nuclei to be put into a single y-scaling format for comparison. Effects of beam particle, beam energy, target, and scattering angle would be subsumed into a single response in an ideal but uninteresting world. The data below will test the assumptions made in prescribing this y-scaling. Three types of scaling have been defined;

1. Scaling of the first kind is that in which a given beam on a given target at all energies and angles yields the same response. This is found for longitudinal (charge) electron scattering [14,30].
2. Scaling of the second kind is that in which the responses are the same for all nuclei, after an adjustment for their slowly varying Fermi momenta [14,30]. These are the f(Y) defined above. 'Super-scaling' is noted when both these first and second kinds of scaling are realized [14].
3. A third kind of scaling is noted when comparing a range of scattering spectra with pions, kaons, and protons, in which all simple hadrons give the same responses on a given nucleus at some fixed momentum transfer [31].

A large quasi-free peak was noted in continuum electron spectra, as noted early on [32]. Most of this peak is very similar to that expected from scattering from a Fermi gas of non-interacting nucleons, characterized by only two parameters, a binding energy that locates y = 0 (or Y = 0) and a Fermi momentum that determines the width of the peak [32]. These have been accurately determined for electron scattering [15]. The scaling responses below will be compared to parabolic curves for this simple model:

$$f_{FG}(Y) = 0.75\,(1 - Y^2), \tag{8}$$

for values of Y between −1 and +1, or to corresponding curves for F(y).

Four other scaling systems have been proposed to unify measurements made for q and ω [13], but the Nachtmann system was based on conditions inappropriate to intermediate energy beams [33]. The other three scaling systems gave trends and conclusions for carbon very similar to the simple y-scaling system used here for both charge exchange [48] and scattering [33].

With the rules for forming scaling responses now laid out, and the necessary parameters (binding energy SE, Fermi momentum $k_F$, singly-differential cross sections, and $A_{eff}$)

available, we can now examine those responses over a wide range of variables. For cases in which intermediate energy hadron quasi-free spectra do demonstrate scaling consistency, we will then be able to examine those responses for what they tell us about interactions with nucleons within nuclei.

## 3. Interactions and Correlations

Nucleons simply bound in a mean field, without specific interactions among themselves, will have a distribution of momentum given by the Fermi gas model or something very like it. This model will be shown in figures to set off the more interesting regions of the spectra or the responses. The broad quasi-free peak caused by this momentum dominates the continuum spectra from scattered electrons [32] or hadrons. It is the events found at smaller energy losses than the Fermi gas permits that are most interesting, since something other than the mean field has driven them into high momentum or held them in multi-nucleon states.

The several spin transfer $\Delta S = 0, 1$ and isospin $\Delta T = 0, 1$ interactions of a beam hadron with a nucleon, free or bound, can be compared to computed nucleon-nucleon interactions. To form an overall scalar, the properties of the interaction of the beam and the struck nucleon must be the same as those of any interactions that nucleon may be undergoing.

One way that nuclear correlations can influence quasi-free scattering is if the one struck nucleon is partially 'stuck' with another by a long range interaction; this would make the struck nucleon effectively more massive, with less recoil than when free. An example of the momentum transfer dependence of such nucleon-nucleon interactions is found in Ref. [34]. The data shown in this work from intermediate energy hadron quasi-free scattering will cover much the same range as the figures in that work. The longest range and strongest correlations are $\Delta S = \Delta T = 0$, which drive many familiar features of nuclear collectivity. Quasi-free scattering reactions without charge exchange (NCX) are driven by hadron-nucleon scatterings primarily with these quantum number transfers. Hadron charge exchange reactions must be isovector ($\Delta T = 1$), and sense an entirely different set of correlations.

Yet harder collisions from short-range correlations (SRC) among nucleons will scatter hadron (or electron) beams to yet lower values of y or Y. Access to the length scales of SRC is provided by quasi-free scattering at larger momentum transfers, where it is found that neutron-proton SRC are much stronger than those for like nucleons [35]. Recent electron scattering work has demonstrated the role of these SRC at momentum transfers larger than for the work presented here [7].

The next section will present data to test the validity of the quasi-free conditions, and for successful cases, illustrate at least the isoscalar/isovector differences.

## 4. Examples

Scaling is expected to be most readily followed in light nuclei, since the correction term to the impulse approximation is smallest there. A good example might be $^{6}$Li, since it has half the density of $^{12}$C. Figure 1 shows the single-nucleon scaling responses for 950 MeV/c negative pions scattered from $^{6}$Li [36,37].

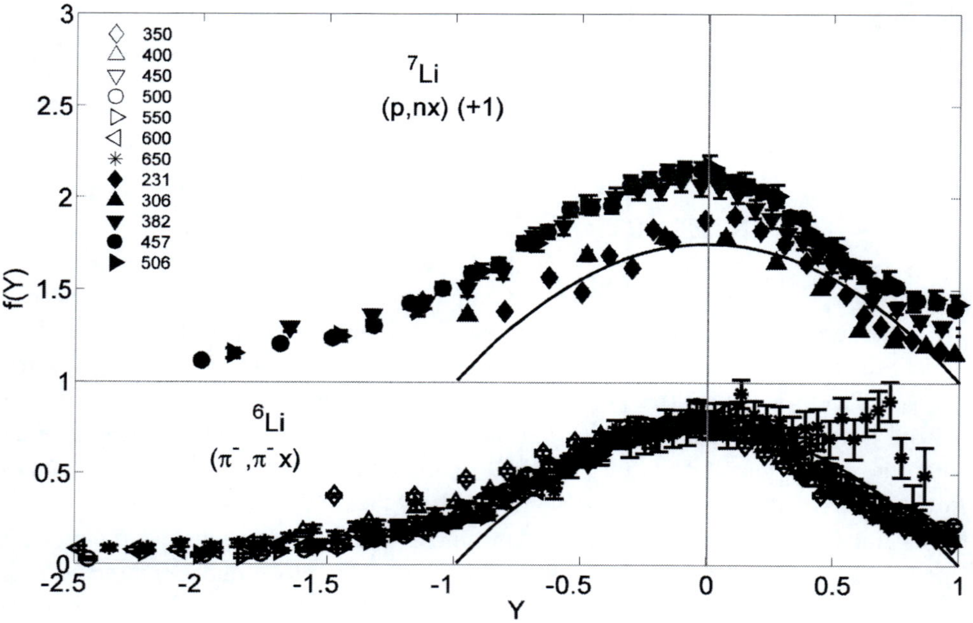


Figure 1. Y-scaling responses, computed from inclusive hadron scattering spectra as described in the text, are presented for pion scattering without charge exchange (NCX) [36, 37] and for proton charge exchange (CX) [38,39] at several momentum transfers for lithium. Each data set is labeled by its 3-momentum transfer q in MeV/c. The pion data below were taken in 50 MeV/c steps of momentum transfer and the charge exchange data at fixed angles of 9, 12, 15, 18 and 20 degrees. The solid curves show the expectation of a non-relativistic Fermi gas (NRFG). Once the momentum transfers reach the values where the quasi-free conditions are met, all of the data points coincide for Y < 0.

The NRFG curve shown is computed for $k_F$ = 165 MeV/c, and an energy of separation of 15 MeV is used [15]. Over a good range of momentum transfers, the scaled data lie on top of one another, and match the simple Fermi gas curve over much of the range of y. Hadron scattering without charge exchange on this light system does indeed follow scaling of the "first kind", as not changing with q once this momentum transfer exceeds 350 MeV/c.

Also shown in Figure 1 are (p,nx) charge exchange data on $^7$Li [38,39]. Above momentum transfers of about 350 MeV/c these data again scale very well. The quasi-free condition described in Section 1 suggested that the kinematic conditions would be met above q = 300 MeV/c. We note that the scattering without charge exchange (NCX) indicates a stronger response from $^6$Li at the lowest value of q, as expected from stronger isoscalar correlations for the struck nucleon. This figure uses the Y-scaling system for better comparison with later figures.

A stronger way to examine the trends with increasing q is to compute the interpolated values of the responses at fixed values of y for carbon (the most commonly used sample) as q increases. Such values are plotted in Figure 2 for sets of experiments that used the same beam energies for both charge exchange (CX) and NCX reactions [38–43]. At y = −300 MeV/c there is little change with q for either class of reaction, for energy losses to the struck nucleon much less than would be permitted by the NRFG, indicating little role of correlations. Charge exchange data at y = −170 MeV/c rise with q, or more properly perhaps, with scattering angle, as events from pion production may contaminate the spectra.

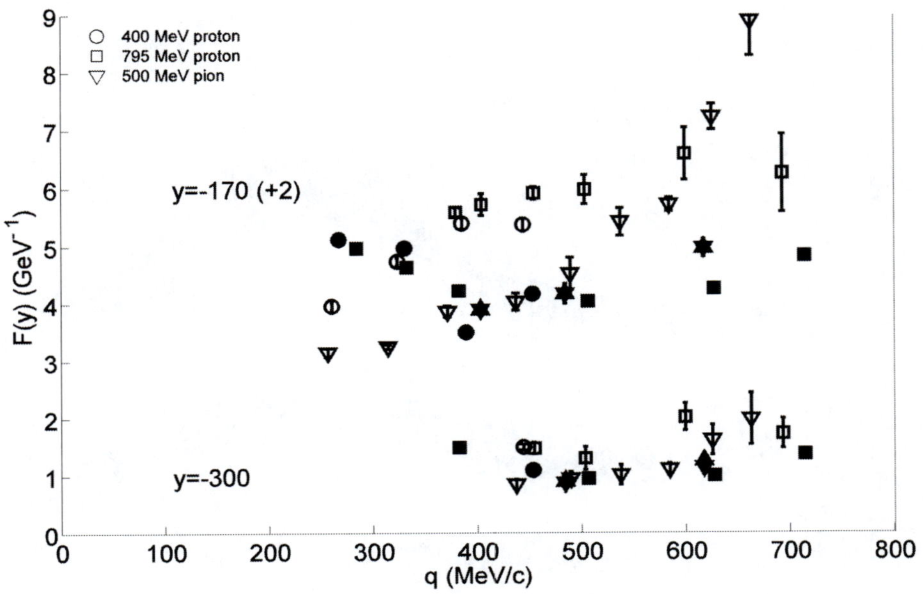


Figure 2. For three inclusive experiments where both CX and NCX data were taken at the same energies for a carbon sample, values of the scaling responses F(y) interpolated for y = −170 and −300 MeV/c are plotted as the three-momentum transfer increases. Open points are for CX, solid points for NCX. Data are for 400 MeV protons (circles) [40], 795 MeV protons [38,39,41] as squares, and 500 MeV pions [42,43], with upward pointing triangles for $\pi^+$ and downward pointing triangles for $\pi^-$. These charge-symmetric data points agree closely for carbon. By y = −300 MeV/c, beyond the range of the NRFG, these responses are constant for both classes of reaction. At y = −170 MeV/c the CX spectra are increasingly contaminated by pion production events. The spectra in the NCX channel show stronger isoscalar correlations as q is lower.

This effect is particularly strong for the pion data, as examined in Ref. [42]. The NCX data at y = −170 MeV/c indicate a stronger role of isoscalar correlations as q decreases.

Scaling of the "second kind" is that in which a given beam and scattering angle yield the same single-nucleon Y responses for all target masses [30]. This is observed for electron charge scattering [14], but for hadrons such scaling is a test of the method used to obtain the effective number of nucleons, those scattered from once and only once. Nuclear samples will each have a Fermi momentum, which determines the shape of the quasi-free peak. This is accounted for by using the dimensionless super-scale variable Y = y/kF, and the response is taken to be $f(Y) = k_F F(y)$, using values of $k_F$ from Ref. [15].

Figure 3 shows data for two NCX studies that covered a good range of target masses. The pion data are for 950 MeV/c beams of negative pions at q = 500 MeV/c [37, 36] and the proton data are for a 795 MeV beam at 20 deg [41]. The free scattering momentum transfer at this proton scattering angle is 503 MeV/c. A momentum transfer near 500 MeV/c meets the criteria for quasi-free scattering. The responses in Figure 3 do indeed agree, indicating that the $A_{eff}$ are correctly computed. The curve shown is just that for a nonrelativistic Fermi gas, of the same shape for all $k_F$.

Figure 4 shows pion charge exchange data over a range of target masses, for 500 MeV negative pions at 50 deg, for q = 487 MeV/c [42].

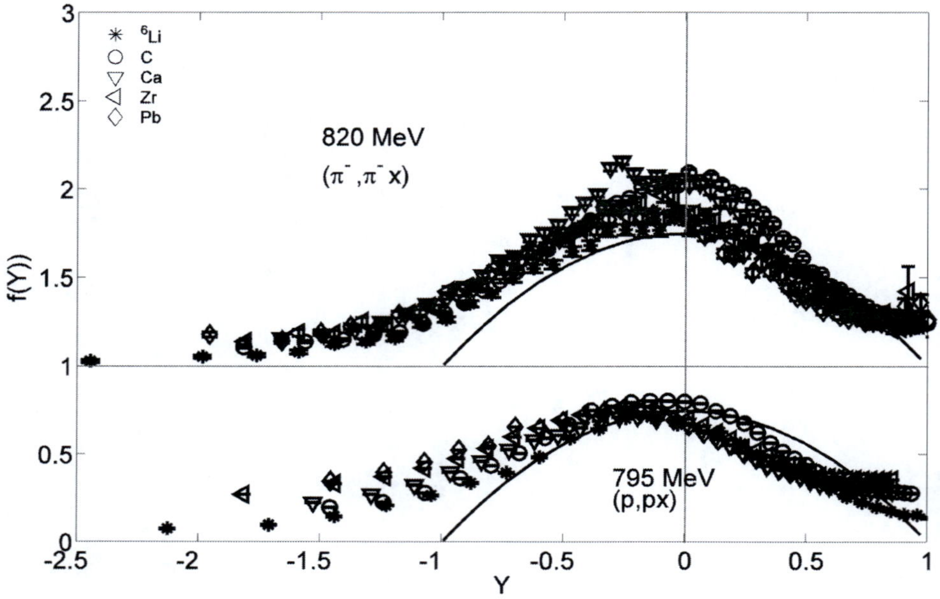


Figure 3. The same two NCX experiments as used for the responses in Figure 1 [37, 41] are used to compute "super-scaling" responses for a range of nuclear masses for momentum transfer near 500 MeV/c. A small peak due to scattering on hydrogen is noted for calcium in the pion data, displaced from Y = 0 by the binding energy used in the scaling analysis.

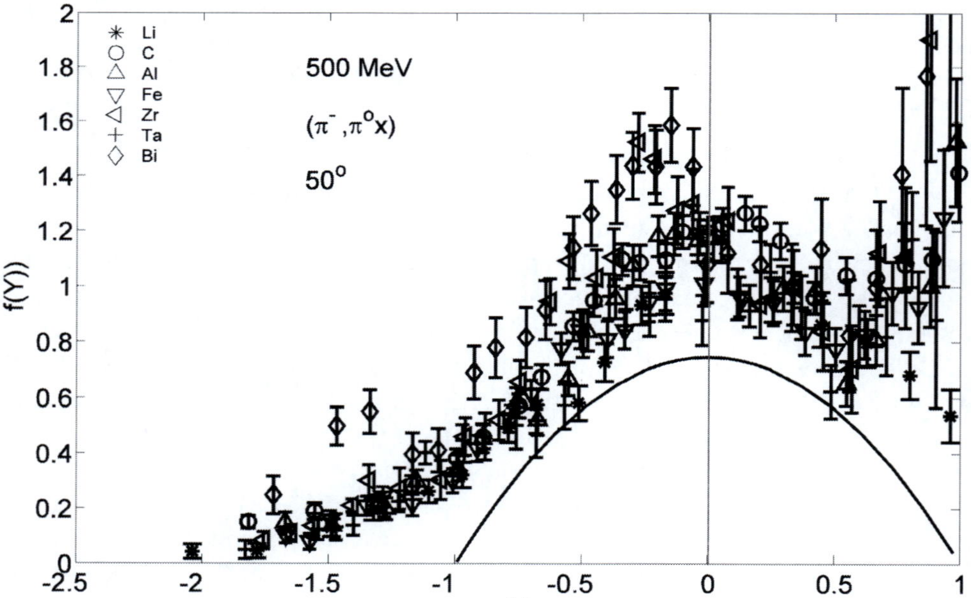


Figure 4. As for Figure 3 but for 500 MeV ($\pi^-,\pi^0$x) charge exchange reactions on a range of nuclei at a momentum transfer near 500 MeV/c. The curve shows the expected NRFG response. Only if the $A_{eff}$ parameter is computed correctly will these data and the data in Figure 3 be found to agree, to give "scaling of the second kind"[30].

This experiment did not have good resolution, but the scaling responses again agree quite well among themselves. For negative pion charge exchange the effective number of protons $Z_{eff}$ is taken to be Z/A times $A_{eff}$. A strong background at positive y arises from pion production. If this background is suitably subtracted the quasi-free data are nearly in agreement with this simple NRFG model over much of the Y range [42–48].

These four figures demonstrate that hadron continuum spectra do scale as the "first" and "second" kinds, and that the methods and parameters used are appropriate. Although much of each spectrum is like the Fermi gas curve which arises from noninteracting nucleons merely bound in a self-consistent potential well, the interesting data are those points found at lower energy losses (more negative y or Y). These scattering events arise from collisions with bound nucleons while they are interacting, by correlations, by hard scattering, or by other mechanisms. It is these points which are of interest for NN interactions within nuclei.

The figures above each used a given hadron species and beam energy. It has also been shown that for a given target (carbon) and a given momentum transfer (q near 500 MeV/c) that all hadron species give the same responses, for a new "scaling of the third kind" [31].

Figure 5 shows y-scaling responses for carbon for nine hadron experiments with q near 500 MeV/c, with data from [37,41,43–46]. Each hadron experiment had a systematic uncertainty near 10%, and most results agree within that uncertainty. Only statistical uncertainties are shown in the figures in this chapter. The 300 MeV case has a beam energy below the criteria noted in Sec. 1, and that response is not in agreement with the other cases. Also shown are responses for charge scattering of electrons from carbon at q = 550 MeV/c [47], also agreeing with the hadron data.

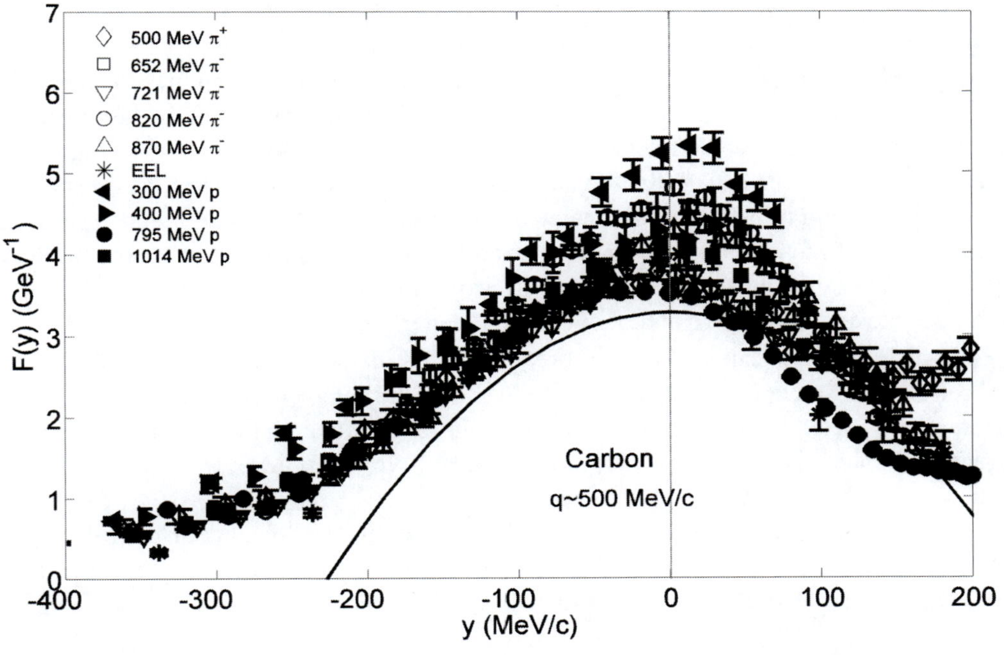


Figure 5. All hadrons scattered from the same nucleus at very similar momentum transfers near 500 MeV/c give much the same responses, for scaling of "the third kind". Hadron data are from Refs. [37,41,43–46]. Also shown are charge (longitudinal EEL) electron scattering data treated exactly as for the hadrons [47].

Quantitative comparisons of several hadron beams found their interpolated responses for NCX [31] and CX [48] to agree to within the systematic uncertainties of the several experiments.

## 5. Spin

Spin observables in the all hadron-nucleon interactions have been vital to specify the partial wave coefficients, with great sensitivity since these observables are the results of interference phenomena. When one of the spin particles is within a nucleus, the interactions may be modified, and spin observables in the quasi-free interactions of hadronic beams can be expected to be more sensitive to 'in-medium' effects than merely cross sections [49]. This idea led to a series of experiments with spin-polarized proton beams and measurements of the spin observables of neutrons in (p,nx) inclusive reactions. Charge exchange reactions were used to select the isovector channel required for sensitivity to pionic degrees of freedom in nuclei.

The momentum transfer used for these reactions was chosen to be sensitive to the "pionic enhancement" regime, where the pionic field within a nucleus would be observable [50]. Calculations in several models to treat hadronic complications were carried out to understand the data [51, 52]. High quality data were taken in several experiments, without seeing the expected signal [27, 38, 53–56]. A momentum transfer of q = 345 MeV/c was used in these studies. This is too low to meet the quasifree conditions. The conclusions of this wide effort led to a famous paper "Where are the Nuclear Pions?" [57], and a rather convoluted answer titled "Where the nuclear pions are" [58]. This issue seems to have evaporated.

Quasi-free electron scattering in the transverse magnetic channel may be selected by varying the beam energy and scattering angle to maintain a fixed momentum transfer, in the Rosenbluth separation. Pion production and a strong excitation of the delta inelastic peak are seen in this channel [14, 59]. While suitable scaling variables can yield common responses in the charge scattering channel, without spin, these same analyses find strong differences with changing momentum transfer q in the magnetic channel, similar for a wide range of nuclei in the same range of momentum transfers examined by (p,n) reactions (Figs. 10 and 11 of Ref. [14]). There are still unresolved issues in the isovector spin-driven responses of nucleons within nuclei, even with the clean tool of electron scattering.

Only hadron beams can access the isoscalar version of spin-driven responses $\Delta S = 1$, $\Delta T = 0$. Proton NCX scattering is driven mostly by isosscalar interactions, but an experiment to measure the expected [60] differences between spin longitudinal and spin transverse transfer also at q = 345 MeV/c did not observe the expected effect [61]. Some of these responses would be driven by smaller isovector terms.

A study of quasi-free scattering of 600 MeV deuterons used beams of both vector and tensor polarized deuterons with measurements of the vector spin transfer to address strictly $\Delta T = 0$ spin responses [62]. It was not possible to unravel the spin responses separately in the nonspin, spin longitudinal ($\boldsymbol{\sigma}.\mathbf{q}$), and spin transverse ($\boldsymbol{\sigma}\mathbf{x}\mathbf{q}$) channels, but since the several spin observables for carbon, calcium and lead were nearly the same as seen for protons and deuterons, it must be inferred that $\Delta T = 0$ spin correlations in complex nuclei are weak.

A more direct comparison between electromagnetic and hadron spin observables can be made with pions. Spin-zero pions may interact with nucleon spin only in the transverse (**σxq**) sense, as do photons. The magnetic moments of neutrons and protons indicate that $\Delta S = 1$ quasi-free electron scattering must be nearly isovector, $\Delta T = 1$. Isovector quasi-free pion charge exchange may then offer access to both $\Delta S = 1$ (transverse) and $\Delta S = 0$ observables, with the calibration check that the $\Delta S = 1$ responses should agree with those from transverse electron scattering.

A transfer of a unit of spin by pion inelastic scattering was noted for transitions to bound states, by changing the beam energy and angle much as is used with electrons with the Rosenbluth separation. The general idea is that spin transfer scattering would be proportional to $\sin^2\theta$ and non-spin to $\cos^2\theta$ [63], and this was demonstrated for a number of known spin excitations of bound states in light nuclei [64].

That separation would be true if there were no strong interaction effects on the integer spin pions, much as the case for spin-1/2 electrons resulting in the $\sin^2\theta/2$ and $\cos^2\theta/2$ that form the basis of the Rosenbluth separation. For spin-zero pions, including the actual strong interactions, a better separation technique is provided using the actual non-spin pi-nucleon scattering amplitudes $f(T,\theta)$ and (transverse) spin amplitudes $g(T,\theta)$. These amplitudes can be obtained from the SAID analysis [23]. Spin transfer cross sections must be zero at reaction angles of zero and 180 degrees, by parity conservation.

The hard part is finding a pion beam energy where the spin amplitudes change quickly enough with angle to permit a clean separation, especially since effects other than just spin terms may change with angle [65].

A fortunate case is found for a pion beam momentum of 750 MeV/c, where the spin-driven cross section is also small near 65 degrees [23]. Only if it was the spin terms, not other reaction complications, that change smoothly with scattering angle would a plot suitable to extract the spin and non-spin terms be a single straight line, as is the case for a Rosenbluth analysis for electrons.

A pion charge exchange experiment with the nearly-4π Crystal Ball detector at 750 MeV/c was able to select charge exchange cross sections resulting in one and only one outgoing pion, avoiding the strong background of produced pions [66]. Poor statistics required use of only the maximum of the quasi-free peak, $F(y = 0)$. The slope of a plot of $(1 + x)\,F(y = 0)$ vs. x, with $x = |g|^2/|f|^2$, gave a slope measuring the spin transverse response for C, Al, and Cu that agreed with the response to transverse electron scattering at the same momentum transfer, giving a verification of the method. The intercept of this plot gave the isovector non-spin quasi-free response, which was found to be somewhat stronger than the spin response [66].

So yes, quasi-free hadron experiments on complex nuclei have been able to provide indications of the spin-dependence of hadron-nucleon couplings when the target nucleons are subject to in-medium effects. Quasi-free electron scattering gives access to only two combinations of spin and isospin, with non-spin scattering from protons and (largely) isovector magnetic scattering. Hadron quasi-free scattering with spin observables and a range of beam probes can give access to all six combinations of isospin ($\Delta T = 0$ and $\Delta T = 1$) and spin (non-spin, **σ.q**, and **σxq**). The results to date indicate that some of the expected or possible in-medium effects are not seen. But at least, now we know that from spin-dependent hadron quasi-free scattering.

## 6. Deltas in Nuclei

The energy density within a nucleus could conceivably excite nucleons to the 3-3 or Δ(1232) resonance, with J = T = 3/2 [67]. Or, one could use pion beams at resonance energies near 170 MeV to create delta-hole states in nuclei. These states then serve as "doorways" to following stages, as described in Sec. 7.4 of Ref. [68]. The delta will sense some complex nuclear mean field during its brief lifetime, with the imaginary potential of about 40 MeV adding to the natural width of the free delta when it is observed in nuclei. A variety of analyses indicate a real attractive potential of about 30 MeV [68] for these deltas during their fleeting existence within nuclei for pion beam energies near the resonance. These real potentials are much smaller at the intermediate energies used for quasi-free analyses [5].

Or perhaps, with the right set of quantum numbers a stable nuclear state may contain fossil traces of deltas. One striking observation comes from an excitation function for pion inelastic scattering exciting the T = 0 12.7 MeV and T = 1 15.1 MeV $1^+$ states in $^{12}C$ [69–73]. These experiments varied the beam energy and scattering angle so as to maintain a fixed momentum transfer q = 125 MeV/c, near the top of the diffraction pattern differential cross section to these states. The data are shown in Figure 6. Data points that average $\pi^+$ and $\pi^-$ data will have cancelled out any effects of isospin mixing between the two $1^+$ states.

These data show the T = 0 cross section to rise as expected from the $\sin^2\theta$ dependence for a pion scattering spin excitation, as also computed from DWIA calculations [73].

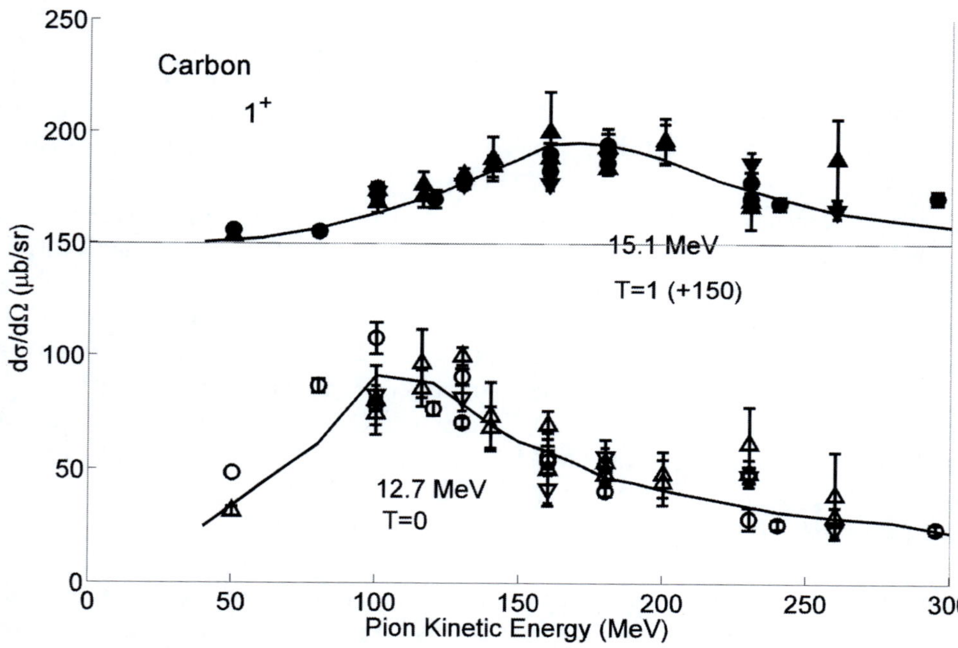


Figure 6. Excitation functions for pion inelastic scattering to the 12.7 MeV (T = 0) and 15.1 MeV (T = 1) $1^+$ states of $^{12}C$ at q = 125 MeV/c show very different trends. The shape of the DWIA prediction for scattering to a $1^+$ nuclear state matches the data for the T = 0 transition, but it is the shape of the energy-dependent pi-nucleon differential cross section that matches the T = 1 data. Upward triangles are for $\pi^+$ data, downward for $\pi^-$ data, and circle points show their average. Data are from Refs. [69-73].

The T = 1 peak cross section, however, shows an energy dependence closely following the pi-nucleon total cross section shape across the delta resonance, as shown. Only the T = 1 $1^+$ state has the quantum numbers required for an explicit delta-hole state, perhaps indicating some role or presence of this mode in the well-known 15.1 MeV state of $^{12}C$.

Direct scattering [74] or charge exchange [75] to nuclear excitations corresponding to excitation of a delta have also been observed in complex nuclei. A quasi-free analysis similar to that of this work for nucleonic responses was been carried out, similar to that used for electron scattering to this excitation [59].

Direct quasi-free pion double charge may uniquely de-excite a delta found within a nucleus, as $\Delta^-(\pi^+,\pi^-)p$. The results of such an experiment with 500 MeV pions found a few percent of deltas instead of nucleons within a wide range of nuclei [76]. This fraction is a bit less than initially predicted [67].

These two classes of experiment, excitation functions through the delta region with pion kinetic energies near 160 MeV and quasi-free double charge exchange at higher energies, have given direct evidence that delta excitations of nucleons are indeed actors within complex nuclei.

## 7. "Swollen Nucleons"?

One very direct idea is that the nuclear medium, saturated with strongly interacting entities, could change the size of a bound nucleon. It has been shown that quasi-free electron scattering did not measure such an effect [77, 78], but perhaps some effect could be found in a strong interaction with bound nucleons.

Their large strong interaction cross sections will not allow hadronic beams to sense nucleons within the surface region. For instance, a 800 MeV proton, with an average proton-proton/ proton-neutron total cross section of 42.34 mb [23] will sense one-and-only-one collision with a bound nucleon no further into the medium than only 23% of the central density of carbon and 7.7% of the central density of lead. All of the points made in this chapter for protons and pions will thus be relevant only to the low density regions of nuclei.

At these low densities NN interactions and correlations will not be the same as those for nuclear interiors or even nuclear averages. For instance, the nucleon effective mass, less than the free mass M in the interior, is expected to increase in the nuclear surface, even greater than the free mass due to surface tension [79]. The masses of mesons that mediate strong interactions are also dependent on density [80], and more phenomenological calculations also predict NN forces that vary with $k_F$ or density [81]. These are only a few examples of the many calculations on NN interactions within the interior, but are enough to open the possibilities of significant changes in the properties or interactions of bound nucleons as seen in quasi-free scattering.

The one hadron that is able to penetrate to somewhat greater densities is the $K^+$ meson, protected by its anti-strange quark from the simplest quark annihilation processes that limit propagation. This beam probe is thus a unique probe of hadron-nucleon interactions within nuclei up to a beam momentum of about 900 MeV/c [82]. With an average total cross exaction of 12.5 mb, this hadron can penetrate to 90% of the central density of carbon, 64% of the central density of calcium and 55% of the central density of lead [83].

When early data for $K^+$-nucleus elastic scattering were presented [84], they exceeded what were considered to be reliable model calculations [85, 86], leading to the suggestion that these beam particles encountered "swollen nucleons" as they reached within the nuclear surface. More reliable data confirmed the observation.

Observables include precise measurements of the total and reaction cross sections of $K^+$ on nuclei [87,88] and elastic scattering [89, 90]. These observables may be related to fundamental $K^+$-nucleon interactions by way of the distorted wave impulse approximation (DWIA), as also commonly used for pion-nucleus interactions [91]. At these low momenta, the $K^+$-nucleon coupling is largely isoscalar and S-wave and may be reliably dealt with in the DWIA; a relevant code has been presented [92]. Quasi-free scattering data for $K^+$ on nuclei have been presented, but the two quantities required to form the response, $A_{eff}$ and $d\sigma/d\Omega$, are both influenced by any changes in the interaction, in opposite directions. The $K^+$ quasi-free data [93] have been analyzed, with the conclusion that such quasi-free scattering shows no evidence for an change in $A_{eff}$ [94]. A larger $K^+$-nucleon total cross section would lead to a smaller $A_{eff}$.

The simple S-wave isoscalar interaction between a $K^+$ and a nucleon permits a very simple parameterization of the nuclear size. A repulsive hard core interaction varied a single parameter (the S-wave phase shift) in the DWIA for fits to reaction, total, and elastic differential cross sections for symmetric nuclei $^6$Li, $^{12}$C, $^{28}$Si, and $^{40}$Ca over a range of $K^+$ momenta [95]. The slopes of these fitted phase shifts with momentum yielded the hard core radii, which were found to vary with the density of the four nuclei.

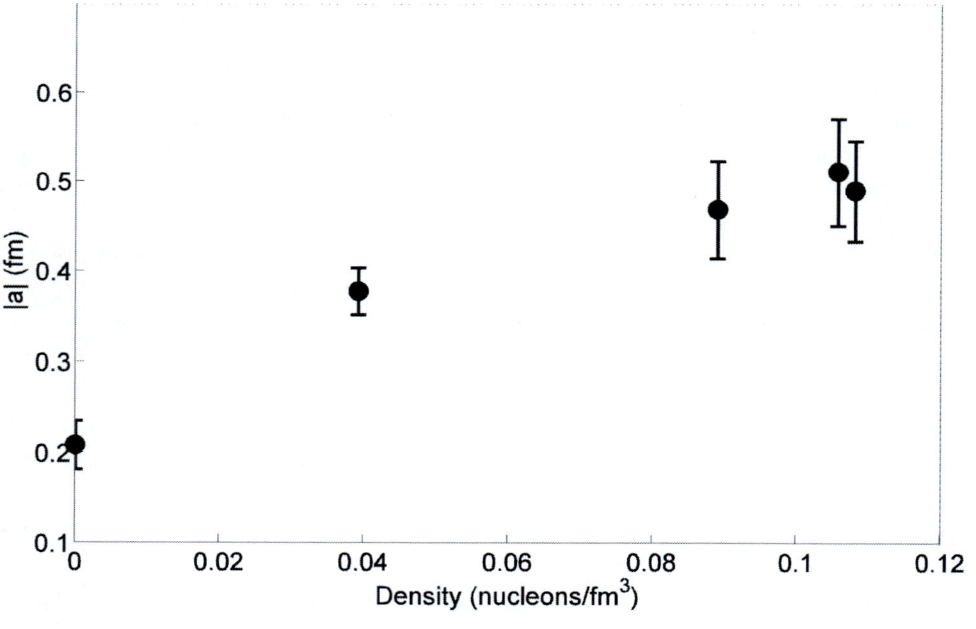


Figure 7. $K^+$ data for elastic scattering, total, and reaction cross sections have been analyzed simultaneously in the DWIA, using an $S_{11}$ phase shift derived from a hard sphere of radius a. This radius was varied to match all the data points for complex nuclei, and shows a smooth trend with nuclear density for symmetric nuclei. See Ref. [95] for details.

Using a density A/9.0128 $<r^2>^{3/2}$ with the known rms radii [96] and the hard core radii gives the results shown in Figure 7, including a point for a similar analysis of free $K^+$-nucleon scattering. The accuracy of the fitted data even permitted uncertainties to be determined for the hard core radii.

Figure 7 shows a definite trend, with the nucleon hard core radius increasing with nuclear density, possibly linearly. This result with a strongly interacting beam differs from the conclusions reached on the electromagnetic size of nucleons, described above.

## CONCLUSION

The simplest and most direct way to measure strong interactions between a hadronic projectile and bound nucleons is the direct look that can be provided by quasi-free scattering, with the kinematic length scale of the interaction matching the size and spacing of nucleons within nuclei. Many intermediate energy experiments have provided a relevant array of data, but the conditions under which these data can be used to reveal hadron-nucleon interactions are at least in some doubt due to the many complications arising from the strong interactions. A simple scaling system can be used to test the conditions and the parameters of a quasi-free analysis that would permit reliable conclusions on the fundamental hadron-nucleon interactions. The methods and successes of such a system have been presented above, and show that there are regions of kinematics and nuclei where the methods work, at least to the usual accuracy of strong interaction physics. The much simpler case of electron scattering analysis has been the guide to these efforts.

Among the successes of such analyses, the relative role of isoscalar and isovector interactions, driving both the nuclear scattering and the correlations that they sense, has been the most obvious. This is not a surprise, because the comparison between charge and magnetic scattering of electrons had already made this effect clear [14]. Perhaps the main offering of hadron quasi-free scattering is access to spin degrees of freedom. A number of sensitive studies with spin-polarized beams and ejectiles found no signs of certain isoscalar [60] or expected isovector [57] correlations. A fortunate choice of beam energy allowed an exclusive pion charge exchange reaction to measure the isovector, non-spin nuclear response; this is a channel unavailable to electron scattering [66].

The strongly interacting neighborhood in which bound nucleons live could create occasional excitations to the 3-3 delta state. A quasi-free pion double charge exchange reaction found deltas to compromise a few percent of nuclear hadrons [72]. Quasi-free studies are akin to the "direct" reactions used to measure such nuclear features as spectroscopic factors. An alternative excitation function method, akin to resonant nuclear reactions, found symptoms of a nuclear delta configuration in the familiar 15.1 MeV T=1 $1^+$ state of $^{12}C$, as evidence of a manifestation of deltas even in coherent nuclear states [73].

Although the $K^+$ meson is an ideal probe of nuclei in many ways, its short lifetime has limited the experiments to apply this, the least strongly interacting of all hadrons. When experiments failed to agree with what should have been reliable models, it was suggested [85, 86] that the $K^+$ had encountered "swollen nucleons", altered within nuclei by their neighbors. An analysis of three types of data on light nuclei found that a one-parameter size analysis was able to correlate that "size" with nuclear density [95].

So yes, over at least some range, intermediate energy hadron beams can directly examine interactions of hadrons (nucleons, pions, and kaons) with bound nucleons, but perhaps the restriction to low density regions of nuclei and strong distortion effects are masking the desired access to interesting degrees of freedom [97]. It is unfortunate that most of the facilities that provided the relevant data are no longer operating, and the methods described above are no longer possible to pursue.

## ACKNOWLEDGMENT

Deep thanks are due to the many colleagues and students I have worked with over the years on many of the experiments used for this work, with inventive experimental methods and late nights of toil. Much of this work involved my own efforts, strongly supported by the US Department of Energy. I also express deep thanks to experimenters who retrieved tables of old data to allow reliable use of their work.

## REFERENCES

[1] Yearian, M. R. and Hofstadter,R. *Phys. Rev.*, 110, 532 (1958).
[2] Goldberger, M. L. and Watson, K. M. Collision Theory, John Wiley, New York, NY, 1964.
[3] Kofoed-Hansen ,O. *Nucl. Phys. B,* 54, 42 (1973).
[4] DiGiacomo, N. J. et al. *Phys. Rev*. C, 19, 1132 (1979).
[5] Hong, S. W. and Kim, B. T. *J. Phys. G,* 25, 1065 (1999).
[6] Gelderloos, C. J. et al. *Phys. Rev*. C, 62, 24612 (2000).
[7] Fomin, N. et al. *Phys. Rev. Lett*., 108 092502 (2012).
[8] van Niekerk, D. D., van der Ventel, B. I. S., Titus, N. P., and Hillhouse, G. C. *Phys. Rev*. C, 83, 044607 (2011).
[9] Baturin, V. N. et al. *Nucl. Phys. A,* 736, 283 (2004).
[10] Abbot, D. et al. *Phys. Rev. Lett*., 80, 5072 (1998).
[11] Rohe,D. et al. *Phys. Rev*. C, 72, 054602 (2005).
[12] Ryckebusch, J.,Cosyns, W. and Vanalst,M. *Phys. Rev*. C, 83, 054601 (2011).
[13] Day, D. B., McCarthy, J. S., Donnelly, T., W., and Sick,I. *Ann. Rev. Nucl. Part. Sci*., 40, 3570 (1990).
[14] Donnelly, T. W. and Sick, I. *Phys. Rev*. C, 60, 065502 (1999).
[15] Maieron, C., Donnelly, T. W. and Sick, I. *Phys. Rev*. C, 65, 025502 (2002).
[16] Alam, Ghosh, S. K., Roy, P. and Safkar, S. *J Phys. G,* 30, 1967 (2004).
[17] Barbaro, M. B. et al. *Nucl. Phys. A,* 643, 137 (1998).
[18] Jourdan, J., *Nucl. Phys. A,* 603, 117 (1996).
[19] Kim, K.,S. ,B. G. Yu, B. G. and L. E. Wright, L. E. *Phys. Rev*. C, 74, 067601 (2006).
[20] Williamson, C. T. et al. *Phys. Rev*. C, 56, 3152 (1997).
[21] Ouyang, J. ,Hoibraten, S. and Peterson, R. J. *Phys. Rev*. C, 47, 2809 (1993).
[22] Patterson, J. D. and Peterson, R. J. *Nucl. Phys. A,* 717, 235 (2003).

[23] SAID, Scattering Analysis Interactive Dial-in, www.gwdac.physics.gwu.edu
[24] Fuchs, C., Faessler, A. and el-Shabshiry, M. *Phys. Rev*. C, 64, 024003 (2001).
[25] Peterson, R. J. *Nucl. Phys. A,* 740, 119 (2004).
[26] Gurvitz, S. A. *Phys. Rev*. C, 33, 422 (1986).
[27] Chen, X. Y. et al. *Phys. Rev*., 1993, C 47, 2159 (1993).
[28] Kishimoto, T et al. *Phys. Rev*. C, 43, 1454 (1991).
[29] Bystricki, J., Lehar, F. and Winteritz, P. *J. Phys. (Paris),* 39, 1 (1978).
[30] Antonov, A. N. et al. *Phys. Rev*. C, 71, 014317 (2005).
[31] Peterson, R. J. *Nucl. Phys. A,* 769, 115 (2006).
[32] Moniz, E. J. et al. *Phys. Rev. Lett.,* 26, 445 (1971).
[33] Peterson, R. J. *Phys. Rev*. C, 85, 064616 (2012).
[34] Shigehara, T., Shimizu K. and Arima, A. *Nucl. Phys. A,* 492, 388 (1989).
[35] Subedi, R. et al. *Science,* 320, 1476 (2008).
[36] Fujii, Y. et al. *Phys. Rev*. C, 64, 034608 (2001).
[37] Fujii, Y. Ph D thesis, Tohoku University 1998.
[38] Prout, D. L. et al. *Phys. Rev*. C, 52, 228 (1995).
[39] Prout, D. L. PhD thesis, University of Colorado `1992.
[40] Otsu, H. PhD thesis, University of Tokyo, 1995.
[41] Chrien, R. E. et al. *Phys. Rev*. C, 21, 1014 (1980).
[42] Ouyang, J. PhD thesis, University of Colorado, 1992.
[43] Zumbro, J. D. et al. *Phys. Rev. Lett.,* 71, 1796 (1993).
[44] Kin, T. et al., *Phys. Rev*. C, 72, 014606 (2005).
[45] Esbensen, H. and Bertsch, G. F., *Phys. Rev*. C, 34, 1419 (1986).
[46] Corley, D. M. et al., *Nucl. Phys. A,* 184, 437 (1972).
[47] Finn, J. M., Lourie, R. W. and Cottman, B. H., *Phys. Rev*. C, 29, 2230 (1984).
[48] Peterson, R. J. *Nucl. Phys. A,* 920, 20 (2013).
[49] Maris,Th. A. J., Theodoro, M. R., and Vasconcellos, C. A. Z. *Nucl. Phys. A,* 322, 461 (1979).
[50] Alberico, W. et al. *Phys. Rev*. C, 38, 109 (1988).
[51] De Pace, A. *Phys. Rev. Lett*., 75, 29 (1995).
[52] Ichimura, M. and Kagahigashi, K. *Phys. Rev*. C, 45, 1822 (1992).
[53] McClelland, J. B. et al. *Phys. Rev. Lett*., 69, 582 (1992).
[54] Taddeucci, T. et al. *Phys. Rev. Lett*., 73, 3516 (1994).
[55] Wakasa, T. et al. *Phys. Rev*. C, 59, 3177 (1999).
[56] Wakasa, T. et al. *Phys. Rev*. C, 69, 054609 (2004).
[57] Bertsch, G. F., Frankfurt, L. and Strikman, M. *Science,* 259, 773 (1993).
[58] Brown,G. E., Buballa,M., Li, Z. B. and Wambach, J. *Nucl. Phys. A,* 593, 295 (1995).
[59] Ivanov, M. V. et al., *Phys. Rev*. C, 77, 034612 (2008).
[60] Shigehara,T., Shimizu, K. and Arima, A. *Nucl. Phys. A,* 510, 106 (1990).
[61] Carey, T. A. et al. *Phys. Rev. Lett*., 53, 144 (1984).
[62] Holcomb, M. D. et al. *Phys. Rev*. C, 57, 1778 (1998).
[63] Siciliano, E. R. and Walker, G. E. *Phys. Rev*. C, 23, 2661 (1981).
[64] Holtkamp, D. B. et al. *Phys. Rev*. C, 31, 957 (1985).
[65] Peterson, R. J., *Int. Jour. Modern Physics,* 20, 686 (2005).

[66] Peterson, R. J. et al. *Phys. Rev.* C, 69, 064612 (2004).
[67] Friman, B. L., Pandharipande, V.R. and Wiringa, R. B. *Phys. Rev. Lett.,* 51, 703 (1983).
[68] Ericson, T. and Weise, W. Pions and Nuclei, Clarendon Press, Oxford 1988.
[69] Peterson, R. J. et al. *Phys. Rev.* C, 21, 1030 (1980).
[70] Cottingame, W. B. et al. Phys. Rev. C, 36, 230 (1987).
[71] Ritchie, B. G. et al. *Phys. Rev.* C, 37, 1347 (1988).
[72] Morris, C. L. et al. *Phys. Lett. B,* 108, 172 (1982).
[73] Oakley, D. S. et al. *Phys. Rev.* C, 38, 2978 (1988).
[74] McGill, J. A. et al. *Phys. Rev.* C, 29, 204 (1984).
[75] Bergqvist, I. et al. *Nucl. Phys. A,* 469, 648 (1987).
[76] Morris, C. L. et al. *Phys. Lett. B,* 419, 25 (1998).
[77] Kumano, S. and Moniz, E. J. *Phys. Rev.* C, 37, 2088 (1988).
[78] Kim, K. S. and Wright, L. E. *Phys. Rev.* C, 68, 027601 (2003).
[79] Ma, Z. Y. and Wambach, J. *Nucl. Phys. A,* 402, 275 (1983).
[80] Shen,H., Tamenaga, S. and Toki, H. *Nucl. Phys. A,* 745, 121 (2001).
[81] Ray, L. *Phys. Rev.* C, 41, 2816 (1990).
[82] Particle Data Group *Phys. Lett. B,* 667, 1 (2008).
[83] Kormanyos, C. et al. *Phys. Rev. Lett.,* 71, 2571 (1993).
[84] Marlow, D. et al. *Phys. Rev.* C, 25, 2619 (1982).
[85] Jiang, P. B., Ernst, D. J. and Chen, C. M. *Phys. Rev.* C, 51, 857 (1995).
[86] Siegel, P. B., Kaufmann, W. B. and Gibbs, W. R. *Phys. Rev.* C, 31, 2184 (1985).
[87] Weiss, R. et al. *Phys. Rev.* C, 49, 2569 (1994).
[88] Friedman, E. et al.,*Phys. Rev.* C, 55, 1304 (1997).
[89] Chrien, R. E. et al. Nucl. Phys. *A,* 625, 251 (1997).
[90] Michael, R. et al. *Phys. Lett. B,* 582, 29 (1995).
[91] Ebrahim, A. A. and Peterson, R. J. *Phys. Rev.* C, 54, 2499 (1996).
[92] Peterson, R. J., Ebrahim, A. A. and Bhang, H. C. *Nucl. Phys. A,* 625, 261 (1997).
[93] Kormanyos, C. et al. *Phys. Rev.* C 51, 669 (1995).
[94] De Pace, A, Garcia-Recio, C and Oset E. *Phys. Rev.*, 1997 C 55, 1394 (1997).
[95] Peterson, R. J. *Phys. Rev.* C, 60, 022201 (1999).
[96] de Vries, H., de Jager, C. W. and de Vries, C. *Atomic Data and Nuclear Data Tables,* 36, 495 (1987).
[97] De Pace, A and Viviani M. *Phys. Lett. B,* 254, 20 (1991).

# About the Authors

## Editors

*Leonid Blokhintsev*
Moscow State University, Russia
Email: blokh@srd.sinp.msu.ru

*Igor Strakovsky*
The George Washington University, US
Email: igor@gwu.edu

## Contributing Authors

*Claudio Ciofi degli Atti*
INFN, Italy
Email: Claudio.ciofi@pg.infn.it

*Sergei Dubovichenko*
Fessenkov Astrophysical Inst., Kazakhstan
Email: dubovichenko@gmail.com

*Franz Gross*
Jefferson Lab, US
Email: gross@jlab.org

*Vasily Kulikov*
Moscow State Univ., Russia
Email: kulikov@nucl-th.sinp.msu.ru

*Ruprecht Machleidt*
Idaho Univ., US
Email: machleid@uidaho.edu

*Pieter Maris*
Iowa State Univ., US
Email: pmaris@iastate.edu

*Roy Jerome Peterson*
Univ. of Colorado Boulder, US
Email: Jerry.Peterson@Colorado.EDU

*Misak Sargsian*
Florida State Univ., US
Email: sargsian@fiu.edu

*Andrey Shirokov*
Moscow State Univ., Russia
Email: shirokov@iastate.edu

*Alfred Stadler*
Universidade de Évora, Portugal
Email: alfred.stadler@tecnico.ulisboa.pt

*James Vary*
Iowa State Univ., US
Email: jvary@iastate.edu

*Colin Wilkin*
University College London, UK
Email: c.wilkin@ucl.ac.uk

*Yrii Yzikov*
JINR, Russia
Email: uzikov@jinr.ru

# INDEX

## F



## H



## I



## J



## K



## L



## M



## N



## P